Roger Z. Ríos-Mercado • Yasmín A. Ríos-Solís
Editors

Just-in-Time Systems

Springer

Editors
Roger Z. Ríos-Mercado
Department of Mechanical and Electrical Engineering
Graduate Program in Systems Engineering
Universidad Autónoma de Nuevo León
66450 San Nicolás de los Garza
Nuevo León
Mexico
roger.rios@uanl.edu.mx

Yasmín A. Ríos-Solís
Department of Mechanical and Electrical Engineering
Graduate Program in Systems Engineering
Universidad Autónoma de Nuevo León
66450 San Nicolás de los Garza
Nuevo León
Mexico
yasmin@yalma.fime.uanl.mx

ISSN 1931-6828
ISBN 978-1-4614-1122-2 e-ISBN 978-1-4614-1123-9
DOI 10.1007/978-1-4614-1123-9
Springer New York Dordrecht Heidelberg London

Library of Congress Control Number: 2011938811

Mathematics Subject Classification (2010): 37M10, 92B20, 68M20, 68Q15, 68Q25, 82C32, 9002, 9008, 90B05, 90B25, 90B30, 90B35, 90B36, 90B50, 90B90, 90C10, 90C11, 90C27, 90C39, 90C59, 90C60, 90C90, 91A80, 91B50

Printed on acid-free paper

Springer is part of Springer Science+Business Media (www.springer.com)

Preface

Just-in-time (JIT) philosophy was introduced in the 1950s. Its aim was to reduce inventory and its associated costs. Nowadays, JIT is associated with efficiency, continuous improvement, quality, and optimal flow. JIT started in industries but it quickly attracted the interest of the researchers since its implementation involves difficult problems that have been, and will be, challenging the scientific community.

There is a large community dealing with JIT scheduling. As we show in the first part of this book, the scheduling models, their analysis, and their resolution methods are vast and are continuously expanding. Nevertheless, JIT is covering and integrating other parts of the supply chain as presented in the second part of the book. If we define a system as a set of interacting components forming an integrated whole, then the title of this book emerges: "JIT systems". Indeed, not only one of the main objectives for us is to present state-of-the-art articles in JIT scheduling but also is to show that JIT includes lot-sizing, forecasting, game theory, mass production, and transfer lines.

The book is comprised of invitation-only articles written by experts in the JIT field. Each manuscript has been strictly peer-reviewed to ensure the quality of the work. The aim of the book is to present models, methods, algorithms, applications, or case studies of JIT systems. The target audience of this book will be professionals, researchers, and graduate students from diverse fields such as Operations Research, Management Science, Computer Science, and Discrete Mathematics.

Part I of the book deals with a surprising variety of JIT scheduling models. First, a survey of scheduling to maximizing the number of JIT jobs is presented by Shabtay and Steiner. It deals from the single machine case to the shop cases. Then, Tanaka contributed an article dealing with the exact resolution of the single machine earliness–tardiness scheduling. Indeed, earliness penalties can be considered as inventory costs while tardiness ones are related to the displeasure of the customer when a delay occurs. Yang and Yang present another field studied in scheduling that arises when the jobs or tasks have deteriorating and learning effects on a single machine.

In the scheduling literature, the single machine problem has received most of the attention. Nevertheless, the parallel machine case is interesting and closer to the real problems industries have. Vallada and Ruiz present a genetic algorithm for the parallel machine earliness–tardiness problem with setup times. Moreover, Ronconi and Birgin deal with the exact resolution of the flowshop earliness–tardiness scheduling case. Soukhal and Hyunh Toung analyse the particular but challenging case of having jobs with equal processing times. Last article of Part I is by Oulamara and focus on scheduling problems where the jobs must be processed in batches as it often happens in the real word.

Part II of this book is intended to problems that include or integrate other parts of the production system. First, Józefowska deals with the mass production environment where it is too costly to define and control due dates for individual items. Instead, the objective is to construct schedules with minimum deviation from an ideal product rate. Absi, Dauzère-Pérez, and Kedad-Sidhoum show that JIT systems integrating the lot-sizing and the scheduling phases are actually interesting and challenging. Gourvès, Monnot, and Telelis examine an algorithmic game-theoretic approach to optimizing the performance of distributed systems utilized by autonomous self-interested users. They focus on scheduling with parallel machines with setup times.

For a JIT system to work properly, accurate forecasts are sought. In particular, neural networks can be extremely useful. This subject is treated by Cabrera-Ríos, Salazar-Aguilar, Villareal-Marroquín, and Anaya Salazar. The paper of Low and Jayawickrama present a real life building project that is managed with the JIT system. Finally, Gökçe, Dinçer, and Örnek study the throughput rates of transfer lines with pull systems.

Collectively, this book describes recent advances in JIT systems. We acknowledge that there are many topics that are not covered due to the limitations of space. Nevertheless, we trust that those that are included here will provide information and motivation to explore this research area.

We thank foremost the authors for their outstanding contributions and their cooperation for this project. Furthermore, we thank Professor Panos Pardalos, editor of the Springer Series on Optimization and Its Applications, for accepting and encouraging this project. We also thank Vaishali Damle and Meredith Rich from Springer for their support throughout the edition process of this book. We specially thank the anonymous reviewers who have greatly improved the quality of the chapters. We are also grateful to Omar Ibarra-Rojas and Mónica Elizondo-Amaya from UANL for their help during the edition of the book.

San Nicolás de los Garza, Mexico

Yasmín A. Ríos-Solís
Roger Z. Ríos-Mercado

Contents

Contributors

Nabil Absi
CMP, Ecole des Mines de Saint-Étienne, 880 route de Mimet, F-13541 Gardanne, France, e-mail: absi@emse.fr

Ángela Patricia Anaya Salazar
Industrial Engineering Department, Universidad de San Buenaventura-Cali, Cali, Valle, Colombia, e-mail: angelitanaya@gmail.com

Ernesto G. Birgin
Department of Computer Science, IME-USP, University of São Paulo, Rua do Matão, 1010, Cidade Universitária, 05508-090 São Paulo SP, Brazil, e-mail: egbirgin@ime.usp.br

Mauricio Cabrera-Ríos
Industrial Engineering Department, University of Puerto Rico at Mayagüez, Puerto Rico, e-mail: Mauricio.cabrera1@upr.edu

Stéphane Dauzère-Pérès
CMP, Ecole des Mines de Saint-Étienne, 880 route de Mimet F-13541 Gardanne, France, e-mail: dauzere-peres@emse.fr

M. Cemali Dinçer
Department of Industrial Engineering, Istanbul Bilgi University, Kazim Karabekir Cad. No:2/13, Eyup Istanbul, Turkey, e-mail: cemal.dincer@bilgi.edu.tr

Mahmut Ali Gökçe
Department of Industrial Systems Engineering, Izmir University of Economics, Sakarya Cad. No: 156 Balcova Izmir, Turkey, e-mail: ali.gokce@ieu.edu.tr

Laurent Gourvès
LAMSADE, CNRS FRE 3234, Université de Paris-Dauphine, 75775 Paris, France, e-mail: laurent.gourves@lamsade.dauphine.fr

Nguyen Huynh Toung
Faculty of Computer Science and Engineering, University of Technology of Ho Chi Minh City, Ho Chi Minh City, Vietnam, e-mail: nguyen.huynh@yahoo.com

Thilini Shiromani Jayawickrama
Department of Building, National University of Singapore, Singapore, Singapore, e-mail: a0066405@nus.edu.sg

Joanna Józefowska
Poznań University of Technology, Poznań, Poland, e-mail: joanna.jozefowska@cs.put.poznan.pl

Safia Kedad-Sidhoum
LIP6, Université Pierre et Marie Curie, 4 Place Jussieu, F-75252 Paris Cedex, France, e-mail: safia.kedad-sidhoum@lip6.fr

Low Sui Pheng
Department of Building, National University of Singapore, Singapore, Singapore, e-mail: bdglowsp@nus.edu.sg

Jérôme Monnot
LAMSADE, CNRS FRE 3234, Université de Paris-Dauphine, 75775 Paris, France, e-mail: monnot@lamsade.dauphine.fr

M. Arslan Örnek
Department of Industrial Systems Engineering, Izmir University of Economics, Sakarya Cad. No: 156 Balcova Izmir, Turkey, e-mail: arslan.ornek@ieu.edu.tr

A. Oulamara
LORIA laboratory, Nancy University, Ecole des Mines de Nancy, Parc de Saurupt, 54042 Nancy, France, e-mail: Ammar.Oulamara@loria.fr

Débora P. Ronconi
Department of Production Engineering, EP-USP, University of São Paulo, Av. Prof. Almeida Prado, 128, Cidade Universitária, 05508-900 São Paulo SP, Brazil, e-mail: dronconi@usp.br

Rubén Ruiz
Grupo de Sistemas de Optimización Aplicada, Instituto Tecnológico de Informática (ITI), Ciudad Politécnica de la Innovación, Edificio 8G, Acceso B, Universidad Politécnica de Valencia, Camino de Vera s/n, 46022 Valencia, Spain, e-mail: rruiz@eio.upv.es

María Angélica Salazar-Aguilar
CIRRELT–HEC, Université de Montréal, Quebec, Canada, e-mail: angelica.salazar@cirrelt.ca

Dvir Shabtay
Department of Industrial Engineering and Management, Ben-Gurion University of the Negev, Beer-Sheva, Israel, e-mail: dvirs@bgu.ac.il

Ameur Soukhal
Laboratoire d'Informatique, Université François Rabelais, 64 Avenue Jean Portalis, 37200 Tours, France, e-mail: ameur.soukhal@univ-tours.fr

George Steiner
Operations Management Area, DeGroote School of Business, McMaster University, Hamilton, ON, Canada, e-mail: steiner@mcmaster.ca

Shunji Tanaka
Department of Electrical Engineering, Kyoto University, Kyotodaigaku-Katsura, Nishikyo-ku, Kyoto 615-8510, Japan, e-mail: tanaka@kuee.kyoto-u.ac.jp

Orestis A. Telelis
Center for Mathematics and Computer Science (CWI) Amsterdam, The Netherlands, e-mail: telelis@cwi.nl

Eva Vallada
Grupo de Sistemas de Optimización Aplicada, Instituto Tecnológico de Informática (ITI), Ciudad Politécnica de la Innovación, Edificio 8G, Acceso B, Universidad Politécnica de Valencia, Camino de Vera s/n, 46022 Valencia, Spain, e-mail: evallada@eio.upv.es

María Guadalupe Villarreal-Marroquín
Integrated Systems Engineering Department, The Ohio State University, Columbus, OH, USA, e-mail: lupita.villarreal@gmail.com

Dar-Li Yang
Department of Information Management, National Formosa University, Yun-Lin 632, Taiwan, e-mail: dlyang@nfu.edu.tw

Suh-Jenq Yang
Department of Industrial Engineering and Management, Nan Kai University of Technology, Nan-Tou 542, Taiwan, e-mail: sjyang@nkut.edu.tw

Part I
Just-in-Time Scheduling Systems

Chapter 1
Scheduling to Maximize the Number of Just-in-Time Jobs: A Survey

Dvir Shabtay and George Steiner

Abstract In just-in-time (*JIT*) scheduling, the usual objective is to minimize a cost function which includes a penalty for both the early and tardy completion of jobs. In this paper, we survey results for a cost function that is related to the *number* of early and tardy jobs rather than the actual earliness and tardiness values. More specifically, we study the problem of maximizing the weighted number of jobs which are completed exactly on their due date (i.e., in JIT mode). Our survey covers the literature for various scheduling environments both with fixed and controllable processing times. We also describe several new algorithms for certain flow-shop problems.

1.1 Introduction

In just-in-time (*JIT*) scheduling, the usual objective is to minimize a cost function which includes a penalty for both the early and tardy completion of jobs. A frequently used cost function sums the penalties due to earliness and tardiness, and the resulting problem is usually referred to as the *earliness–tardiness scheduling problem*. Lann and Mosheiov [21] introduced a different cost function that is related to the *number* of early and tardy jobs rather than the earliness and tardiness values. More specifically, they study the problem of maximizing the weighted number of jobs which are completed exactly on their due date (i.e., in *JIT* mode). We will refer to this problem as the *JIT scheduling problem*.

Dvir Shabtay
Department of Industrial Engineering and Management, Ben-Gurion University of the Negev, Beer-Sheva, Israel, e-mail: dvirs@bgu.ac.il

George Steiner
Operations Management Area, DeGroote School of Business, McMaster University, Hamilton, ON, Canada, e-mail: steiner@mcmaster.ca

R.Z. Ríos-Mercado and Y.A. Ríos-Solís (eds.), *Just-in-Time Systems*, Springer Optimization and Its Applications 60, DOI 10.1007/978-1-4614-1123-9_1,

Baker and Scudder [3] survey the literature for earliness–tardiness scheduling problems up to 1990, and a more up-to-date survey is given by Józefowska [17]. We aim to provide an up-to-date survey for *JIT* scheduling problems in which completing the jobs exactly at their due date is crucial and early or tardy jobs are equally penalized independent of the duration of their earliness or tardiness. Lann and Mosheiov [21] mention several applications for which such a cost structure can arise. Among them are applications in chemical and hi-tech industries, space shuttle scheduling, production of perishable items under deterministic demand, and scheduling production with no capacity for keeping inventory.

The *JIT* scheduling problem can be formally stated as follows: a set of n independent, non-preemptive jobs, $J = \{J_1, ..., J_n\}$, is available for processing at time zero and is to be processed on m machines $M = \{M_1, M_2, ..., M_m\}$. The machines are arranged in a specific technological configuration, which can be a single machine ($m = 1$), machines in parallel, or machines in a flow shop, job shop, or open shop. d_j represents the due date of job J_j, w_j represents the gain (income) for completing job J_j in *JIT* mode (i.e., exactly at time d_j) and p_{ij} represents the processing time of job J_j on machine M_i for $j = 1, ..., n$ and $i = 1, 2, ..., m$ (in the single-machine case, we omit the machine index so that, e.g., p_j is the processing time of job j on the single machine). A partition of set J into two disjoint subsets E and T is considered to be a *feasible partition* (or a *feasible schedule*) if it is possible to schedule the jobs belonging to set E in *JIT* mode. In a feasible schedule, we may assume, without loss of generality, that the jobs belonging to set T are rejected and thus do not need to be processed at all. Alternatively, if the jobs in set T also have to be processed, they can be processed arbitrarily on the machines *after* the entire schedule on set E is completed. The objective is to find a feasible schedule with a maximal weighted number of jobs in set E, that is, to find a feasible schedule that maximizes $\sum_{J_j \in E} w_j$. If all weights are identical, the objective is simply to maximize $|E|$ instead of $\sum_{J_j \in E} w_j$. Note that the criterion in the classical notation usually has to be minimized, but in the *JIT* scheduling problem we have to maximize it.

In the original *JIT* scheduling problem, job processing times are assumed to be fixed positive parameters. However, today it is accepted that in many cases, the scheduler can control processing times by allocating additional resources (such as fuel, gas, catalyzer, or manpower) to compress the job processing times. This control becomes even more important when the objective is to finish as many jobs as possible in *JIT* mode. Janiak [14] describes an interesting application of a scheduling problem with controllable processing times in steel mills, where batches of ingots have to be preheated before being hot-rolled in a blooming mill, and both the preheating time and the rolling time are inversely proportional to the gas flow intensity. Another interesting application arises from scheduling in a machine tooling environment where the job processing time is a function of the feed rate and the spindle speed used for each operation (see, e.g., Trick [37], Kaspi and Shabtay [18], and Kayan and Akturk [19]). Shakhlevich and Strusevich [34] point out that scheduling with controllable processing times could also be used for modeling outsourcing in supply chains. In this case, the compression of job processing times corresponds to the portion of jobs that is outsourced to subcontractors and the "resource consumption costs" represent the subcontracting costs. Due to the large variety of applications, there is extensive literature on the subject of scheduling with controllable processing times (e.g., Vickson [38], Alidaee and Ahmadian [1], Janiak and Kovalyov [15], Cheng et al. [6], Shabtay [28], Shabtay and Kaspi [29], Ng et al. [23], and Shabtay et al. [30]). A survey of results on this subject up to 1990 is provided by Nowicki and Zdrzalka [24]. More recent surveys appear in Chudzik et al. [9], Shabtay and Steiner [31], and Janiak et al. [16].

For a *JIT* scheduling problem with controllable processing times, we assume that the processing time functions (commonly known as the *resource consumption functions*) satisfy the following properties:

Property 1. Each job processing time, $p_{ij}(u_{ij})$, is either a discrete or a continuous, bounded function of the resource allocated to the job, u_{ij}, where $\underline{u}_{ij}, \overline{u}_{ij}$ are the lower and upper bounds of the resource that can be assigned to process job J_j on machine M_i for $j = 1, ..., n$ and $i = 1, ..., m$, respectively.

Property 2. Each $p_{ij}(u_{ij})$ is a decreasing function of the amount of resource allocated, and its inverse, $(p_{ij})^{-1}(x)$, exists and can be evaluated in constant time for any $x \in [p_{ij}(\overline{u}_{ij}), p_{ij}(\underline{u}_{ij})]$.

If job J_j does not get processed on machine M_i, then $u_{ij} = 0$ by definition. A solution S, for a *JIT* scheduling problem with controllable processing times is defined by a partition of set J into $E \cup T$ and by a resource allocation vector $(u_{1j}, u_{2j}, ..., u_{mj})$ for each $J_j \in E$. Such a solution is considered feasible if it is possible to complete all jobs in set E in *JIT* mode, given the processing times $(p(u_{1j}), p(u_{2j}), ..., p(u_{mj}))$ for each $J_j \in E$. The quality of a feasible solution S for a *JIT* scheduling problem with controllable processing times is measured by two criteria. The first one

$$W(S) = \sum_{J_j \in E} w_j$$

is the total gain (income), while the second one

$$U(S) = \sum_{i=1}^{m} \sum_{J_j \in E} c_{ij} u_{ij}$$

is the total resource consumption cost, where c_{ij} is the possibly machine-dependent unit resource allocation cost for job J_j on machine M_i. Since a *JIT* problem with controllable processing times has two criteria, four different optimization problems can arise:

- The first one, which we denote by P1, is to maximize the net profit, that is, to maximize $\sum_{J_j \in E} w_j - \sum_{i=1}^{m} \sum_{J_j \in E} c_{ij} u_{ij}$.
- The second one, denoted by P2, is to find a schedule S maximizing $W(S) - \sum_{J_j \in E} w_j$ subject to $\sum_{i=1}^{m} \sum_{J_j \in E} c_{ij} u_{ij} \leq U$, where U is a limitation (budget) on the total resource consumption cost.

- The third one, which we denote by P3, is to find a schedule S minimizing $U(S) = \sum_{i=1}^{m} \sum_{J_j \in E} c_{ij} u_{ij}$ subject to $\sum_{J_j \in E} w_j \geq K$, where K is a given lower bound (target) on the total targeted income.
- The last one, denoted by P4, is to identify each *Pareto-optimal point* (W,U) and associated *Pareto optimal schedule*, where a schedule S with $W = W(S)$ and $U = U(S)$ is called *Pareto-optimal* (or *efficient*) if there does not exist another schedule S' such that $W(S') \geq W(S)$ and $U(S') \leq U(S)$ with at least one of these inequalities being strict. The corresponding pair (W,U) is referred to as a Pareto optimal point.

It should be noted that solving problem P4, also solves problems P1–P3 as a by-product. Note also that the decision versions of P2 and P3 are identical and it is defined below:

Definition 1.1. *DVP:* Given a *JIT* scheduling problem and parameters K and U, is there a feasible job schedule with $\sum_{J_j \in E} w_j \geq K$ and $\sum_{i=1}^{m} \sum_{J_j \in E} c_{ij} u_{ij} \leq U$?

The fact that both P2 and P3 share the same decision version implies that either both or neither of them is $\mathcal{NP}$-hard. Studying the bicriteria versions P2–P4 becomes especially important for situations where the two criteria are not compatible and adding them together makes no sense, for example, when $c_{ij} = 1$ for all i and j and the second criterion represents the actual amount of the resource used.

In order to specify each problem, we generalize the classical 3-field notation $\alpha|\beta|\gamma$ introduced in [11]. The α field describes the machine environment. We consider the case of a single machine when $\alpha = 1$, the case of m identical parallel machines for which $\alpha = Pm$, the case of unrelated machines in parallel for which $\alpha = Rm$, the case of a flow shop for which $\alpha = Fm$, the case of a job shop for which $\alpha = Jm$ and the case of an open shop for which $\alpha = Om$. The β field exhibits the job processing characteristics and constraints and may contain no entry, a single entry, or multiple entries. For example, if $p_{ij} = p_{ij}(u_{ij})$ appears in this field, it means that the processing times are resource dependent (controllable) and satisfy Properties 1 and 2. Otherwise, the processing times are assumed to be constant parameters. For P2- and P3-type problems with controllable processing times, we also include in β the resource limitation or income target, respectively. If *dscr* appears in the β field it means that the resource can be allocated only in *discrete* quantities. Otherwise, we can use *continuous* quantities for allocating the resource. The γ field contains the optimizing criteria. For a P4-type problem, we include both criteria in this field. For example, $1||\sum_{J_j \in E} w_j$ denotes the scheduling problem of maximizing the weighted number of *JIT* jobs on a single machine, while $1|p_j = p_j(u_j)|\left(\sum_{J_j \in E} w_j, \sum_{J_j \in E} u_j\right)$ denotes the single-machine bicriteria problem of constructing the $\left(\sum_{J_j \in E} w_j, \sum_{J_j \in E} u_j\right)$ trade-off curve (the efficient points), where the job processing times are controllable (and satisfy Properties 1 and 2) and the resource can be allocated in continuous quantities.

1.2 The *JIT* Problem on a Single Machine

1.2.1 Constant Processing Times

In what follows, we denote by $[i]$ the job scheduled in the ith position in a fixed sequence for $i = 1, 2, ..., n$. Let $E = \{[1], [2], ..., [|E|]\}$ be a set of *JIT* jobs sequenced according to their processing order on the single machine. Then we have the following easy-to-prove lemma.

Lemma 1.1. *Set E is feasible only if*

$$p_{[i]} \leq d_{[i]} - d_{[i-1]}, \tag{1.1}$$

for $i = 1, ..., |E|$ *where* $d_{[0]} \stackrel{def}{=} 0$. *Otherwise, job* $[i]$ *cannot be completed in JIT mode given that job* $[i-1]$ *is completed in JIT mode.*

The next corollary is a straightforward consequence of Lemma 1.1.

Corollary 1.1. *The JIT set E is always sequenced in earliest due date (EDD) order.*

Let $S_{[i]}$ be the starting time of job $[i]$ that guarantees its completion in *JIT* mode. Then according to Lemma 1.1 we have,

$$d_{[i-1]} \leq S_{[i]} \stackrel{\text{def}}{=} d_{[i]} - p_{[i]}. \tag{1.2}$$

Lann and Mosheiov [21] present the following procedure that solves the $1||\,|E|$ problem in $O(n \log n)$ time.

Algorithm 1: Lann and Mosheiov's optimization algorithm for the $1||\,|E|$ problem.

Step 0. Calculate $S_j = d_j - p_j$ for $J_j \in J$ and renumber the jobs in a non-increasing order of S_j. Set $j = 0$, $t = \max_{j=1,...,n} d_j$ and $E = \emptyset$.

Step 1. Set $j = j + 1$. If $j > n$ then Go to step 2.
 If $d_j \leq t$ (i.e., the inclusion of job J_j in set E yields a feasible partial schedule) then set $E = E \cup \{J_j\}$ and $t = S_j$. Go to Step 1.

Step 2. Output: E is the set of *JIT* jobs and $T = J \setminus E$ is the set of rejected jobs.

Lann and Mosheiov [21] also present an $O(n^2)$ dynamic programming (*DP*) algorithm to solve the weighted problem $1||\sum_{J_j \in E} w_j$. They also mention that the problem can be solved by transforming it to a longest path problem in a directed acyclic graph (acyclic digraph). We describe this approach in more detail next: The acyclic digraph $G = (V, A)$ includes $n+2$ vertices numbered from 0 to $n+1$, with vertices $1, 2, ..., n$ corresponding to the n jobs in *EDD* order. Every arc $(i, j) \in A$, connecting vertices $i, j \in V$ with $0 \leq i < j \leq n+1$, is directed from i to j. Vertex 0, representing a fictitious job at the start of the schedule, is connected to every vertex in $\{1,2,...,n\}$, and every vertex in $\{1,2,...,n\}$ is connected to $n+1$, which corresponds to another fictitious job at the end of the *JIT* set. The arc (i,j) exists if it is feasible to schedule job J_j in *JIT* mode immediately following the *JIT* job J_i, that is, according to Lemma 1.1, they satisfy the condition that $p_j \leq d_j - d_i$. The length of (i,j) is

$$L_{ij} = \begin{cases} w_j \text{ for } & 0 \leq i < j \leq n \\ 0 \text{ for } & 0 \leq i < j = n+1 \end{cases} \tag{1.3}$$

corresponding to the profit for completing job J_j in *JIT* mode (right after job J_i).

A path $(0,[1]),([1],[2]),...,([k],n+1)$ from vertex 0 to vertex $n+1$ implies that it is possible to schedule jobs $[1],[2],...,[k]$, in this order, to be just in time, that is, it represents a sequence for a feasible set E. Therefore, the length of any path in the graph equals the total profit for scheduling the corresponding set E in *JIT* mode. Hence, finding a longest path in G is equivalent to finding a feasible schedule and set E with maximum profit. The longest path in the graph can be determined in $O(n^2)$ time by applying the following simple recursion:

$$G_l = \max_{k=0,...,l-1} (G_k + L_{kl}), \tag{1.4}$$

where $G_0 = 0$ is the initial condition, and G_l is the length of a longest path that ends in l in the subgraph that includes the vertices $0,...,l$. Thus, G_{n+1} corresponds to the length of the longest path in the graph. To summarize, Algorithm 2 below can be used to optimally solve the $1||\sum_{J_j \in E} w_j$ problem in $O(n^2)$ time.

Finally, we mention that Hiraishi et al. [13] study an extension of the $1|||E|$ problem in which the gain of w_j corresponds to the completion of job J_j within a *time window* $[\underline{d}_j, \overline{d}_j]$ (rather than at a specific due-date) and show that this problem is $\mathcal{NP}$-hard.

1.2.2 Controllable Processing Times

Since the EDD order is independent of the processing times, Corollary 1.1 holds for all problem types (P1–P4) and resource types (discrete and continuous). Moreover, the feasibility condition in Lemma 1.1 should be replaced with

$$p_{[i]}(\overline{u}_{[i]}) \leq d_{[i]} - d_{[i-1]}. \tag{1.5}$$

The following lemma proved by Leyvand et al. [22] is applicable to all problem types (P1–P4):

Lemma 1.2. *The optimal resource allocation for the JIT set E is*

$$u^*_{[i]} = \begin{cases} \underline{u}_{[i]} & \text{if} \quad p_{[i]}(\underline{u}_{[i]}) < d_{[i]} - d_{[i-1]} \\ p^{-1}_{[i]}(d_{[i]} - d_{[i-1]}) & \text{if } p(\overline{u}_{[i]}) \leq d_{[i]} - d_{[i-1]} \leq p_{[i]}(\underline{u}_{[i]}) \end{cases} \tag{1.6}$$

Algorithm 2: The optimization algorithm for the $1||\sum_{J_j \in E} w_j$ problem.

Input: n, $\mathbf{d} = (d_1, ..., d_n)$, $\mathbf{w} = (w_1, ..., w_n)$, $\mathbf{p} = (p_1, ..., p_n)$ where the jobs are numbered in EDD order.
Step 1. Construct the digraph $G(V,A)$ with $V = \{0, ..., n+1\}$ and $A = \{(i,j) \mid p_j \le d_j - d_i\}$ and arc (i,j) is directed from i to j.
Step 2. Calculate the length of arcs (i,j) for $0 \le i < j \le n$ by (1.3).
Step 3. Find a longest path in the digraph by applying the recursion (1.4) for $l = 1, ..., n+1$ with the initial condition $G_0 = 0$. Trace back the longest path in the graph G to identify its internal vertices $[1], [2], ..., [k]$.
Step 4. Set $E = \{[1], [2], ..., [k]\}$ and $T = J \setminus E$.
Step 5. Set the starting time of job $[j]$ to $S_{[j]} = d_{[j]} - p_{[j]}$ for $j = 1, ..., k$.

for a continuous resource type, while it is

$$u^*_{[i]} = \begin{cases} \underline{u}_{[i]} & \text{if} \quad p_{[i]}(\underline{u}_{[i]}) < d_{[i]} - d_{[i-1]} \\ \left\lceil p^{-1}_{[i]}(d_{[i]} - d_{[i-1]}) \right\rceil & \text{if } p(\overline{u}_{[i]}) \le d_{[i]} - d_{[i-1]} \le p_{[i]}(\underline{u}_{[i]}) \end{cases} \tag{1.7}$$

for a discrete resource type.

The P1 problem on a single machine, that is, the $1|p_j = p_j(u_j)|\sum_{J_j \in E} w_j - \sum_{J_j \in E} c_j u_j$ problem, can be solved in $O(n^2)$ time by modifying the previous transformation of the $1||\sum_{J_j \in E} w_j$ problem to a longest path problem. The graph is constructed in a similar manner with the following modifications:

- There is an arc (i,j) if it is feasible to schedule job J_j in *JIT* mode immediately following the *JIT* job J_i, that is, according to (1.5) they satisfy the condition $p_j(\overline{u}_j) \le d_j - d_i$.
- The length of arc (i,j) is defined as the *net profit* (gain minus resource allocation cost) from selecting job J_j to be part of set E subject to the condition that the last previously selected job for set E is job J_i. It is calculated by

$$L_{ij} = \begin{cases} w_j - c_j u^*_{ij} & \text{for} \quad 0 \le i < j \le n \\ 0 & \text{for } 0 \le i < j = n+1 \end{cases}, \tag{1.8}$$

where u^*_{ij} is the optimal resource allocation to job J_j *if* it follows job J_i. By Lemma 1.2, u^*_{ij} can be calculated by

$$u^*_{ij} = \begin{cases} \underline{u}_j & \text{if} \quad p(\underline{u}_j) < d_j - d_i \\ p^{-1}_j(d_j - d_i) & \text{if } p_j(\overline{u}_j) \le d_j - d_i \le p_j(\underline{u}_j) \\ \infty & \text{if} \quad p_j(\overline{u}_j) > d_j - d_i \end{cases} \quad \text{for } 0 \le i < j \le n, \tag{1.9}$$

if the resource type is continuous and by

$$u_{ij}^* = \begin{cases} \underline{u}_j & \text{if} \quad p(\underline{u}_j) < d_j - d_i \\ \left\lceil p_j^{-1}(d_j - d_i) \right\rceil & \text{if } p_j(\overline{u}_j) \leq d_j - d_i \leq p_j(\underline{u}_j) \\ \infty & \text{if} \quad p_j(\overline{u}_j) > d_j - d_i \end{cases} \quad \text{for } 0 \leq i < j \leq n, \quad (1.10)$$

if the resource type is discrete (where $d_0 \stackrel{def}{=} 0$).

Note that if $u_{ij}^* = \infty$ results in the above calculations, then arc (i, j) will not exist in the corresponding graph. To summarize, Algorithm 3 below can be used to optimally solve the $1|p_j = p_j(u_j)|\sum_{J_j \in E} w_j - \sum_{J_j \in E} c_j u_j$ problem in $O(n^2)$ time.

In contrast with the polynomial-time solvability of the P1 problem, the following theorem of Leyvand et al. [22] shows that the P2–P4 problems on a single machine are all $\mathcal{NP}$-hard.

Theorem 1.1 ([22]). *DVP (see Definition 1.1) is $\mathcal{NP}$-complete on a single machine for both discrete and continuous resource type even if $c_j = 1$ for $j = 1, \ldots, n$.*

1.3 The *JIT* Scheduling Problem on Parallel Machines

Usually, three different systems of parallel machines are considered in the scheduling literature. In each of these, every job must be processed on any one of the m machines.

- *Identical machines* ($\alpha = Pm$) where the processing times are machine independent, that is, $p_{ij} = p_j$.
- *Uniform machines* ($\alpha = Qm$) where $p_{ij} = p_j/\tau_i$ and τ_i denotes the uniform speed of machine M_i.
- *Unrelated machines* ($\alpha = Rm$) where each machine has a different speed for each job.

1.3.1 Constant Processing Times

Kovalyov et al. [20] show that the *JIT* scheduling problem on a set of m unrelated parallel machines, that is, the $Rm||\sum_{J_j \in E} w_j$ problem, is equivalent to a maximum-weight coloring problem in a given interval graph, which they call the *fixed interval scheduling problem*. Kovalyov et al. [20] observe that various formulations and special cases of the fixed interval scheduling problem have been considered in the literature under different names (e.g., interval selection, scheduling with discrete starting times, channel assignment (reservation), k-track assignment, k-coloring of

Algorithm 3: [22] The optimization algorithm for the $1|p_j = p_j(u_j)|$ $\sum_{J_j \in E} w_j - \sum_{J_j \in E} c_j u_j$ problem for both resource types (continuous and discrete).

Input: n, $\mathbf{d} = (d_1, ..., d_n)$, $\mathbf{w} = (w_1, ..., w_n)$, $\mathbf{p}(\mathbf{u}) = (p_1(u_1), ..., p_n(u_n))$ with the jobs numbered in EDD order.
Step 1. Construct the digraph $G(V,A)$ with $V = \{0, ..., n+1\}$ and $A = \left\{(i, j) \mid p_j(\overline{u}_j) \le d_j - d_i\right\}$ and arc (i, j) is directed from i to j.
Step 2. Calculate the length of arcs (i, j) for $0 \le i < j \le n$ by (1.8), where u^*_{ij} is calculated by (1.9) if the resource is continuous and by (1.10) if the resource is discrete.
Step 3. Find a longest path in the digraph by applying the recursion (1.4) for $l = 1, ..., n+1$ with the initial condition $G_0 = 0$. Trace back the longest path in the graph G to identify its internal vertices $[1], [2], ..., [k]$.
Step 4. $E = \{[1], [2], ..., [k]\}$ and $T = J \setminus E$. The optimal resource allocation for job $[j]$ is $u^*_{[j-1][j]}$ for $j = 1, ..., k$ (calculated in Step 2 for the corresponding arc) and no resources are allocated to the jobs in T.
Step 5. Set the starting time of job $[j]$ to $S_{[j]} = d_{[j]} - p_{[j]}(u^*_{[j-1][j]})$ for $j = 1, ..., k$.

intervals, and maximizing the number of on-time jobs). Thus, many results for several variants of the fixed interval scheduling problem have been reinvented or obtained independently for equivalent formulations of the same problem. Carlisle and Lloyd [5] present an algorithm that solves the problem with unit gains on m identical parallel machines, that is, the $Pm|| \, |E|$ problem, in $O(n \log n)$ time. Other polynomial time algorithms for the same problem (in many cases under different titles and more general problems) appear in Frank [10], Arkin and Silverberg [2], Yannakakis and Gavril [39], Hsiao et al. [12], and Čepek and Sung [8]. It seems, however, that the algorithm by Carlisle and Lloyd has the best computational complexity.

Arkin and Silverberg [2] present an $O(n^2 \log n)$ time optimization algorithm for the weighted case $Pm|| \sum_{J_j \in E} w_j$ on m identical parallel machines, which is based on reformulating the problem as a minimum cost flow problem. Bouzina and Emmons [4] and Carlisle and Lloyd [5] give more efficient minimum cost flow algorithms for the same problem by modeling it on a network which has only $O(n)$ arcs. Both algorithms solve the problem in $O(mn \log n)$ time. Additional polynomial time algorithms, which are less efficient or have unspecified polynomial complexity, are given by Steiner [35], Sarrafzadeh and Lou [27], Pal and Bhattacharjee [25], Saha and Pal [26] and Hiraishi et al. [13]. Hiraishi et al. [13] show that the problem can be solved in polynomial time even if a non-negative setup time is required between every two consecutive jobs on the same machine.

Both Arkin and Silverberg [2] and Sung and Vlach [36] show that the weighted problem $Rm|| \sum_{J_j \in E} w_j$ on m unrelated parallel machines is solvable in $O(mn^{m+1})$ time, which is polynomial for a fixed number of machines. They also prove, however, that the problem becomes $\mathcal{NP}$-hard in the strong sense if the number of machines is part of the input.

1.3.2 Controllable Processing Times

Leyvand et al. [22] study the *JIT* scheduling problem with controllable processing times on m parallel machines. They show that the algorithm presented by Hiraishi et al. [13] can be extended to solve the more general $Pm|p_j = p_j(u_j)|\sum_{J_j \in E} w_j - \sum_{J_j \in E} c_j u_j$ problem in $O(mn^2)$ time. Moreover, they show that the P1 problem $Rm|p_{ij} = p_{ij}(u_{ij})|\sum_{J_j \in E} w_j - \sum_{J_j \in E} c_{ij} u_{ij}$ is also solvable in polynomial time with a fixed number of unrelated machines by providing an $O(mn^{m+1})$ time optimization algorithm. Note that the $Rm|p_{ij} = p_{ij}(u_{ij})|\sum_{J_j \in E} w_j - \sum_{J_j \in E} c_{ij} u_{ij}$ problem is strongly $\mathcal{NP}$-hard if the number of machines is part of the input, due to the strong $\mathcal{NP}$-hardness of the less general $Rm||\sum_{J_j \in E} w_j$ problem ([2] and [36]). Leyvand et al. [22] also present a DP based optimization algorithm to solve the $\mathcal{NP}$-hard $Rm|p_{ij} = p_{ij}(u_{ij})|\left(\sum_{J_j \in E} w_j, \sum_{J_j \in E} c_{ij} u_{ij}\right)$ P4-type problem in $O(mn^{m+1} U_{\max})$ time, where $U_{\max} \overset{\text{def}}{=} \sum_{j=1}^{n} \max_{i=1,\ldots,m} c_{ij} \bar{u}_{ij}$. This time complexity is pseudo-polynomial for a fixed number of machines proving that this version of the problem is only ordinary $\mathcal{NP}$-hard. Moreover, they show that the algorithm can be converted into an FPTAS for finding an approximate Pareto optimal solution in $O\left(\frac{m}{\varepsilon} n^{m+2}\right)$ time, which is polynomial for a fixed number of machines.

1.4 The *JIT* Scheduling Problem in Other Multi-machine Systems

Here we present the known results for flow shops, job shops, and open shops. These different systems are briefly described below.

- *Flow shop* $(\alpha = Fm)$: In a flow shop, the machines are linearly ordered and the jobs all follow the same processing route (from the first machine to the last machine).
- *Job shop* $(\alpha = Jm)$: In a job shop, each job has its own predetermined processing route to follow between the machines. Usually, it is assumed that each job visits each machine at most once. Note that the flow shop is a special case of the job shop.
- *Open shop* $(\alpha = Om)$: In an open shop, each job needs to be processed exactly once on each of the machines, but the scheduler has to determine the processing route of each job in any order of the machines, and different jobs may have different routes.

1.4.1 Flow Shops

Permutation schedules are schedules in which the jobs get processed in the same sequence on every machine. Since for the $Fm||\sum_{J_j \in E} w_j$ problem, we are restricted to schedule the jobs on the last machine according to the EDD rule, any permutation schedule is restricted to the EDD permutation (order). The following lemma shows that this permutation schedule is indeed optimal for the $F2||\sum_{J_j \in E} w_j$ problem.

Lemma 1.3 ([7]). *There exists an optimal schedule for the $F2||\sum_{J_j \in E} w_j$ problem in which the JIT jobs are scheduled in an increasing order of due dates (i.e., according to the EDD order) on both machines.*

It is interesting to note, however, that the above property does not extend to the case of more than two machines, as a permutation (EDD) schedule would not necessarily be an optimal schedule. This result was obtained by Choi and Yoon [7] by providing a counter example of three jobs on three machines. We give a different counter example below, which shows that permutation schedules are not optimal even if we have only two jobs:

Example 1.1. Consider an instance of the $Fm|||E|$ problem with $m = 3$ machines, $n = 2$ jobs and processing times and due dates as given in Table 1.1.

Table 1.1: A small counter example.

j	1	2
p_{1j}	3	1
p_{2j}	1	3
p_{3j}	1	1
d_j	6	7

By processing the two jobs in the EDD sequence on all three machines, we can see that the earliest completion time of job J_2 on the last machine is at time 8 and thus, a permutation (EDD) schedule yields a schedule with $|E| = 1$ (see Fig. 1.1). However, by processing job J_2 earlier than job J_1 on machines M_1 and M_2, we can construct a schedule with $|E| = 2$ (see Fig. 1.2).

The following two lemmas are proved in [32].

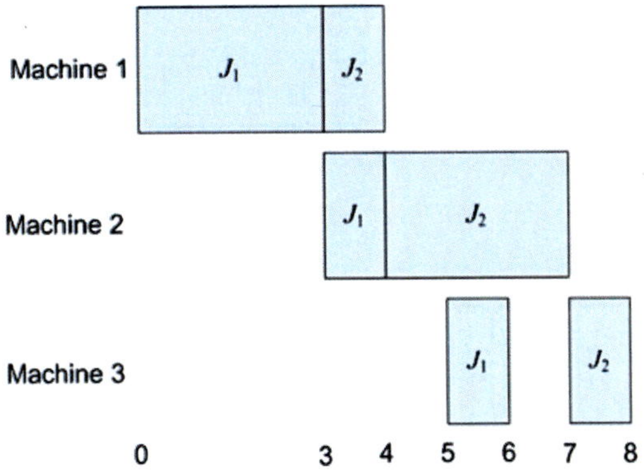

Fig. 1.1: A non-optimal permutation schedule

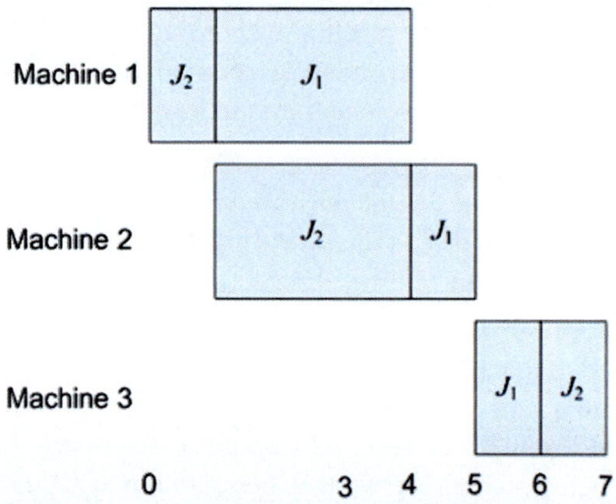

Fig. 1.2: An optimal non-permutation schedule

Lemma 1.4. *There exists an optimal schedule for the $F2||\sum_{J_j \in E} w_j$ problem in which there are no idle times on the first machine.*

Let us denote by $F(l,D)$ the maximum obtainable profit over any feasible *JIT* subset of jobs from $\{J_1,...,J_l\}$ that contains J_l, with D being the total processing time of this subset on the first machine. Let $1 \le i < j \le n$, let $E(i) \subseteq \{J_1,...,J_i\}$ be a feasible set of *JIT* jobs such that $J_i \in E(i)$ and $E(j) \subseteq \{J_1,...,J_j\}$ be a feasible set of *JIT* of jobs with $J_j \in E(j)$, and let $S(i)$ and $S(j)$ be the corresponding subschedules, respectively. Let $D_i = \sum_{J_l \in E(i)} p_{1l}$ and $D_j = \sum_{J_l \in E(j)} p_{1l}$ be the completion time of subschedules $S(i)$ and $S(j)$, respectively, on the first machine.

Lemma 1.5 ([32]). *If $D_i \le D_j$ and $F(i,D_i) \ge F(j,D_j)$, then subschedule $S(i)$ dominates subschedule $S(j)$, that is, every feasible continuation of $S(j)$ is also a feasible continuation of $S(i)$, which results in at least as high profit as the continuation of $S(j)$.*

Choi and Yoon [7] show that the $F2|||E|$ problem can be solved in $O(n^4)$ time by exploiting the above three lemmas. They propose an iterative algorithm, which at iteration k finds the entire set of non-dominated subschedules whose *JIT* subset has size k. The algorithm ends at the first iteration k^*+1 in which it is not possible to construct a feasible subset of size k^*+1. Then, any constructed subschedule of size k^* is optimal. We present next a faster (and less complicated) optimization algorithm for the $F2|||E|$ problem.

Let $\mathscr{E}_j(k) = \{E_j(k)\}$ be the set of all feasible subsets of *JIT* jobs from among jobs $\{J_1,...,J_j\}$ with $J_j \in E_j(k)$ and $|E_j(k)| = k$ for $1 \le k \le j \le n$. Moreover, let $D_j(k) = \sum\limits_{J_l \in E_j(k)} p_{1l}$ be the total processing time of set $E_j(k)$ on machine M_1 and $\mathscr{D}_j(k) = \min\limits_{E_j(k) \in \mathscr{E}_j(k)} \{D_j(k)\}$ be the minimal total processing time on machine M_1 among all feasible subsets $E_j(k) \in \mathscr{E}_j(k)$ of *JIT* jobs. Then we can compute all $\mathscr{D}_j(k)$ values for $1 \le k \le j = 1,...,n$ by applying the following recursive relation:

$$\mathscr{D}_j(k) = \min_{i=k-1,\dots,j-1} \begin{cases} \mathscr{D}_i(k-1)+p_{1j} & \text{if } \max\left(\mathscr{D}_i(k-1)+p_{1j},d_i\right)+p_{2j} \le d_j \\ \infty & \text{if } \max\left(\mathscr{D}_i(k-1)+p_{1j},d_i\right)+p_{2j} > d_j \end{cases} \tag{1.11}$$

with the initial conditions

$$\mathscr{D}_0(0)=0, \mathscr{D}_j(0)=\infty \text{ for } 1 \le j \le n \text{ and } \mathscr{D}_j(k)=\infty \text{ for any } 1 \le k \le j \le n. \tag{1.12}$$

We note that if job J_i is the last job which was included in a *JIT* set $E_i(k-1)$, then job J_j $(j > i)$ cannot start its processing on the second machine before $\max(\mathscr{D}_i(k-1)+p_{1j},d_i)$ and therefore, it cannot be completed before $\max(\mathscr{D}_i(k-1)+p_{1j},d_i)+p_{2j}$. Thus, the condition in (1.11), namely $\max\left(\mathscr{D}_i(k-1)+p_{1j},d_i\right)+p_{2j} \le d_j$, is a necessary and sufficient condition for the resulting subschedule to be feasible.

The maximal number of *JIT* jobs is then given by the maximum k value for which $\mathscr{D}_j(k)$ is finite among all $\mathscr{D}_j(k)$ values for $j \in \{1,...,n\}$ and $j \ge k$, that is,

$$k^* = \max_{1\le k\le j\le n} \left\{k \mid \mathscr{D}_j(k) < \infty\right\}. \tag{1.13}$$

To summarize, we suggest using the following optimization algorithm to solve the $F2||\,|E|$ problem.

Algorithm 4: The optimization algorithm for the $F2||\,|E|$ problem.

Input: n, $\mathbf{d} = (d_1,...,d_n)$, $\mathbf{P} = (p_{ij})$ for $i = 1,2$, and $j = 1,...,n$.
Step 1. Renumber the jobs according to the EDD order and initialize $k^* = 0$.
Step 2. Apply the recursion defined by (1.11) with the initial conditions in (1.12) for $1 \le k \le j = 1,...,n$. Whenever a finite $\mathscr{D}_j(k)$ value is obtained with $k > k^*$, reset $k^* = k$.
Step 3. Determine the optimal schedule by tracing back the solution from the end.

Theorem 1.2. *Algorithm 4 solves the $F2||\,|E|$ problem in $O(n^3)$ time.*

Proof. According to Lemma 1.5, for any combination of j and k, all feasible subschedules are dominated by the one with the smallest $D_j(k)$ value. Thus, Algorithm 4 finds the set of all (finite) $D_j(k)$ values for the entire set of non-dominated subschedules. The maximum number of *JIT* jobs is given by (1.13), which is clearly obtained in Step 2. Step 1 requires a sorting operation and thus takes $O(n\log n)$ time. Since computing every $\mathscr{D}_j(k)$ value by (1.11) requires $O(n)$ time, and since there are at most $O(n^2)$ such $\mathscr{D}_j(k)$ values, Step 2 overall takes at most the $O(n^3)$ time. Thus, the complexity of Algorithm 4 is indeed $O(n^3)$. □

The following theorem of Choi and Yoon [7] shows that the weighted version of the flow-shop problem on two machines cannot be solved in polynomial time, unless $\mathscr{P} = \mathscr{N}\mathscr{P}$.

Theorem 1.3. *The $F2||\sum_{J_j \in E} w_j$ problem is $\mathcal{NP}$-hard.*

Choi and Yoon prove Theorem 1.3 by a reduction from the *Partition* problem, leaving the open question whether the $F2||\sum_{J_j \in E} w_j$ is strongly or ordinarily $\mathcal{NP}$-hard. Shabtay and Bensoussan [32] answer this open question by providing a pseudo-polynomial time list construction algorithm to solve the $F2||\sum_{J_j \in E} w_j$ problem in $O(n^2 P_\Sigma)$ time, where $P_\Sigma = \sum_{i=1}^{n} p_{1i}$. In the following, we present an alternative algorithm which is based on a DP formulation and has the same $O(n^2 P_\Sigma)$ complexity.

Let $F(j,D)$ be the maximum obtainable profit for the *JIT* scheduling of any subset of jobs from $\{J_1,...,J_j\}$, subject to the condition that the last *JIT* job is job J_j and the total processing time of the subset of *JIT* jobs on the first machine is D. Thus, the value $F(j,D)$ can be computed as the maximal profit that can be obtained by extending a partial solution with value $F(i,D-p_{1j})$ for $0 \le i \le j-1$ by adding to it the profit for appending job J_j in *JIT* mode to the corresponding partial schedule:

$$F(j,D) = \max_{i=0,...,j-1} Q(i,j,D), \tag{1.14}$$

where

$$Q(i,j,D) = \begin{cases} F(i,D-p_{1j}) + w_j & \text{if } \max(d_i,D) + p_{2j} \le d_j \\ -\infty & \text{otherwise,} \end{cases} \tag{1.15}$$

for any $p_{1j} \le D \le \sum_{k=1}^{j} p_{1k}$ with $d_0 \overset{\text{def}}{=} 0$. We note that if job J_j is included in a *JIT* set E right after job J_i, that is, job J_j is processed right after job J_i on both machines in a given feasible schedule S, then the second machine will be idle during time interval $[d_i, d_j - p_{2j}]$ in schedule S. Using the recursion defined by (1.14) and (1.15), and the following initial conditions

$$F(0,D) = \begin{cases} 0 & \text{for } D = 0 \\ -\infty & \text{for } D > 0 \end{cases},$$

$$F(j,D) = -\infty, \text{ for } D < p_{1j} \text{ and } D > \sum_{k=1}^{j} p_{1k}, \tag{1.16}$$

all possible state values can be determined. Then, the value of the optimal schedule can be found by

$$F^* = \max_{\substack{j=1,...,n, \\ p_{1j} \le D \le \sum_{k=1}^{j} p_{1k}}} F(j,D), \tag{1.17}$$

and the optimal schedule can be determined by tracing the solution back from the end. In summary, the following optimization algorithm solves the $F2||\sum_{J_j \in E} w_j$ problem in a pseudo-polynomial ($O(n^2 P_\Sigma)$) time.

Algorithm 5: The optimization algorithm for the $F2||\sum_{J_j \in E} w_j$ problem.

Input: n, $\mathbf{d} = (d_1, ..., d_n)$, $\mathbf{w} = (w_1, ..., w_n)$, $\mathbf{P} = (p_{ij})$ for $i = 1, 2$, and $j = 1, ..., n$.
Step 1. Renumber the jobs according to the EDD order.
Step 2. Apply the recursion defined by (1.14) and (1.15) with the initial conditions in (1.16) for $j = 1, ..., n$ and $p_{1j} \leq D \leq \sum_{k=1}^{j} p_{1k}$ with $d_0 \overset{def}{=} 0$.
Step 3. The maximum weighted number of *JIT* jobs is given by (1.17) and the optimal schedule can be determined by tracing back the solution from the end.

Shabtay et al. [33] show that Algorithm 5 can be extended to solve the more general bicriteria $F2|p_{ij} = p_{ij}(u_{ij})|\left(\sum_{J_j \in E} w_j, \sum_{i=1}^{2} \sum_{J_j \in E} c_{ij}(u_{ij})\right)$ problem with controllable processing times. It requires $O(n^3 P_\Sigma C_\Sigma)$ pseudo-polynomial time, where $C_\Sigma = \sum_{i=1}^{2} \sum_{k=1}^{n} c_{ik}(\overline{u}_{ik})$ and $u_{2\max} = \max_{j=1,...,n} u_{2j}$. Moreover, they show that the pseudo-polynomial time algorithm can be converted into an FPTAS to find an approximate Pareto optimal solution in $O\left(n^6/\varepsilon\right)$ time.

In contrast with the above results for various versions of the flow-shop problem on two machines, we mention the following theorem.

Theorem 1.4 ([7]). *The* $F3|| |E|$ *problem is strongly* $\mathcal{NP}$*-hard.*

1.4.2 Job Shops and Open Shops

Shabtay and Bensoussan [32] prove that even the unweighted version of the *JIT* problem is intractable on two machines in both job-shop and open-shop machine environments.

Theorem 1.5 ([32]). *The* $J2|| |E|$ *problem is strongly* $\mathcal{NP}$*-hard.*

Theorem 1.6 ([32]). *The* $O2|| |E|$ *problem is strongly* $\mathcal{NP}$*-hard.*

1.5 Summary

We survey the up-to-date literature on certain *JIT* scheduling problems. The main complexity results for problems with constant processing times are summarized in Table 1.2, while Table 1.3 summarizes the known complexity results for problems with controllable processing times.

Finally, we present some challenging problems for future research in this area:

Table 1.2: Computational complexity results for the JIT scheduling problem with constant processing times.

Problem	Complexity	Reference
$1\|\|\sum_{J_j\in E} w_j$	$O(n^2)$**	[21]
$1\|[\underline{d}_j,\bar{d}_j]\|\,\|E\|$	$\mathcal{NP}$-hard	[13]
$Pm\|\|\,\|E\|$	$O(n\log n)$**	[5]
$Pm\|\|\sum_{J_j\in E} w_j$	$O(mn\log n)$**	[4], [5]
$Rm\|\|\sum_{J_j\in E} w_j$	$O(mn^{m+1})$* and **	[2], [36]
$F2\|\|\sum_{J_j\in E} w_j$	Ordinary $\mathcal{NP}$-hard**	[7] and [32]
$F2\|\|\,\|E\|$	$O(n^3)$**	Theorem 1.2
$F3\|\|\,\|E\|$	strongly $\mathcal{NP}$-hard	[7]
$J2\|\|\,\|E\|$	strongly $\mathcal{NP}$-hard	[7]
$O2\|\|\,\|E\|$	strongly $\mathcal{NP}$-hard	[7]
* Polynomial for a fixed number of machines and strongly $\mathcal{NP}$-hard if the number of machines is a part of the input		
** This result can be extended to solve the more general case in which sequence-dependent setup time is required between each pair of consecutive jobs J_i and J_j		

Table 1.3: Computational complexity results for the JIT scheduling problem with controllable processing times.

Problem	Complexity	Reference
$1\left\|p_{ij}=p_{ij}(u_{ij})\right\|\sum_{J_j\in E} w_j-\sum_{J_j\in E} v_j u_j$	$O(n^2)$*** and ****	[22]
$1\left\|p_{ij}=p_{ij}(u_{ij}),\sum_{J_j\in E} u_j\le U\right\|\sum_{J_j\in E} w_j$	$\mathcal{NP}$-hard****	[22]
$1\|p_{ij}=p_{ij}(u_{ij}),\ \sum_{J_j\in E} w_j\ge K\|\sum_{j=1}^n u_j$	$\mathcal{NP}$-hard****	[22]
$1\|p_{ij}=p_{ij}(u_{ij})\|\left(\sum_{J_j\in E} w_j,\sum_{j=1}^n v_j u_j\right)$	$\mathcal{NP}$-hard****	[22]
$Pm\left\|p_{ij}=p_{ij}(u_{ij})\right\|\sum_{J_j\in E} w_j-\sum_{i=1}^m\sum_{J_j\in E} v_{ij}u_{ij}$	$O(n^4\log n)$*** and ****	[22]
$Rm\left\|p_{ij}=p_{ij}(u_{ij})\right\|\sum_{J_j\in E} w_j-\sum_{i=1}^m\sum_{J_j\in E} v_{ij}u_{ij}$	$O(mn^{m+1})$*, *** and ****	[22]
$Rm\|p_{ij}=p_{ij}(u_{ij}),dscr\|\left(\sum_{J_j\in E} w_j,\sum_{i=1}^m\sum_{J_j\in E} v_{ij}u_{ij}\right)$	$O(mn^{m+1}U_{\max})$** and ***	[22]
$F2\|p_{ij}=p_{ij}(u_{ij})\|\left(\sum_{J_j\in E} w_j,\sum_{i=1}^2\sum_{J_j\in E} c_{ij}(u_{ij})\right)$	$O(n^3P_\Sigma C_\Sigma)$ ***,**** and *****	[33]
* Polynomial for a fixed number of machines and strongly $\mathcal{NP}$-hard if the number of machines is a part of the input		
** Pseudo polynomial for a fixed number of machines and strongly $\mathcal{NP}$-hard if the number of machines is a part of the input		
*** This result can be extended to solve the more general case in which sequence-dependent setup time is required between each pair of consecutive jobs J_i and J_j		
**** The results are applicable for both discrete and continuous type of resource		
***** $P_\Sigma=\sum_{i=1}^2 p_{1j}$, $C_\Sigma=\sum_{i=1}^2\sum_{k=1}^n c_{ik}(\bar{u}_{ik})$ and $u_{2\max}=\max_{j=1,\dots,n} u_{2j}$		

- Determine whether the $Qm||\sum_{J_j \in E} w_j$ and the $Rm|||E|$ problems can be solved in polynomial time if the number of machines is part of the input.
- Find approximation algorithms for the strongly $\mathcal{NP}$-hard problems ($F3|||E|$, $J2|||E|$ and $O2|||E|$ problems).
- Study special cases of the *JIT* scheduling problem, including identical jobs (equal processing times).
- Determine whether the P1-type version of the unweighted two-machine flow-shop problem with controllable processing times is solvable in polynomial time.

References

1. Alidaee, B., Ahmadian, A.: Two Parallel Machine Sequencing Problems Involving Controllable Job Processing Times. European Journal of Operational Research **70**, 335-341 (1993)
2. Arkin, E.M., Silverberg, E.L.: Scheduling Jobs with Fixed Start and Finish Times. Discrete Applied Mathematics **18**, 1-8 (1987)
3. Baker, K.R., Scudder, G.D.: Sequencing with Earliness and Tardiness Penalties: A Review. Operations Research **38**, 22-36 (1990)
4. Bouzina, K.I., Emmons, H.: Interval Scheduling on Identical Machines. Journal of Global Optimization **9**, 379-393 (1996)
5. Carlisle, M.C., Lloyd, E.L.: On the *k*-coloring of Intervals. Discrete Applied Mathematics **59**, 225-235 (1995)
6. Cheng, T.C.E., Janiak, A., Kovalyov, M.Y.: Bicriterion Single Machine Scheduling with Resource Dependent Processing Times. SIAM Journal on Optimization **8**(2), 617-630 (1998)
7. Choi, B.C., Yoon, S.H.: Maximizing the Weighted Number of Just-in-Time Jobs in Flow Shop Scheduling. Journal of Scheduling **10**, 237-243 (2007)
8. Čepek, O., Sung, S. C.: A Quadratic Time Algorithm to Maximize the Number of Just-in-Time Jobs on Identical Parallel Machines. Computers and Operations Research **32**, 3265-3271 (2005)
9. Chudzik, K., Janiak, A., Lichtenstein, M.: Scheduling Problems with Resource Allocation. In A. Janiak, ed., Scheduling in Computer and Manufacturing Systems, WK L, Warszawa (2006)
10. Frank, A.: On Chains and Antichains Families of a Partially Ordered Set. Journal of Combinatorial Theory Series B **29**, 176-184 (1980)
11. Graham, R.L., Lawler, E.L., Lenstra, J.K., Rinnooy Kan, A.H.G.: Optimization and Approximation in Deterministic Sequencing and Scheduling: A Survey. Annals Discrete Mathematics **5**, 287-326 (1979)
12. Hsiao, J.Y., Tang, C.Y., Chang, R.S.: An Efficient Algorithm for Finding a Maximum Weight 2-independent Set of Interval Graphs. Information Processing Letters **43**, 229-235 (1992)
13. Hiraishi, K., Levner, E., Vlach, M.: Scheduling of Parallel Identical Machines to Maximize the Weighted Number of Just-in-Time Jobs. Computers and Operations Research **29**(7), 841-848 (2002)
14. Janiak, A.: One-Machine Scheduling with Allocation of Continuously-Divisible Resource and with No Precedence Constraints. Kybernetika **23**(4), 289-293 (1987)
15. Janiak, A., Kovalyov, M.Y.: Single Machine Scheduling Subject to Deadlines and Resource Dependent Processing Times. European Journal of Operational Research **94**, 284-291 (1996)
16. Janiak, A., Janiak, W., Lichtenstein, M.: Resource Management in Machine Scheduling Problems: A Survey. Decision Making in Manufacturing and Services **1**, 59-89 (2007)
17. Józefowska, J.: Just-in-time Scheduling: Models and Algorithms for Computer and Manufacturing Systems, Springer-Verlag New York Inc (2007)
18. Kaspi, M., Shabtay, D.: Optimization of Machining Economics Problem for a Multi-Stage Transfer Machine Under Failure, Opportunistic and Integrated Replacement Strategies. International Journal of Production Research **41**(10), 2229-2248 (2003)

19. Kayan, R.K., Akturk, M.S.: A New Bounding Mechanism for the CNC Machine Scheduling Problem with Controllable Processing Times. European Journal of Operational Research **167**, 624-643 (2005)
20. Kovalyov, M.Y., Ng, C.T., Cheng, T.C.E.: Fixed Interval Scheduling: Models, Applications, Computational Complexity and Algorithms. European Journal of Operational Research **178**, 331-342 (2007)
21. Lann, A., Mosheiov, G.: Single Machine Scheduling to Minimize the Number of Early and Tardy Jobs. Computers and Operations Research **23**, 765-781 (1996)
22. Leyvand Y., Shabtay, D., Steiner, G., Yedidsion, L.: Just-in-Time Scheduling with Controllable Processing Times on Parallel Machines. Journal of Combinatorial Optimization **19**(3), 347-368 (2010)
23. Ng, C.T.D., Cheng, T.C.E., Janiak, A., Kovalyov, M.Y.: Group Scheduling with Controllable Setup and Processing Times: Minimizing Total Weighted Completion Time. Annals of Operations Research **133**, 163-174 (2005)
24. Nowicki, E., Zdrzalka, S.: A Survey of Results for Sequencing Problems with Controllable Processing Times. Discrete Applied Mathematics **26**, 271-287 (1990)
25. Pal, M., Bhattacharjee, G.P.: A Sequential Algorithm for Finding a Maximum Weight K-independent Set on Interval Graphs. International Journal of Computer Mathematics **60**, 205-214 (1996)
26. Saha, A., Pal, M.: Maximum Weight k-independent Set Problem on Permutation Graphs. International Journal of Computer Mathematics **80**, 1477-1487 (2003)
27. Sarrafzadeh, M., Lou, R.D.: Maximum k-Covering of Weighted Transitive Graph with Applications. Algorithmica **9**, 84-100 (1993)
28. Shabtay, D.: Single and a Two-Resource Allocation Algorithms for Minimizing the Maximal Lateness in a Single Machine-Scheduling Problem. Computers and Operations Research, **31**(8), 1303-1315 (2004)
29. Shabtay, D., Kaspi, M.: Minimizing the Total Weighted Flow Time in a Single Machine with Controllable Processing Times. Computers and Operations Research **31**(13), 2279-2289 (2004)
30. Shabtay, D., Kaspi, M., Steiner, G.: The No-Wait Two-Machine Flow-Shop Scheduling Problem with Convex Resource-Dependent Processing Times. IIE Transactions **39**(5), 539–557 (2007)
31. Shabtay, D., Steiner, G.: A Survey of Scheduling with Controllable Processing Times. Discrete Applied Mathematics **155**(13), 1643-1666 (2007)
32. Shabtay, D., Bensoussan, Y.: Maximizing the Weighted Number of Just-in-Time Jobs in a Two Machine Flow and Open Shop Scheduling Systems. Journal of Scheduling, DOI: 10.1007/s10951-010-0204-y
33. Shabtay, D., Bensusan, Y., Kaspi, M.: A Bicriteria Approach to Maximize the Weighted Number of Just-In-Time Jobs and to Minimize the Total Resource Consumption Cost in a Two-Machine Flow-Shop Scheduling System, International Journal of Production Economics, to appear
34. Shakhlevich, N.V., Strusevich, V.A.: Pre-emptive Scheduling Problems with Controllable Processing Times. Journal of Scheduling **8**, 233-253 (2005)
35. Steiner, G.: Finding the Largest Suborder of Fixed Width. Order **9**, 357-360 (1992)
36. Sung, S.C., Vlach, M.:Maximizing Weighted Number of Just-in-Time Jobs on Unrelated Parallel Machines, Journal of Scheduling **8**, 453-460 (2005)
37. Trick, M.: Scheduling Multiple Variable-Speed Machines. Operations Research **42**, 234-248 (1994)
38. Vickson, R.G.: Two Single Machine Sequencing Problems Involving Controllable Job Processing Times. AIIE Transactions **12**(3), 258-262 (1980)
39. Yannakakis, M., Gavril, F.: The Maximum k-Colorable Subgraph Problem for Chordal Graphs. Information Processing Letters **24**, 133-137 (1987)

Chapter 2
An Exact Algorithm for the Single-Machine Earliness–Tardiness Scheduling Problem

Shunji Tanaka

Abstract This paper introduces our exact algorithm for the single-machine total weighted earliness–tardiness scheduling problem, which is based on the Successive Sublimation Dynamic Programming (SSDP) method. This algorithm starts from a Lagrangian relaxation of the original problem and then constraints are successively added to it until the gap between lower and upper bounds becomes zero. The relaxations are solved by dynamic programming, and unnecessary dynamic programming states are eliminated in the course of the algorithm to suppress the increase of states caused by the addition of constraints. This paper explains the methods employed in our algorithm to construct the Lagrangian relaxations, to eliminate states and to compute an upper bound together with some other improvements. Then, numerical results for known benchmark instances are given to show the effectiveness of our algorithm.

2.1 Introduction

This paper introduces our exact algorithm [29] for the single-machine total weighted earliness–tardiness scheduling problem. Let us consider that n jobs (job $1,\ldots,$ job n) are to be processed on a single machine that can process at most one job at a time. No preemption is allowed and once the machine starts processing a job, it cannot be interrupted. After the machine finishes processing a job, it can be idle even when there exist unprocessed jobs. Each job i is given a processing time p_i, due date d_i and release date r_i, where $d_i \geq r_i + p_i$. It is also given a earliness weight α_i and a tardiness weight β_i. The earliness E_i and the tardiness T_i are defined by

$$E_i = \max(d_i - C_i, 0), \quad T_i = \max(C_i - d_i, 0), \tag{2.1}$$

Shunji Tanaka
Department of Electrical Engineering, Kyoto University, Kyotodaigaku-Katsura, Nishikyo-ku, Kyoto 615-8510, Japan, e-mail: tanaka@kuee.kyoto-u.ac.jp

R.Z. Ríos-Mercado and Y.A. Ríos-Solís (eds.), *Just-in-Time Systems*, Springer Optimization and Its Applications 60, DOI 10.1007/978-1-4614-1123-9_2,

where C_i denotes the completion time of job i. Our objective is to find a schedule that minimizes $\sum_{1 \le i \le n}(\alpha_i E_i + \beta_i T_i)$.

This problem, the single-machine total weighted earliness–tardiness problem ($1||\sum(\alpha_i E_i + \beta_i T_i)$ and $1|r_i|\sum(\alpha_i E_i + \beta_i T_i)$ according to the standard classification of scheduling problems [11]) includes the single-machine total weighted tardiness problem ($1||\sum w_i T_i$) as a special class[1] that appears frequently in the literature as a typical strongly NP-hard scheduling problem [17, 18]. Therefore, it is also strongly NP-hard. Nonetheless, many researchers have tackled this problem and constructed exact algorithms [3,6,7,10,13,16,22,25,26,33,34] because of its importance in JIT scheduling. Almost all the existing algorithms are branch-and-bound algorithms. To the best of the author's knowledge, the best algorithm so far is that by Sourd [26] and it can solve instances with up to 50 or 60 jobs within 1,000 s.

On the other hand, our exact algorithm proposed in [29] is based on the Successive Sublimation Dynamic Programming (SSDP) method [14]. The SSDP method is a dynamic-programming-based exact algorithm that starts from a relaxation of the original problem and then repeats the following procedure until the gap between lower and upper bounds becomes zero:

1. The relaxation is solved by dynamic programming and a lower bound is computed.
2. Unnecessary states are eliminated.
3. A relaxation with more detailed information of the original problem, which is referred to as sublimation, is constructed.

In [15], the SSDP method was applied to $1||\sum(\alpha_i E_i + \beta_i T_i)$ without idle time by utilizing the relaxations proposed in [1] as sublimations. Their algorithm could solve instances with up to 35 jobs and hence was better than the branch-and-bound algorithm in [1] that could solve only those with up to 25 jobs. However, they concluded that it is not easy to apply the algorithm to larger instances because of its heavy memory usage for storing dynamic programming states.

Things have changed much from those days and more powerful computers with several gigabytes memory are now readily accessible. It motivated us to improve and extend their algorithm for the total weighted earliness–tardiness problem. First, we showed in [28] that their algorithm with several significant improvements can solve instances with up to 300 jobs of $1||\sum w_i T_i$ and $1||\sum(\alpha_i E_i + \beta_i T_i)$ without idle time. Next, in [29], we extended the algorithm to the problem with idle time and showed that it can solve benchmark instances of $1||\sum(\alpha_i E_i + \beta_i T_i)$ and $1|r_i|\sum(\alpha_i E_i + \beta_i T_i)$ with even 200 jobs. Moreover, the framework is not restricted to these problems, but is applicable to a wide class of single-machine scheduling problems.

The purpose of this paper is to introduce this algorithm, but only its essential part will be given because of limited space. For details, please refer to our papers [28, 29]. Up-to-date information will be available on the web page:

`http://turbine.kuee.kyoto-u.ac.jp/~tanaka/SiPS/.`

---

[1] When α_i are chosen as zero, the total weighted earliness-tardiness problem is reduced to the total weighted tardiness problem.

This paper is organized as follows. In Sect. 2.2, our problem will be given as so-called a time-indexed formulation and then its network representation will be introduced. In Sect. 2.3, Lagrangian relaxations of the problem, which are utilized as the relaxation and sublimations in our algorithm (3. in the SSDP method), will be expressed in terms of the network representation. Dynamic programming recursions to solve them (1. in the SSDP method) will also be explained. Next, Sect. 2.4 will state how to reduce the networks corresponding to the Lagrangian relaxations (2. in the SSDP method). Then, Sect. 2.5 will summarize our algorithm. Section 2.6 will describe how to obtain a tight upper bound from a solution of the Lagrangian relaxation in this algorithm. Section 2.7 will present results of numerical experiments and show the effectiveness of our algorithm. Finally, in Sect. 2.8, the contents of this paper and future research directions will be summarized.

2.2 Problem Formulation

This section will give a time-indexed formulation [8, 21, 27, 31] of our problem and its network representation. It is well-known that a tight lower bound can be obtained by a linear programming (LP) relaxation of a time-indexed formulation. However, there is a major drawback that the number of decision variables is very large. In our algorithm, an alternative way to apply the Lagrangian relaxation technique is taken to obtain a lower bound, which will be explained in the next section by the network representation.

2.2.1 *Time-Indexed Formulation*

Let us assume that the processing times p_i, duedates d_i and release dates r_i are all integral. Then, we can assume without loss of optimality that the job completion times C_i are also integral. Thus, we introduce binary decision variables x_{it} ($1 \le i \le n$, $1 \le t \le T_E$) such that

$$x_{it} = \begin{cases} 1 & \text{if } C_i = t, \\ 0 & \text{otherwise,} \end{cases} \tag{2.2}$$

where T_E is a scheduling horizon and can be chosen, e.g., as

$$T_E = \max_{1 \le i \le n} d_i + \sum_{1 \le i \le n} p_i. \tag{2.3}$$

We also introduce a cost function $f_i(t)$ ($1 \le i \le n$, $1 \le t \le T_E$) defined by

$$f_i(t) = \begin{cases} \infty & t < r_i + p_i, \\ \max\left(\alpha_i(d_i - t), \beta_i(t - d_i)\right) & t \ge r_i + p_i. \end{cases} \tag{2.4}$$

Then, the problem can be formulated as follows:

$$(\mathrm{P}):\min_{\mathbf{x}} \sum_{\substack{1\le i\le n\\ 1\le t\le T_{\mathrm{E}}}} f_i(t)x_{it}, \tag{2.5}$$

$$\text{s.t.} \sum_{\substack{1\le i\le n\\ t\le s\le \min(t+p_i-1,T_{\mathrm{E}})}} x_{is} \le 1, \quad 1\le t\le T_{\mathrm{E}}, \tag{2.6}$$

$$\sum_{1\le t\le T_{\mathrm{E}}} x_{it} = 1, \quad 1\le i\le n, \tag{2.7}$$

$$x_{it}\in\{0,1\}, \quad 1\le i\le n,\ 1\le t\le T_{\mathrm{E}}. \tag{2.8}$$

In (P), the constraints (2.6) require that at most one job can be processed in the interval $[t-1,t)$, and the constraints (2.7) ensure that each job is processed exactly once. This type of formulation is referred to as time-indexed formulation.

For convenience hereafter, let us assume that a dummy job is processed when no (ordinary) job is processed and the machine is idle. For this purpose, job 0 with $p_0=1$ and $f_0(t)=0\,(1\le t\le T_{\mathrm{E}})$ is introduced. This job, which is referred to as idle job, is processed $(T_{\mathrm{E}}-\sum_{1\le i\le n} p_i)$ times because it is the total length of idle time. By using the idle job, the problem (P) can be reformulated as follows.

$$(\mathrm{P}'):\min_{\mathbf{x}} \sum_{\substack{0\le i\le n\\ 1\le t\le T_{\mathrm{E}}}} f_i(t)x_{it}, \tag{2.9}$$

$$\text{s.t.} \sum_{\substack{0\le i\le n\\ t\le s\le \min(t+p_i-1,T_{\mathrm{E}})}} x_{is} = 1, \quad 1\le t\le T_{\mathrm{E}}, \tag{2.10}$$

$$\sum_{1\le t\le T_{\mathrm{E}}} x_{it} = 1, \quad 1\le i\le n, \tag{2.11}$$

$$x_{it}\in\{0,1\}, \quad 0\le i\le n,\ 1\le t\le T_{\mathrm{E}}. \tag{2.12}$$

In this formulation, the constraint on the number of occurrences is not imposed on the idle job because it is automatically satisfied. Please also note that the inequality constraints (2.6) in (P) become the equality constraints (2.10) in (P').

2.2.2 Network Representation

The problem (P') formulated in the preceding subsection can be converted to a constrained shortest path problem on a network. In this subsection, this network representation will be introduced as a preparation for the next section.

To construct the network, that is, an acyclic weighted directed graph $G=(V,A)$, we allocate a node to every decision variable x_{it}. More specifically, the node set V is defined by

$$V=\{v_{n+1,0}\}\cup V_{\mathrm{O}}\cup\{v_{n+1,T+1}\}, \tag{2.13}$$

$$V_{\mathrm{O}}=\{v_{it}\,|\,0\le i\le n,\ 1\le t\le T_{\mathrm{E}}\}. \tag{2.14}$$

Here, another dummy job $n+1$ with $p_{n+1} = 1$, $f_{n+1}(t) = 0$ is introduced. This job is assumed to be completed at 0 and $T_E + 1$, and $v_{n+1,0}$ and $v_{n+1,T+1}$ denote the source and sink nodes, respectively. The arc set A is defined by

$$A = \{(v_{j,t-p_i}, v_{it}) \,|\, v_{j,t-p_i}, v_{it} \in V\}. \tag{2.15}$$

In addition, the length (weight) of an arc $(v_{j,t-p_i}, v_{it}) \in A$ is given by $f_i(t)$.

Let us consider a path from $v_{n+1,0}$ to $v_{n+1,T+1}$ on this network. Then, its length is equal to the objective value (2.9) of (P') if we choose $x_{it} = 1$ when v_{it} $(0 \le i \le n)$ is on the path. These decision variables satisfy the resource constraints (2.10) but do not always satisfy the constraints (2.11). To satisfy (2.11), v_{it} $(1 \le t \le T_E)$ should be visited exactly once for any i $(1 \le i \le n)$. Therefore, (P') is equivalent to the problem to find a shortest path from $v_{n+1,0}$ to $v_{n+1,T+1}$ on G under the constraints that v_{it} $(1 \le t \le T_E)$ should be visited exactly once for any i $(1 \le i \le n)$.

Here, some notation and definitions are introduced. Let us denote by $\mathscr{P}$ a set of nodes visited by a path from $v_{n+1,0}$ to $v_{n+1,T+1}$ on G. A node set $\mathscr{P}$ and the corresponding path are both referred to as "path $\mathscr{P}$" unless there is ambiguity. Let $L(\mathscr{P})$ be the length of a path $\mathscr{P}$ defined by

$$L(\mathscr{P}) = \sum_{\substack{v_{it} \in \mathscr{P} \\ 0 \le i \le n}} f_i(t) = \sum_{\substack{v_{it} \in \mathscr{P} \\ 1 \le i \le n}} f_i(t). \tag{2.16}$$

We also define by $\mathscr{V}_i(\mathscr{P})$ $(1 \le i \le n)$ the number of occurrences of v_{it} $(1 \le t \le T_E)$ in $\mathscr{P}$. That is,

$$\mathscr{V}_i(\mathscr{P}) = |\{v_{it} \,|\, v_{it} \in \mathscr{P}\}|. \tag{2.17}$$

Then, the constraints that v_{it} $(1 \le t \le T_E)$ should be visited exactly once on a path $\mathscr{P}$ for any i $(1 \le i \le n)$ can be written by

$$\mathscr{V}_i(\mathscr{P}) = 1, \quad 1 \le i \le n. \tag{2.18}$$

Accordingly, define a set of all the feasible paths by

$$\mathscr{Q} = \{\mathscr{P} \,|\, \mathscr{V}_i(\mathscr{P}) = 1,\ 1 \le i \le n\}. \tag{2.19}$$

Then, our problem on the network, which is referred to as (N), can be described simply by

$$(\mathrm{N}): \ \min_{\mathscr{P}} L(\mathscr{P}) \quad \text{s.t. } \mathscr{P} \in \mathscr{Q}. \tag{2.20}$$

2.3 Lagrangian Relaxation

As already mentioned in the preceding section, the LP relaxation of (P') obtained by removing the integrity constraints (2.12) yields a tight lower bound. However, it is not easy to solve when the number of jobs n becomes large because (P') has $O(nT_{\mathrm{E}})$ decision variables. To avoid this difficulty, the Lagrangian relaxation technique is employed instead. There are two types of relaxations for (P'), that is, the relaxations of (2.10) and (2.11), respectively. One of the advantages of the former relaxation is that (P') can be decomposed into trivial n subproblems corresponding to the n jobs. Therefore, it is sometimes referred to as Lagrangian decomposition. There is an early attempt by Fisher [9] to apply this relaxation for lower bound computation in a branch-and-bound algorithm. On the other hand, the primary advantage of the latter relaxation is that it gives an easy way to obtain a tighter lower bound than that by the LP relaxation. It is also referred to as state-space relaxation, which originates in the study by Christofides et al. [4] for routing problems. It was first applied to single-machine scheduling by Abdul-Razaq and Potts [1] and following their study, Ibaraki and Nakamura [15] proposed an exact algorithm based on the SSDP method [14]. Our exact algorithm also utilizes this type of relaxation. It also appears in the context of the column generation approach [32], or branch-and-bound algorithms [19,26] for single-machine scheduling problems.

In the following subsections, the Lagrangian relaxation of (2.11) in (P') will be explained for its counterpart (N) in the network representation. Then, three types of constraints will be introduced and imposed on it to improve the lower bound.

2.3.1 *Lagrangian Relaxation of the Number of Job Occurrences*

To begin with, the violation of the constraints on the number of job occurrences (2.11) in (P') are penalized by Lagrangian multipliers μ_i $(1 \leq i \leq n)$. It corresponds to the relaxation of the constraints (2.18) in (N), and its objective function $L(\mathscr{P})$ becomes

$$\begin{aligned}
L(\mathscr{P}) &+ \sum_{1\leq i\leq n} \mu_i(1-\mathscr{V}_i(\mathscr{P})) \\
&= \sum_{\substack{v_{it}\in\mathscr{P} \\ 1\leq i\leq n}} f_i(t) + \sum_{1\leq i\leq n} \mu_i - \sum_{1\leq i\leq n} \mu_i |\{v_{it} \mid v_{it} \in \mathscr{P}\}| \\
&= \sum_{\substack{v_{it}\in\mathscr{P} \\ 1\leq i\leq n}} (f_i(t)-\mu_i) + \sum_{1\leq i\leq n} \mu_i \\
&= L_{\mathrm{R}}(\mathscr{P};\,\mu) + \sum_{1\leq i\leq n} \mu_i, \qquad (2.21)
\end{aligned}$$

where $L_{\mathrm{R}}(\mathscr{P};\ \mu)$ is defined by

$$L_{\mathrm{R}}(\mathscr{P};\ \mu) = \sum_{\substack{v_{it} \in \mathscr{P} \\ 1 \le i \le n}} (f_i(t) - \mu_i). \tag{2.22}$$

Equations (2.21) and (2.22) imply that the relaxation for a fixed set of multipliers is equivalent to the problem to find a shortest unconstrained path from $v_{n+1,0}$ to $v_{n+1,T+1}$ on G where the length of an arc $(v_{j,t-p_i}, v_{it}) \in A$ is given not by $f_i(t)$ but by $f_i(t) - \mu_i$ (we assume that $\mu_0 = \mu_{n+1} = 0$). This relaxation is denoted by (LR$_0$), that is,

$$(\mathrm{LR}_0):\ \min_{\mathscr{P}} L_{\mathrm{R}}(\mathscr{P};\ \mu) + \sum_{1 \le i \le n} \mu_i. \tag{2.23}$$

Clearly,

$$\min_{\mathscr{P}} L_{\mathrm{R}}(\mathscr{P};\ \mu) + \sum_{1 \le i \le n} \mu_i \le \min_{\mathscr{P} \in \mathscr{Q}} L(\mathscr{P}) \tag{2.24}$$

holds and (LR$_0$) gives a lower bound of the original problem (N).

The relaxation (LR$_0$) is easy to solve in $O(nT_{\mathrm{E}})$ time by dynamic programming [1]. If we denote the partial path from $v_{n+1,0}$ to v_{it} by $\mathscr{P}_{\mathrm{P}}(v_{n+1,0}, v_{it})$, the forward dynamic programming recursion is expressed by

$$\min_{\mathscr{P}} L_{\mathrm{R}}(\mathscr{P};\ \mu) = h_0(T_{\mathrm{E}} + 1;\ \mu), \tag{2.25}$$

$$\begin{aligned} h_0(t;\ \mu) &= \min_{v_{it} \in V} L_{\mathrm{R}}(\mathscr{P}_{\mathrm{P}}(v_{n+1,0}, v_{it});\ \mu) \\ &= \min_{v_{it} \in V} (h_0(t - p_i;\ \mu) + f_i(t) - \mu_i). \end{aligned} \tag{2.26}$$

We can also formulate the backward dynamic programming recursion in a similar manner.

To improve the lower bound more, the following three types of constraints are imposed on this relaxation. The first and the third were proposed in [4] and were applied in [1,15]. The second constraints were proposed in [26,28], and were applied in our previous algorithm [28] together with the other two.

2.3.2 Constraints on Successive Jobs

The first constraints are to forbid job duplication in successive jobs of a solution. In the network representation, these are interpreted as constraints on successively visited nodes on a path. More specifically, they are described as follows.

> For any i ($1 \le i \le n$), nodes corresponding to job i, that is, v_{it} ($1 \le t \le T_{\mathrm{E}}$) should not be visited more than once in any $\lambda + 1 > 0$ successive nodes on a path.

Therefore, these constraints forbid $v_{i,t-p_i} \to v_{it}$ when $\lambda = 1$ and $v_{i,t-p_i-p_j} \to v_{j,t-p_i} \to v_{it}$ when $\lambda = 2$, and so on. A subset of paths satisfying these constraints is denoted by $\mathscr{Q}_\lambda$ ($\mathscr{Q} \subseteq \cdots \subseteq \mathscr{Q}_2 \subseteq \mathscr{Q}_1$), and the relaxation with the constraints is denoted by

$$(\mathrm{LR}_\lambda): \min_{\mathscr{P}} L_{\mathrm{R}}(\mathscr{P};\, \mu) + \sum_{1\le i\le n} \mu_i \quad \text{s.t. } \mathscr{P} \in \mathscr{Q}_\lambda. \tag{2.27}$$

Clearly, (LR_λ) gives a better (or at least not worse) lower bound as λ increases. However, the time complexity also increases because it is given by $O(n^\lambda T_{\mathrm{E}})$ [1, 19].

It might not be intuitive that the time complexity of (LR_1) is $O(nT_{\mathrm{E}})$ because that of (LR_0) is also $O(nT_{\mathrm{E}})$, but the following recursion confirms this fact:

$$\min_{\mathscr{P}\in\mathscr{Q}_1} L_{\mathrm{R}}(\mathscr{P};\, \mu) = y_1(T_{\mathrm{E}}+1;\, \mu), \tag{2.28}$$

$$y_1(t;\, \mu) = \min_{v_{it}\in V} h_1(v_{it};\, \mu), \tag{2.29}$$

$$m_1(t;\, \mu) = \arg\min_{\substack{i \\ v_{it}\in V}} h_1(v_{it};\, \mu), \tag{2.30}$$

$$z_1(t;\, \mu) = \min_{\substack{v_{it}\in V \\ i\neq m_1(t;\,\mu)}} h_1(v_{it};\, \mu), \tag{2.31}$$

$$h_1(v_{it};\, \mu) = \begin{cases} y_1(t-p_i;\, \mu) + f_i(t) - \mu_i & \text{if } i=0 \text{ or } i \neq m_1(t-p_i;\, \mu), \\ z_1(t-p_i;\, \mu) + f_i(t) - \mu_i & \text{otherwise}, \end{cases} \tag{2.32}$$

$$y_1(0;\, \mu) = 0, \quad z_1(0;\, \mu) = +\infty, \quad m_1(0;\, \mu) = n+1. \tag{2.33}$$

Roughly speaking, the shortest path $y_1(t;\, \mu)$ and the second shortest path $z_1(t;\, \mu)$ among those from $v_{n+1,0}$ to v_{it} are stored in the above recursion. When the path is expanded ($h_1(v_{it};\, \mu)$), the shortest path is used if no successive job duplication occurs and otherwise the second shortest is used. It can be shown in a similar way that the time complexity of (LR_λ) for $\lambda \ge 2$ is $O(n^\lambda T_{\mathrm{E}})$.

In our algorithm, only (LR_1) and (LR_2) are considered. The relaxation (LR_1) becomes more tractable if we introduce a subnetwork $G_{\mathrm{S}} = (V, A_{\mathrm{S}})$, where A_{S} is defined by

$$A_{\mathrm{S}} = A \backslash \{(v_{i,t-p_i}, v_{it}) \,|\, v_{i,t-p_i}, v_{it} \in V_{\mathrm{O}},\ 1 \le i \le n\}, \tag{2.34}$$

and the length of an arc $(v_{j,t-p_i}, v_{it}) \in A_{\mathrm{S}}$ is given by $f_i(t) - \mu_i$. Indeed, (LR_1) is equivalent to the unconstrained shortest path problem on G_{S}. On the other hand, (LR_2) is equivalent to the constrained shortest path problem even on G_{S}, under the constraints on three successive nodes.

2.3.3 Constraints on Adjacent Pairs of Jobs

The second constraints derive from the dominance theorem of dynamic programming [20] for adjacent pairs of jobs. For example, consider that two jobs i and j ($0 \leq i, j \leq n$, $i \neq j$) are successively processed and completed at t. The total completion cost of the two jobs is $f_i(t - p_j) + f_j(t)$ when they are sequenced as $i \to j$, and $f_j(t - p_i) + f_i(t)$ when $j \to i$. It follows that $i \to j$ never occurs at t in an optimal solution if $f_i(t - p_j) + f_j(t) > f_j(t - p_i) + f_i(t)$ because interchanging these jobs decreases the objective value without affecting the other jobs. On the other hand, $j \to i$ never occurs at t if $f_i(t - p_j) + f_j(t) < f_j(t - p_i) + f_i(t)$. Therefore, the processing order of jobs i and j at any t can be restricted by checking the total cost of the two. This also holds even if $f_i(t - p_j) + f_j(t) = f_j(t - p_i) + f_i(t)$, and either (but not arbitrary) processing order can be forbidden without loss of optimality [28]. To summarize, the processing order of adjacent pairs of jobs can be restricted and it is imposed on the relaxation as constraints. Please note that the processing order of an ordinary job and the idle job can also be restricted.

In the network representation, these adjacency constraints eliminate from G_S, those arcs corresponding to the forbidden processing orders. Thus, we define a subnetwork $\widehat{G}_S = (V, \widehat{A}_S)$ of G_S, where

$$\widehat{A}_S = A_S \setminus \{(v_{j,t-p_i}, v_{it}) \mid j \to i \text{ is forbidden at } t\}. \tag{2.35}$$

The relaxations (LR_1) and (LR_2) with the adjacency constraints are equivalent to the unconstrained and constrained shortest path problems on $\widehat{G}_S$, respectively. Since the time complexities of (LR_1) and (LR_2) with the adjacency constraints are both $O(n^2 T_E)$ [26,28], only (LR_2) with the adjacency constraints, which is, denoted by ($\widehat{\text{LR}}_2$), is used as in our previous algorithm. Let $\widehat{\mathscr{Q}}_2$ denote a subset of $\mathscr{Q}_2$ composed of paths on $\widehat{G}_S$ that satisfy the constraints on three successive nodes.

2.3.4 Constraints on State-Space Modifiers

The last constraints are described in terms of state-space modifiers: Each ordinary job i ($1 \leq i \leq n$) is given a value $q_i \geq 0$ called state-space modifier and the constraint that the total modifier in a solution should be $\sum_{1 \leq i \leq n} q_i$ is imposed on ($\widehat{\text{LR}}_2$). In our algorithm, the modifiers are chosen so that $q_i = 1$ for some i and $q_j = 0$ for $j \neq i$ ($1 \leq j \leq n$). In this case, the constraint simply requires that job i should be processed exactly once and hence is equivalent to $\mathscr{V}_i(\mathscr{P}) = 1$, that is, the constraint (2.18) for job i. It follows that all the constraints (2.18) are once relaxed, but one of them is recovered to improve the lower bound.

Let us consider that not the constraint (2.18) for a single job i but those for a subset of jobs $\mathscr{M}$ are recovered to ($\widehat{\text{LR}}_2$). Hereafter, ($\widehat{\text{LR}}_2$) with the constraints

$$\mathscr{V}_i(\mathscr{P}) = 1, \quad \forall i \in \mathscr{M} \tag{2.36}$$

is denoted by $(\widehat{\mathrm{LR}}_2^m)$, where $m = |\mathscr{M}|$. Clearly, an optimal solution of $(\widehat{\mathrm{LR}}_2^m)$ is also optimal for the original problem (N) when $m = n$.

The network representation of $(\widehat{\mathrm{LR}}_2^m)$ is a little complicated. Let us define an m-dimensional vector $\mathbf{q}_i^m$ of state-space modifiers for job i by $\mathbf{q}_i^m = (q_{i1}, \ldots, q_{im})$, where

$$q_{ij} = \begin{cases} 1, & \text{if the } j\text{th element of } \mathscr{M} \text{ is } i, \\ 0, & \text{otherwise.} \end{cases} \tag{2.37}$$

Let us also define m-dimensional vectors $\mathbf{q}_0^m$ and $\mathbf{q}_{n+1}^m$ by $\mathbf{q}_0^m = \mathbf{q}_{n+1}^m = (0, \ldots, 0)$. Next, a weighted directed graph $\widehat{G}_\mathrm{S}^m = (V^m, \widehat{A}_\mathrm{S}^m)$ is introduced. The node set V^m is defined by

$$V^m = \{v_{n+1,0}^{\mathbf{0}_m}\} \cup V_\mathrm{O}^m \cup \{v_{n+1,T+1}^{\mathbf{1}_m}\}, \tag{2.38}$$

$$V_\mathrm{O}^m = \{v_{it}^{\mathbf{b}} \mid v_{it} \in V_\mathrm{O},\ \mathbf{q}_i^m \le \mathbf{b} \le \mathbf{1}_m\}, \tag{2.39}$$

where $\mathbf{0}_m$ and $\mathbf{1}_m$ denote m-dimensional vectors whose elements are all zero and all one, respectively. The arc set $\widehat{A}_\mathrm{S}^m$ is defined by

$$\widehat{A}_\mathrm{S}^m = \{(v_{j,t-p_i}^{\mathbf{b}-\mathbf{q}_i^m}, v_{it}^{\mathbf{b}}) \mid (v_{j,t-p_i}, v_{it}) \in \widehat{A}_\mathrm{S},\ \mathbf{q}_i^m + \mathbf{q}_j^m \le \mathbf{b} \le \mathbf{1}_m\}, \tag{2.40}$$

and the length of an arc $(v_{j,t-p_i}^{\mathbf{b}-\mathbf{q}_i^m}, v_{it}^{\mathbf{b}})$ is given by $f_i(t) - \mu_i$. Then, $(\widehat{\mathrm{LR}}_2^m)$ is equivalent to the shortest path problem from $v_{n+1,0}^{\mathbf{0}_m}$ to $v_{n+1,T+1}^{\mathbf{1}_m}$ on $\widehat{G}_\mathrm{S}^m$ under the constraints on three successive nodes. The set of paths from $v_{n+1,0}^{\mathbf{0}_m}$ to $v_{n+1,T+1}^{\mathbf{1}_m}$ on $\widehat{G}_\mathrm{S}^m$ that satisfy the constraints on three successive nodes is denoted by $\widehat{\mathscr{Q}}_2^m$. $(\widehat{\mathrm{LR}}_2^m)$ is solvable by dynamic programming in $O(n^2 2^m T_\mathrm{E})$ time.

2.4 Network Reduction

Our exact algorithm utilizes $(\widehat{\mathrm{LR}}_2)$ and $(\widehat{\mathrm{LR}}_2^m)$ as sublimations of (LR_1). More specifically, it first solves (LR_1), next $(\widehat{\mathrm{LR}}_2)$ and then $(\widehat{\mathrm{LR}}_2^m)$ with jobs added to $\mathscr{M}$ (with m increased). As already explained in the preceding section, all these are solvable by dynamic programming. In the SSDP method, unnecessary dynamic programming states are eliminated in the course of the algorithm. The efficiency of this state elimination determines the efficiency of the SSDP method and hence is very important because it enables us to reduce both computational efforts and memory usage.

The state elimination is interpreted as the removal of unnecessary nodes and arcs from G_S, $\widehat{G}_\mathrm{S}$ and $\widehat{G}_\mathrm{S}^m$ in the network representation. This section will give two types of network reductions utilized in our algorithm.

2.4.1 Network Reduction by Upper Bound

The first network reduction [15] utilizes an upper bound and is applied to all the relaxations (LR_1), ($\widehat{\text{LR}}_2$) and ($\widehat{\text{LR}}_2^m$). Here, only the reduction for (LR_1) will be described because it does not differ much from those for ($\widehat{\text{LR}}_2$) and ($\widehat{\text{LR}}_2^m$).

Let us denote by $h_1(v_{it};\ \mu)$ the shortest path length from $v_{n+1,0}$ to v_{it} on G_S (see (2.32)). Let us also denote by $H_1(v_{it};\ \mu)$ the shortest path length from v_{it} to $v_{n+1,T+1}$ on G_S. Clearly,

$$h_1(v_{it};\ \mu)+H_1(v_{it};\ \mu) = \min_{\substack{\mathscr{P}\in\mathscr{Q}_1 \\ v_{it}\in\mathscr{P}}} L_\text{R}(\mathscr{P};\ \mu) \tag{2.41}$$

holds. In other words, the lefthand side of (2.41) gives the shortest path length from $v_{n+1,0}$ to $v_{n+1,T+1}$ on G_S under the additional constraint that v_{it} should be passed through. Therefore, it can be said from (2.21) that if an upper bound UB of (N) satisfies

$$\text{UB} < h_1(v_{it};\ \mu)+H_1(v_{it};\ \mu)+\sum_{1\le i\le n}\mu_i, \tag{2.42}$$

any optimal path for (N) never passes through v_{it}. Hence, v_{it} can be eliminated from G_S.

$h_1(v_{it};\ \mu)$ appears in the forward recursion of dynamic programming, while $H_1(v_{it};\ \mu)$ in the backward recursion. Therefore, this reduction can be performed by applying dynamic programming in both the directions.

2.4.2 Network Reduction by Dominance of Successive Jobs

To reduce the size of $\widehat{G}_\text{S}^m$ for ($\widehat{\text{LR}}_2^m$) more, dominance of n_D ($n_\text{D}\ge 3$) successive jobs is utilized to eliminate unnecessary arcs [28]. Let us define a set of paths $\mathscr{Q}^m$ by

$$\mathscr{Q}^m = \{\mathscr{P}\,|\,\mathscr{P}\in\widehat{\mathscr{Q}}_2^m,\ \mathscr{V}_i(\mathscr{P})=1\ (1\le i\le n)\}. \tag{2.43}$$

More specifically, $\mathscr{Q}^m$ is a set of paths on $\widehat{G}_\text{S}^m$ that correspond to the paths belonging to $\mathscr{Q}$ on G or, equivalently, feasible solutions of (N). Let us also define a path $\mathscr{P}_\text{opt}^m$ corresponding to an optimal solution of (N) by

$$\mathscr{P}_\text{opt}^m = \arg\min_{\mathscr{P}\in\mathscr{Q}^m} L(\mathscr{P}). \tag{2.44}$$

If, for every $\mathscr{P}\in\mathscr{Q}^m$ that passes through the arc $(v_{j,t-p_i}^{\mathbf{b}-\mathbf{q}_i^m}, v_{it}^{\mathbf{b}})$, there exists a dominating path $\mathscr{P}'\in\mathscr{Q}^m$ such that

$$L(\mathscr{P}') < L(\mathscr{P}), \tag{2.45}$$

$\mathscr{P}^m_{\mathrm{opt}}$ never passes through the arc and hence it can be eliminated. To check this, only n_{D} nodes visited just before $v^{\mathbf{b}}_{it}$ (including $v^{\mathbf{b}}_{it}$) in $\mathscr{P}$ are considered in forward dynamic programming. That is, $(n_{\mathrm{D}}! - 1)$ paths are checked for one $\mathscr{P}$ as a candidate for a dominating path $\mathscr{P}'$, where the visiting order of the n_{D} nodes is interchanged. Similarly, in backward dynamic programming n_{D} nodes visited just after $v^{\mathbf{b}-\mathbf{q}^m_i}_{j,t-p_i}$ (including $v^{\mathbf{b}-\mathbf{q}^m_i}_{j,t-p_i}$) are considered to eliminate $(v^{\mathbf{b}-\mathbf{q}^m_i}_{j,t-p_i}, v^{\mathbf{b}}_{it})$.

For example, suppose that part of $\widehat{G}^1_{\mathrm{S}}$ $(\mathscr{M} = \{3\})$ is given by Fig. 2.1, where $p_1 = 1$, $p_2 = 3$ and $p_3 = 2$. Let us check whether the arc $(v^1_{3,21}, v^1_{2,24})$ can be eliminated in forward dynamic programming by setting $n_{\mathrm{D}} = 3$. In this case, the following three types of paths passing through the arc $(v^1_{3,21}, v^1_{2,24})$ should be considered:

$$\mathscr{P}_{\mathrm{A}} = (\ldots, v^0_{0,19}, v^1_{3,21}, v^1_{2,24}, \ldots), \tag{2.46}$$

$$\mathscr{P}_{\mathrm{B}} = (\ldots, v^0_{1,19}, v^1_{3,21}, v^1_{2,24}, \ldots), \tag{2.47}$$

$$\mathscr{P}_{\mathrm{C}} = (\ldots, v^0_{2,19}, v^1_{3,21}, v^1_{2,24}, \ldots). \tag{2.48}$$

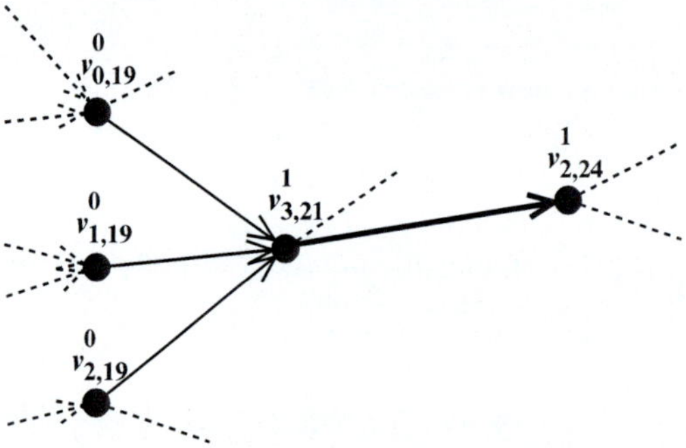

Fig. 2.1: An example of network reduction by dominance of three successive jobs $(n_{\mathrm{D}} = 3)$

Since nodes corresponding to job 2 appear twice in $\mathscr{P}_{\mathrm{C}}$, only $\mathscr{P}_{\mathrm{A}}$ and $\mathscr{P}_{\mathrm{B}}$ belong to $\mathscr{Q}^m$. Here, we are to check whether there exist dominating paths for these two types of paths. First, $\mathscr{P}_{\mathrm{A}}$ is considered and the following five candidates are generated by interchanging the visiting order of $v^0_{0,19}$, $v^1_{3,21}$ and $v^1_{2,24}$:

$$\mathscr{P}'_{\mathrm{A1}} = (\ldots, v^0_{0,19}, v^0_{2,22}, v^1_{3,24}, \ldots), \tag{2.49}$$

$$\mathscr{P}'_{\mathrm{A2}} = (\ldots, v^1_{3,20}, v^1_{0,21}, v^1_{2,24}, \ldots), \tag{2.50}$$

$$\mathscr{P}'_{\mathrm{A3}} = (\ldots, v^1_{3,20}, v^1_{2,23}, v^1_{0,24}, \ldots), \tag{2.51}$$

$$\mathscr{P}'_{\mathrm{A4}} = (\ldots, v^0_{2,21}, v^0_{0,22}, v^1_{3,24}, \ldots), \tag{2.52}$$

$$\mathscr{P}'_{\mathrm{A5}} = (\ldots, v^0_{2,21}, v^1_{3,23}, v^1_{0,24}, \ldots). \tag{2.53}$$

Then, we search for k ($1 \leq k \leq 5$) satisfying $L(\mathscr{P}'_{\mathrm{A}k}) < L(\mathscr{P}_{\mathrm{A}})$. This is not difficult because $L(\mathscr{P}_{\mathrm{A}}) - L(\mathscr{P}'_{\mathrm{A}k})$ depends only on the interchanged three nodes (eg. $L(\mathscr{P}_{\mathrm{A}}) - L(\mathscr{P}'_{\mathrm{A3}}) = f_3(21) + f_2(24) - f_3(20) - f_2(23)$). In addition, we need not consider $k = 1,2$ because $L(\mathscr{P}'_{\mathrm{A1}}) \geq L(\mathscr{P}_{\mathrm{A}})$ and $L(\mathscr{P}'_{\mathrm{A2}}) \geq L(\mathscr{P}_{\mathrm{A}})$ hold from the constraints on adjacent pairs of jobs in Sect. 2.3.3.

Here, assume that $L(\mathscr{P}'_{\mathrm{A3}}) < L(\mathscr{P}_{\mathrm{A}})$ holds. Then, $\mathscr{P}_{\mathrm{B}}$ is checked next and if there also exists a dominating path, the arc $(v^1_{3,21}, v^1_{2,24})$ can be eliminated.

This reduction becomes more effective as n_{D} becomes larger, but both the number of paths passing through the target arc (cf. (2.46)–(2.48)) and the number of permutations of nodes that should be checked (cf. (2.49)–(2.53)) increase exponentially. Therefore, n_{D} is chosen as $n_{\mathrm{D}} = 4$ in our algorithm.

2.5 Algorithm Overview

To summarize, our exact algorithm is composed of the following three stages that correspond to (LR_1), $(\widehat{\mathrm{LR}}_2)$ and $(\widehat{\mathrm{LR}}_2^m)$, respectively. If the gap between lower and upper bounds becomes zero in Stages 1 and 2, the algorithm is terminated[2] and the solution yielding the current upper bound UB is outputted as an optimal solution.

Stage 1 An initial upper bound UB is computed by the algorithm in Sect. 2.6.3. Lagrangian multipliers μ are adjusted by applying the subgradient algorithm to the following Lagrangian dual corresponding to (LR_1):

$$\max_{\mu} \left(\min_{\mathscr{P} \in \mathscr{Q}_1} L_{\mathrm{R}}(\mathscr{P};\ \mu) + \sum_{1 \leq i \leq n} \mu_i \right). \tag{2.54}$$

Then, the network reduction in Sect. 2.4.1 is performed.

Stage 2 Multipliers μ are re-adjusted by applying the subgradient algorithm to the Lagrangian dual corresponding to $(\widehat{\mathrm{LR}}_2)$. In the course of the algorithm, an upper bound is computed by the method in Sect. 2.6 and UB is updated if necessary. The network reduction in Sect. 2.4.1 is applied every time when the best lower bound or UB is updated.

Stage 3 Let the current best lower bound LB be $\mathrm{LB} = \min_{\mathscr{P} \in \widehat{\mathscr{Q}}_2} L_{\mathrm{R}}(\mathscr{P};\ \mu) + \sum_{1 \leq i \leq n} \mu_i$. Then, the subprocedure is repeated by increasing the tentative upper bound $\mathrm{UB}^{\mathrm{tent}}$. It is terminated if $\mathrm{UB}^{\mathrm{tent}} = \mathrm{UB}$ at the end of the subprocedure.

Subprocedure: Let $\mathrm{LB}^{\mathrm{sub}} = \mathrm{LB}$ and $m = |\mathscr{M}| = 0$. Starting from $\widehat{G}^0_{\mathrm{S}} = \widehat{G}_{\mathrm{S}}$, the relaxation $(\widehat{\mathrm{LR}}_2^m)$ for μ is solved with $\mathscr{M}$ increased. When solving $(\widehat{\mathrm{LR}}_2^m)$, all the network reductions in Sect. 2.4 are performed. In the course of the algorithm, an upper bound is computed by the method in Sect. 2.6, and $\mathrm{UB}^{\mathrm{tent}}$ and UB are updated if necessary.

[2] To be more precise, the algorithm can be terminated when the gap becomes less than one because the objective function is integral.

In Stage 3 of the algorithm, a tentative upper bound UB^{tent} is introduced and used for the network reduction in Sect. 2.4.1 in place of the current upper bound UB. Since the effectiveness of the network reduction in Sect. 2.4.1 depends highly on the tightness of an upper bound, shortage of memory space for storing the network structure may occur in Stage 3 unless a tight upper bound is obtained. To reduce this dependence on the tightness, UB^{tent} is chosen as $UB^{tent} \leq UB$, so that the network reduction is performed by UB^{tent}. If UB^{tent} is a valid upper bound, it does not cause any problems and we need not repeat the subprocedure. If otherwise, UB^{tent} is increased and the subprocedure is applied again. It can be easily verified that UB^{tent} is proved to be a valid upper bound if $UB^{tent} = UB$ holds at the end of the subprocedure.

It is also important in Stage 3 which jobs should be added to $\mathcal{M}$. To suppress the increase of memory usage as much as possible, the job whose corresponding nodes appear less frequently in $\widehat{G}_S^m$ is chosen first from $\mathcal{N} \setminus \mathcal{M}$ [28].

2.6 Upper Bound Computation

This section will describe the upper bound computation method in the algorithm of Sect. 2.5. It is composed of two parts: A solution of a relaxation is first converted to a feasible solution of the original problem by some heuristic algorithm, and then it is improved by a neighborhood search. In the following, they will be explained one by one. A tight lower bound is important also for the network reduction in Sect. 2.4.1.

2.6.1 Lagrangian Heuristic

The first part is an extension of the heuristic proposed in [15]. Since it exploits dynamic programming and thus is time-consuming, its greedy version is also applied.

Let us assume that some solution of a relaxation ($(\widehat{\mathrm{LR}}_2)$ or $(\widehat{\mathrm{LR}}_2^m)$) is given as a sequence of jobs that possibly includes duplicated ordinary jobs. Our heuristic converts it to a feasible solution of the original problem by the following two steps:

1. A partial job sequence is generated from the solution by removing the idle job and duplicated jobs (only one job is kept in the sequence for every duplicated ordinary jobs). The number of jobs in the partial sequence is denoted by n_1.
2. The unscheduled $n_2 (= n - n_1)$ ordinary jobs are inserted optimally into the partial sequence without changing the precedence relations of the n_1 jobs, where idle times are taken into account. In other words, when finding optimal job positions, the objective value is evaluated after idle time is optimally inserted into the sequence. An optimal sequence can be obtained by dynamic programming in $O(n_2(n_1+1)2^{n_2}T_E)$ time.

The dynamic programming in 2. is time- (and space-) consuming because its time complexity is multiplied by T_E. Therefore, the method in [23] is adopted, which was originally proposed to improve the efficiency of dynamic programming for optimal idle time insertion. In this method, the objective function is assumed to be piecewise linear and it is evaluated only at the endpoints of linear segments. If the cost function $f_i(t)$ is piecewise linear with few segments, it is much helpful to reduce computational efforts. Nevertheless, it is hard to apply this heuristic when the number of jobs to be inserted, n_2, is large because the time complexity also depends on it exponentially. Hence, a greedy version of the heuristic is applied when n_2 is large. In this case, unscheduled n_2 jobs are inserted one by one according to the shortest processing time (SPT) order into their optimal positions. That is, the following procedure is used in place of 2.:

2'. The unscheduled n_2 ordinary jobs are inserted one by one according to the SPT order into their optimal positions of the partial sequence. Here, the precedence relations of the n_1 jobs are kept unchanged and idle time is taken into account.

2.6.2 *Improvement by Neighborhood Search*

To improve a solution obtained by the heuristics, the dynasearch is applied. The dynasearch is a powerful neighborhood search and was first proposed for the single-machine scheduling problem without idle time [5]. Then, the enhanced dynasearch was proposed in [12] to improve its search ability by enlarging the neighborhood. Another extension [24] was for the problem with idle time and based on the results in [23]. In our algorithm this extended dynasearch is employed.

2.6.3 *Initial Upper Bound*

To obtain an initial upper bound, the greedy version of the heuristic in Sect. 2.6.1 is first employed, where all the jobs are assumed to be unscheduled ($n_1 = 0$ and $n_2 = n$), and not only the SPT order but also the longest processing time (LPT), earliest duedate (EDD) and latest duedate (LDD) orders are considered when jobs are inserted. Then, the extended dynasearch is applied to the best of the four solutions.

2.7 Numerical Experiments

Our algorithm is applied to two sets of benchmark instances: the instance set with equal (zero) release dates ($1||\sum(\alpha_i E_i + \beta_i T_i)$) in [25, 26], and that with distinct release dates ($1|r_i|\sum(\alpha_i E_i + \beta_i T_i)$) in [2]. These are referred to as the Sourd's benchmark set and the Bülbül's benchmark set, respectively. The Sourd's benchmark set is generated in the following way:

1. Processing times p_i are generated from the uniform distribution $U[10,100)$. Let $P = \sum_{i=1}^{n} p_i$.
2. Due dates d_i are generated from $U[d_{\min}, d_{\max}]$, where $d_{\min} = \max(p_i, \lfloor P(\tau - \rho/2) \rfloor)$, $d_{\max} = d_{\min} + \lfloor \rho P \rfloor$.
3. Both tardiness weights α_i and earliness weights β_i are generated from $U[1,5]$.
4. For each combination of (n, τ, ρ), 26 instances are generated.
5. $n \in \{20,30,40,50\}$, $\tau \in \{0.2,0.3,0.4,0.5,0.6,0.7,0.8\}$, and $\rho \in \{0.2,0.3,0.4,0.5,0.6,0.7,0.8\}$.
6. For $n \in \{60,90\}$, only five instances are generated for each combination of (n, τ, ρ), where $\tau \in \{0.2,0.5,0.8\}$ and $\rho \in \{0.2,0.5,0.8\}$.

On the other hand, the generation scheme of the Bülbül's benchmark set is as follows:

1. Processing times p_i are generated from $U[1, p_{\max}]$. Let $P = \sum_{i=1}^{n} p_i$.
2. Release dates are generated from $U[0,P]$.
3. Due dates d_i are generated from $U[d_{\min}, d_{\max}]$, where $d_{\min} = \max(0, \lceil (1-\tau-\rho/2)P \rceil)$ and $d_{\max} = \lceil (1-\tau+\rho/2)P \rceil$.
4. Both earliness weights α_i and tardiness weights β_i are generated from $U[0,100]$.
5. For each combination of (n, $p_{\max}$, τ, ρ), five instances are generated.
6. $\tau \in \{0.2,0.4,0.5,0.6,0.8\}$ and $\rho \in \{0.4,0.7,1.0,1.3\}$. For $n \in \{20,40,60,80\}$, $p_{\max} \in \{10,30,50\}$ and for $n \in \{100,130,170,200\}$, $p_{\max} \in \{50,75,100\}$.

Both the benchmark sets are available from the aforementioned web page.

The algorithm is coded in C (gcc) and we run it on a 3 GHz Intel® Core2 Duo E6850 desktop computer with 4 GB RAM. The maximum memory size for storing the network structure is restricted to 384 MB.

Benchmark results are summarized in Tables 2.1 and 2.2, where the number of optimally solved instances in each stage, the average and maximum CPU times in seconds are shown. We can verify that all the instances are optimally solved. The most efficient algorithm for the problem so far except ours is the branch-and-bound algorithm by Sourd [26]. He reported that he succeeded in solving all the 50 jobs instances in the Sourd's benchmark set within 1,000 s, and all the 60 jobs instances in the Bülbül's benchmark set within 500 s on a 3.2 GHz Intel® Pentium4 desktop computer. On the other hand, our algorithm only takes at most 3.5 and 3 s for these instances, respectively. It is true that our CPU is about twice as fast as Sourd's CPU, but our algorithm is much faster even if this difference is taken into account.

The detailed results are given in Tables 2.3 and 2.4. In these tables, average and maximum CPU times in each stage are given separately. The average and maximum gaps between lower and upper bounds are also shown in percent, where the gap is calculated by $100(\mathrm{UB}-\mathrm{LB})/\mathrm{UB}$. Please note that it is not easy to examine memory usage of our algorithm because the number of jobs added to $\mathscr{M}$ at one iteration of the subprocedure in Stage 3 is changed adaptively. Therefore, it is not shown here. It can be observed from the tables that the gap becomes small in Stage 2. This implies that ($\widehat{\mathrm{LR}}_2$) yields a tight lower bound.

Table 2.1: Computational results for the Sourd's benchmark set of $1||\sum(\alpha_i E_i + \beta_i T_i)$

n	Instances	Optimally solved instances				CPU time (s)	
		Stage1	Stage2	Stage3	Total	Ave.	Max.
20	1,274	133	952	189	1,274	0.05	0.16
30	1,274	22	902	350	1,274	0.22	0.55
40	1,274	13	757	504	1,274	0.58	1.77
50	1,274	2	631	641	1,274	1.33	3.44
60	45	0	22	23	45	2.61	6.09
90	45	0	10	35	45	13.22	30.25

Table 2.2: Computational results for the Bülbül's benchmark set of $1|r_i|\sum(\alpha_i E_i + \beta_i T_i)$

n	Instances	Optimally solved instances				CPU time (s)	
		Stage1	Stage2	Stage3	Total	Ave.	Max.
20	300	37	221	42	300	0.01	0.05
40	300	1	155	144	300	0.18	0.58
60	300	0	63	237	300	1.02	3.05
80	300	0	35	265	300	3.16	11.63
100	300	0	6	294	300	17.83	44.92
130	300	0	2	298	300	51.62	140.26
170	300	0	0	300	300	155.27	379.97
200	300	0	0	300	300	303.27	771.52

Table 2.3: Detailed results for the Sourd's benchmark set of $1||\sum(\alpha_i E_i + \beta_i T_i)$

n	Stage 1					Stage 2					Stage 3		
	Time (s)		Gap (%)		Solved	Time (s)		Gap (%)		Solved	Time (s)		Solved
	Ave	Max	Ave	Max		Ave	Max	Ave	Max		Ave	Max	
20	0.04	0.15	6.17	62.88	133	0.01	0.08	0.61	17.01	952	0.01	0.04	189
30	0.15	0.34	6.45	54.38	22	0.07	0.42	0.49	11.01	902	0.01	0.10	350
40	0.36	1.00	6.14	59.91	13	0.21	1.15	0.39	9.30	757	0.03	0.23	504
50	0.73	1.56	5.59	49.95	2	0.57	2.71	0.37	5.62	631	0.08	0.46	641
60	1.44	2.38	5.12	27.67	0	1.09	4.67	0.35	3.96	22	0.17	0.66	23
90	5.80	9.98	5.77	30.59	0	6.79	19.43	0.33	1.89	10	0.81	3.73	35

Table 2.4: Detailed results for the Bülbül's benchmark set of $1|r_i|\sum(\alpha_i E_i + \beta_i T_i)$

n	Stage 1					Stage 2					Stage 3		
	Time (s)		Gap (%)		Solved	Time (s)		Gap (%)		Solved	Time (s)		Solved
	Ave	Max	Ave	Max		Ave	Max	Ave	Max		Ave	Max	
20	0.01	0.04	3.02	48.09	37	0.00	0.03	0.23	8.29	221	0.00	0.01	42
40	0.09	0.28	3.58	32.19	1	0.08	0.45	0.17	1.79	155	0.01	0.07	144
60	0.36	0.93	3.82	14.77	0	0.61	2.11	0.18	1.71	63	0.06	0.43	237
80	0.90	2.35	3.41	13.77	0	2.07	9.46	0.17	0.93	35	0.22	1.06	265
100	4.56	9.79	3.35	20.70	0	12.09	35.11	0.17	1.20	6	1.21	7.90	294
130	11.28	24.69	3.10	15.02	0	34.81	112.59	0.16	0.75	2	5.57	51.13	298
170	26.41	53.50	3.07	11.65	0	100.01	277.55	0.16	0.59	0	28.85	174.55	300
200	43.48	91.59	2.96	11.95	0	185.44	427.72	0.15	0.48	0	74.35	405.55	300

2.8 Conclusion

This paper introduced our exact algorithm [29] for the single-machine total weighted earliness–tardiness problem. Due to the tightness of lower and upper bounds, this dynamic-programming-based exact algorithm is so efficient that it outperformed the existing best algorithm and could optimally solve 200 jobs instances. In our most recent paper [30] it is shown that a new algorithm after several improvements is also effective for other types of single-machine problems such as $1|r_i|\sum_i w_i C_i$ and $1|r_i|\sum_i w_i T_i$.

The algorithm is based on dynamic programming and hence reduction of memory usage is crucial for further improving the algorithm. Some additional constraints (cuts) to obtain a better lower bound and/or new network reduction methods to reduce memory usage directly would be necessary for this purpose, which we should consider in future research. Another direction of research will be to extend the algorithm to a wider class of problems such as the problems with precedence constraints, setup times and so on.

Acknowledgements This work is partially supported by Grant-in-Aid for Young Scientists (B) 19760273, from the Ministry of Education, Culture, Sports, Science and Technology (MEXT) Japan.

References

1. Abdul-Razaq, T.S., Potts, C.N.: Dynamic programming state-space relaxation for single-machine scheduling. Journal of the Operational Research Society **39**, 141–152 (1988)
2. Bülbül, K., Kaminsky, P., Yano, C.: Preemption in single machine earliness/tardiness scheduling. Journal of Scheduling **10**, 271–292 (2007)
3. Chang, P.C.: A branch and bound approach for single machine scheduling with earliness and tardiness penalties. Computers and Mathematics with Applications **37**, 133–144 (1999)

4. Christofides, N., Mingozzi A, Toth P.: State-space relaxation procedures for the computation of bounds to routing problems. Networks **11**, 145–164 (1981)
5. Congram, R.K., Potts, C.N., van de Velde, S.L.: An iterated dynasearch algorithm for the single machine total weighted tardiness scheduling problem. INFORMS Journal on Computing **14**, 52–67 (2002)
6. Davis, J.S., Kanet, J.J.: Single-machine scheduling with early and tardy completion costs. Naval Research Logistics **40**, 85–101 (1993)
7. Detienne, B., Pinson, É., Rivreau., D.: Lagrangian domain reductions for the single machine earliness-tardiness problem with release dates, European Journal of Operational Research **201**, 45–54 (2010)
8. Dyer, M.E, Wolsey, L.A.: Formulating the single-machine sequencing problem with release dates as a mixed integer problem. Discrete Applied Mathematics **26**, 255–270 (1990)
9. Fisher, M.L.: Optimal solution of scheduling problems using Lagrange multipliers: Part I. Operations Research **21** 1114–27 (1973)
10. Fry, T.D., Armstrong, R.D., Darby-Dowman, K., Philipoom, P.R.: A branch and bound procedure to minimize mean absolute lateness on a single processor. Computers & Operations Research **23**, 171–182 (1996)
11. Graham, R.L., Lawler, E.L., Lenstra, J.K., Rinnooy Kan, A.H.G.: Optimization and approximation in deterministic sequencing and scheduling: A survey. Annals of Discrete Mathematics **5**, 287–326 (1979)
12. Grosso, A., Della Croce, F., Tadei, R.: An enhanced dynasearch neighborhood for the single machine total weighted tardiness scheduling problem. Operations Research Letters **32**, 68–72 (2004)
13. Hoogeveen, J.A., van de Velde, S.L.: A branch-and-bound algorithm for single-machine earliness-tardiness scheduling with idle time. INFORMS Journal on Computing **8**, 402–412 (1996)
14. Ibaraki, T.: Enumerative approaches to combinatorial optimization. Annals of Operations Research **10** and **11** (1987)
15. Ibaraki, T., Nakamura, Y.: A dynamic programming method for single machine scheduling, European Journal of Operational Research **76**, 72–82 (1994)
16. Kim, Y.-D., Yano, C.A.: Minimizing mean tardiness and earliness in single-machine scheduling problems with unequal due dates. Naval Research Logistics **41**, 913–933 (1994)
17. Lawler, E.L.: A "pseudopolynomial" algorithm for sequencing jobs to minimize total tardiness, Annals of Discrete Mathematics **1**, 331–342 (1977)
18. Lenstra, J.K, Rinnooy Kan, A.H.G. and Brucker, P.: Complexity of machine scheduling problems. Annals of Discrete Mathematics **1**, 343–362 (1977)
19. Péridy, L., Pinson, É., Rivreau, D.: Using short-term memory to minimize the weighted number of late jobs on a single machine. European Journal of Operational Research **148**, 591–603 (2003)
20. Potts, C.N., Van Wassenhove, L.N.: A branch and bound algorithm for the total weighted tardiness problem. Operations Research **33**, 363–377 (1985)
21. Pritsker, A.A.B., Watters, L.J., Wolfe, P.M.: Multiproject scheduling with limited resources: A zero-one programming approach. Management Science **16**, 93–108 (1969)
22. Sourd, F., Kedad-Sidhoum, S.: The one-machine problem with earliness and tardiness penalties. Journal of Scheduling **6** 533–549 (2003)
23. Sourd, F.: Optimal timing of a sequence of tasks with general completion cost. European Journal of Operational Research **165**, 82–96 (2005)
24. Sourd, F.: Dynasearch for the earliness-tardiness scheduling problem with release dates and setup constraints. Operations Research Letters **34**, 591–598 (2006)
25. Sourd, F., Kedad-Sidhoum, S.: A faster branch-and-bound algorithm for the earliness-tardiness scheduling problem. Journal of Scheduling **11**, 49–58 (2008)
26. Sourd, F.: New exact algorithms for one-machine earliness-tardiness scheduling. INFORMS Journal on Computing **21**, 167–175 (2009)
27. Sousa, J.P., Wolsey, L.A.: A time indexed formulation of non-preemptive single machine scheduling problems. Mathematical Programming **54**, 353–367 (1992)

28. Tanaka, S., Fujikuma, S., Araki, M.: An exact algorithm for single-machine scheduling without machine idle time. Journal of Scheduling **12**, 575–593 (2009)
29. Tanaka, S., Fujikuma, S.: An efficient exact algorithm for general single-machine scheduling with machine idle time. 4th IEEE Conference on Automation Science and Engineering (IEEE CASE 2008), 371–376 (2008)
30. Tanaka, S., Fujikuma, S.: A dynamic-programming-based exact algorithm for single-machine scheduling with machine idle time. Journal of Scheduling, available online. DOI: 10.1007/s10951-011-0242-0
31. van den Akker, J.M., van Hoesel, C.P.M., Savelsbergh, M.W.P.: A polyhedral approach to single-machine scheduling problems. Mathematical Programming **85** 541–572 (1999)
32. van den Akker, J.M., Hurkens, C.A.J., Savelsbergh, M.W.P.: Time-indexed formulations for machine scheduling problems: column generation. INFORMS Journal on Computing **12**, 111–124 (2000)
33. Yano, C.A., Kim, Y.-D.: Algorithms for a class of single-machine weighted tardiness and earliness problems. European Journal of Operational Research **52**, 167–178 (1991)
34. Yau, H., Pan, Y., Shi, L.: New solution approaches to the general single machine earliness-tardiness problem. IEEE Transactions on Automation Science and Engineering **5**, 349–360 (2008)

Chapter 3
Single-Machine Scheduling Problems Simultaneous with Deteriorating and Learning Effects Under a Deteriorating Maintenance Consideration

Suh-Jenq Yang and Dar-Li Yang

Abstract Machine scheduling problems in just-in-time production environments are important issues in modern operations management to satisfy customer demand for superior service. In this paper we investigate single-machine scheduling problems with simultaneous considerations of the effects of deterioration and learning. Due to the deteriorating effect, maintenance may be performed on the machine to improve its production efficiency. We assume that at most one maintenance is allowed throughout the scheduling horizon. We further assume that the maintenance duration depends on its starting time. Our goal is to find jointly the optimal time to perform the maintenance, the optimal location of the due-window, and the optimal job sequence such that the total cost that includes earliness, tardiness, and due-window size and location penalties is minimized. We also aim to investigate the makespan, the total completion time, and the total absolute deviation of completion times minimization problems. We propose polynomial time algorithms for all the studied problems.

3.1 Introduction

Scheduling problems with deteriorating jobs have been extensively studied in the last two decades in various machine settings and performance measures. In scheduling with the effect of deterioration, the actual processing time of a job is usually

Suh-Jenq Yang
Department of Industrial Engineering and Management, Nan Kai University of Technology, Nan-Tou 542, Taiwan, e-mail: sjyang@nkut.edu.tw

Dar-Li Yang
Department of Information Management, National Formosa University, Yun-Lin 632, Taiwan, e-mail: dlyang@nfu.edu.tw

R.Z. Ríos-Mercado and Y.A. Ríos-Solís (eds.), *Just-in-Time Systems*, Springer Optimization and Its Applications 60, DOI 10.1007/978-1-4614-1123-9_3,

modeled as an increasing function of its starting time. Gupta and Gupta [17] and Browne and Yechiali [5] were the first researchers to introduce the deteriorating effect into scheduling problems. Kunnathur and Gupta [28] and Mosheiov [38] presented several real-life situations where deteriorating jobs might occur such as steel production, resource allocation, fire fighting, maintenance or cleaning, etc., in which any delay in processing a job may result in an increasing effort to perform the job. Since then, related models of deteriorating jobs scheduling problems under different machine environments have been proposed from a variety of perspectives. For details on this stream of research, the reader may refer to the comprehensive surveys by Alidaee and Womer [1], Cheng et al. [7], Janiak and Kovalyov [21], and the recent book by Gawiejnowicz [14].

On the other hand, learning and its effect on productivity is well recognized in industry. In general, workers learn from experience in performing repetitive tasks and the learning effect demonstrates itself in labour productivity improvement. In scheduling with the learning effect, the actual processing time of a job is modeled as a decreasing function if it is scheduled later in a job sequence. Cheng and Kovalyov [9] were probably the first researchers who formally considered the learning effect in scheduling problems. Gawiejnowicz [12] introduced another learning model in scheduling to minimize the makespan. Since then, researchers have presented many different learning models and have applied the concept of the learning effect to many scheduling problems involving various performance measures. Biskup [3] proposed a position-dependent learning model in which the actual time required to perform an operation decreases by the number of repetitions. Cheng and Wang [10] considered a single-machine scheduling problem with a volume-dependent piecewise linear processing time function to model the effect of learning. Mosheiov and Sidney [41] extended Biskup's learning model by the introduction of different learning ratios for jobs, that is, the job-dependent learning model. Kuo and Yang [29] introduced a time-dependent learning model in which the actual time required to perform a job is a function of the total normal processing time of jobs scheduled in front of it. Janiak and Rudek [23] provided an extensive study of a single-machine scheduling with an experience based learning model, where job processing times are described by S-shaped functions that are dependent on the experience of the machine. Yin et al. [59] developed a general model with learning effects where the actual processing time of a job is not only a function of the total normal processing times of jobs already processed, but also a function of the job's scheduled position. For a complete list of studies, the reader may refer to the comprehensive surveys by Bachman and Janiak [2], Janiak and Rudek [22], Biskup [4], and Janiak and Rudek [23].

In addition, the phenomena of deteriorating and learning effects occurring simultaneously can be found in many real-life situations [49]. Lee [31] first considered single-machine scheduling problems with both deteriorating and learning effects. As of now, several studies have investigated scheduling with simultaneous considerations of the effects of deterioration and learning under different machine environments, including Wang [48–51], Wang and Cheng [52], Wang and Cheng [54],

Toksar et al. [47], Wang et al. [53], Sun [46], Huang et al. [20], and Yang and Kuo [56,57].

The majority of the scheduling literature assumed that a machine is continuously available during the scheduling horizon. However, this assumption may be inappropriate in many practical situations. In real production systems, maintenance is important to production as it helps improve the machine efficiency or the product quality. During the maintenance, the machine is not available for processing jobs. Scheduling in such an environment is known as scheduling with availability constraints and it has attracted many researchers. Sanlaville and Schmidt [44], Schmidt [45], and Ma et al. [37] provided extensively surveys related to machine scheduling problems with the maintenance. Moreover, research on scheduling with simultaneous considerations of the deteriorating effect and the maintenance has received increasing attention in recent years. Several studies have investigated this subject in a single-machine environment, which include Wu and Lee [55], Ji et al. [24], Gawiejnowicz [13], Low et al. [36], Lodree and Geiger [35], Gawiejnowicz and Kononov [15], and Hsu et al. [19]. However, most literature on scheduling with the maintenance regarded the maintenance duration constant no matter what conditions of the machine are. Nevertheless, in real production settings, the maintenance duration may depend on its running time. That is, the later the maintenance is started, the worse the machine conditions are, and the longer time is needed to maintain it. This kind of maintenance is known as a deteriorating maintenance. Several recent papers have studied scheduling problems with the deteriorating maintenance [26,27,42,58].

To the best of our knowledge, however, scheduling simultaneous with deteriorating and learning effects under a deteriorating maintenance consideration has never been explored. In this paper, we investigate the scheduling with jointly deteriorating and learning effects under a deteriorating maintenance consideration in the context of the common due-window assignment problem and the maximum completion time, the total completion time, and the total absolute deviation of completion times (TADC) minimization problems, which deal with job scheduling in a just-in-time (JIT) production environment. We will show that all the studied problems are polynomially solvable.

JIT is beneficial for companies engaged in repetitive, customer service, job shop, or process manufacturing. According to the principle of JIT production, jobs should be finished as close as possible to their due-dates. If a job is finished earlier than its due-date, it has to be stored in inventory, which results in an earliness penalty. On the other hand, if a job is finished later than its due-date, it incurs a tardiness penalty as a result of violating contractual obligation with the customer. In practice, however, the supply contract between the supplier and the customer often specifies a time interval such that a job completed within the time interval is considered to be on time and will not incur any penalty. Therefore, the supplier should produce the necessary products in the right quantities and the right quality within the right time interval. This time interval is called the due-window of a job. The left end and the right end of the due-window are, respectively, called the due-window starting time and the due-window finishing time. A job completed on or before the due-window starting time is considered to be early while a job completed after the due-window finishing time is considered to be tardy. Clearly, a wide due-window enhances the supplier's production and delivery flexibility. On the contrary, a large due-window and delaying job completion reduce the supplier's competitiveness and his level of customer service.

The problem of joint scheduling and due-window assignment has received the attention of many researchers. Cheng [6] initiated research on scheduling with the concept of due-window. He studied the earliness and tardiness problem with the assumption that the window is small enough such that at most one job can be completed within this window. Lee [30] considered the problem of minimizing the maximum earliness subject to no tardy jobs. He showed that the problem is NP-hard if the due-window size is arbitrary while the problem is polynomially solvable if the size of the due-window is specified. Liman and Ramaswamy [32] studied the minimization of the weighted sum of earliness and number of tardy jobs. They showed that the problem is NP-hard. Liman et al. [33] considered a single-machine common due-window scheduling problem where the due-window size is given. The objective was to find the optimal job sequence and the optimal common due-window location such that the weighted sum of earliness, tardiness, and due-window location penalties is minimized. Liman et al. [34] further provided a polynomial time solution for the single-machine due-window assignment problem, where both the location and the size of the due-window are decision variables. Later, Mosheiov and Sarig [39] extended the study of Liman et al. [34] to the case with position-dependent job processing times. They assumed that the processing time of a job is a function of its position in the sequence. In the succeeding study, Mosheiov and Sarig [40] considered a single-machine due-window assignment and scheduling with a rate-modifying activity to minimize the total earliness, tardiness, and due-window starting time and size costs. They showed that the rate-modifying activity may be optimally scheduled at time zero, prior to any of the jobs processed after the due-window, or there is no maintenance at all. Recently, Yang et al. [58] investigated the problem of a common due-window assignment and scheduling with a job-dependent aging effect under a deteriorating maintenance consideration on a single-machine. The objective was to find jointly the optimal time to perform the maintenance, the optimal location and size of the due-window, and the optimal job sequence to minimize the total earliness, tardiness, and due-window related costs. They showed that the problem can be optimally solved in a polynomial time algorithm. Cheng et al. [11] studied a common due-window assignment and scheduling of linear time-dependent deteriorating jobs with a deteriorating maintenance on a single-machine. They showed that the problem under study remains polynomially solvable.

In the next section we formulate the problem. We provide a polynomial time algorithm to optimize the common due-window assignment scheduling problem in Sect. 3.3. We introduce polynomial time algorithms for minimizing the makespan, the total completion time, and TADC problems in Sects. 3.4, 3.5, and 3.6, respectively. We conclude the findings in the last section.

3.2 Problem Formulation

In this study, we consider the scheduling of the common due-window assignment problem and the makespan, the total completion time, and TADC minimization problems. The problem under study can be formally described as follows: There are n jobs $J = \{J_1, J_2, \ldots, J_n\}$ to be processed on a single-machine. All the jobs are non-preemptive and available for processing at time zero. The machine can handle at most one job at a time. If job J_j is started at time t and scheduled in position r in a job sequence, then its actual processing time p_{jr} is defined by

$$p_{jr} = (p_j + \lambda t)r^a, \quad j = 1, 2, \ldots, n \text{ and } r = 1, 2, \ldots, n,$$

where $\lambda > 0$ is a common deterioration rate, p_j is the normal processing time of job J_j, $t \geq 0$ is the starting time of job J_j, and $a < 0$ is the learning index. The assumption that all the jobs have a common deterioration rate is reasonable, particularly in the case of scheduling problems with deteriorating machine, when all processing times are increased by a common factor caused by the machine [8].

Due to the effect of deterioration, the machine may need maintenance to improve its production efficiency. We assume that at most one maintenance is allowed throughout the scheduling horizon. The maintenance can be performed immediately after the processing of any job is completed. However, the starting time of the maintenance in the sequence is not known in advance. In order to model the problem more realistic, we further assume that: (1) after the maintenance, the machine deterioration will reset first and then starts anew and (2) the maintenance duration is a linear function of its starting time and is given by $f(t) = \mu + \sigma t$, where $\mu > 0$ is the basic maintenance time, $\sigma \geq 0$ is a maintenance factor, and t is the starting time of the maintenance.

We assume that all the jobs have a common due-window. Denote by $d \geq 0$ and $w \geq d$ the due-window starting time and the due-window finishing time, respectively. Both d and w are decision variables in this study. For a given schedule π, denote by C_j the completion time of job J_j. Then $E_j = \max\{0, d - C_j\}$ and $T_j = \max\{0, C_j - w\}$ represent the earliness and tardiness of job J_j, respectively, and $C_{\max} = \max\{C_j\}$ and TC represent the makespan and the total completion time of all jobs, respectively. We also denote by i the position of the job preceding the maintenance (i.e., position $(i+1)$ is the first position after the maintenance).

We aim to find jointly the optimal due-window location, the optimal time to perform the maintenance, and the optimal job sequence π such that the following cost function

$$f(d, D, \pi) = Z = \sum_{j=1}^{n} (\alpha E_j + \beta T_j + \gamma d + \delta D) \tag{3.1}$$

is minimized, where $\alpha > 0$, $\beta > 0$, $\gamma > 0$, and $\delta > 0$ are the earliness, tardiness, due-window starting time, and due-window size costs per unit time, respectively, and $D = (w - d)$ is the due-window size. We also investigate the makespan, the total completion time, and TADC minimization problems in this study. Using the three-field notation introduced by Graham et al. [16], we, respectively, denote our problems as $1/p_{jr} = (p_j + \lambda t)r^a, ma/\sum_{j=1}^{n}(\alpha E_j + \beta T_j + \gamma d + \delta D)$, $1/p_{jr} = (p_j + \lambda t)r^a, ma/TC$, and $1/p_{jr} = (p_j + \lambda t)r^a, ma/TADC$, where *ma* in the second field denotes the maintenance.

3.3 Problem $1/p_{jr} = (p_j + \lambda t)r^a$, $ma/\sum_{j=1}^{n}(\alpha E_j + \beta T_j + \gamma d + \delta D)$

In this section, we first present several important lemmas for an optimal schedule of the problem, and then show that the problem can be optimally solved in a polynomial time algorithm.

3.3.1 Preliminary Analysis

In Lemma 3.1, we show that an optimal schedule exits with no idle time between consecutive jobs on the machine.

Lemma 3.1. *An optimal schedule π exists in which the machine is not idle between the processing of jobs.*

Proof. Assume that all the jobs are available for processing at time $t_0 \geq 0$. If job J_i is started at time t and scheduled in position r in a job sequence, then its actual processing time is given by $p_{jr} = (p_j + \lambda t)r^a$, $j = 1,2,\ldots,n$ and $r = 1,2,\ldots,n$. Without loss of generality, we assume that the optimal job sequence is $\pi = (J_1, J_2, \ldots, J_n)$. Denote by d_{opt}, an optimal due-window starting time such that (3.1) is minimized. Suppose that there exists an optimal schedule in which the machine is idle between the processing of jobs J_j and J_{j+1} (see Fig. 3.1). Let C_j be the completion time of job J_j and $(C_j + \Delta)$ be the starting time of job J_{j+1}, where $\Delta > 0$. Clearly, we have that $C_j < (C_j + \Delta)$. Then, the maintenance duration is given by $f(t) = \mu + \sigma t = \mu + \sigma C_i$. It should be noted that the maintenance can be performed immediately after the processing of any job is completed.

We prove the result by considering the following cases covering all the possible locations of the optimal due-window starting time d_{opt} and job J_i. Recall that job J_i is the job preceding the maintenance.

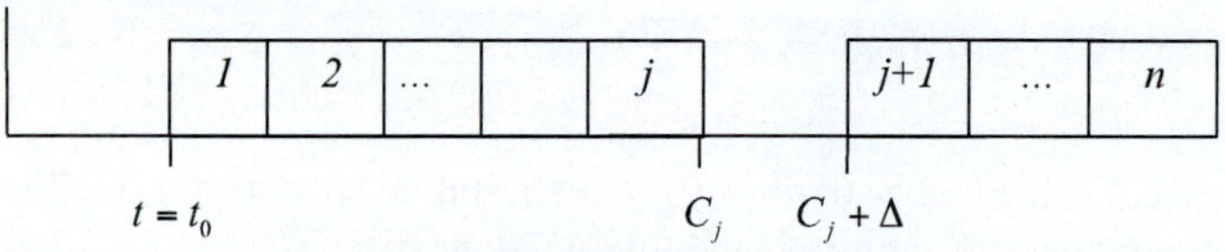

Fig. 3.1: Schematic diagram for Lemma 3.1

Case (i): $d_{\text{opt}} \leq C_j$ and $C_i < C_j$

Let t_i be the completion time of maintenance and t_j be the starting time of job J_j. If $d_{\text{opt}} \geq C_j$ and $C_i < C_j$, then the completion times of jobs J_j and J_{j+1} are, respectively,

$$C_j = t_j + [p_j + \lambda(t_j - t_i)]\, j^a = (1+\lambda j^a)t_j + p_j j^a - \lambda t_i j^a$$

and

$$\begin{aligned} C_{j+1} &= C_j + \Delta + \left[p_{j+1} + \lambda(C_j + \Delta - t_i)\right](j+1)^a \\ &= [1+\lambda(j+1)^a]C_j + [1+\lambda(j+1)^a]\Delta + p_{j+1}(j+1)^a - \lambda t_i (j+1)^a. \end{aligned}$$

If we shift job J_{j+1} to the left such that the starting time of job J_{j+1} coincides with the completion time of job J_j (i.e., C_j), then the completion time of job J_{j+1} becomes

$$\begin{aligned} \overline{C}_{j+1} &= C_j + \left[p_{j+1} + \lambda(C_j - t_i)\right](j+1)^a \\ &= [1+\lambda(j+1)^a]C_j + p_{j+1}(j+1)^a - \lambda t_i (j+1)^a. \end{aligned}$$

Thus, the difference between C_{j+1} and $\overline{C}_{j+1}$ is $C_{j+1} - \overline{C}_{j+1} = [1+\lambda(j+1)^a]\Delta$. Since $\lambda > 0$ and $\Delta > 0$, $(C_{j+1} - \overline{C}_{j+1}) > 0$. That is, if we shift job J_{j+1} to the left, then the tardiness cost of job J_{j+1} decreases. Hence, the total cost decreases.

Case (ii): $d_{\text{opt}} \leq C_j$ and $C_i \geq C_{j+1}$

Let t_j be the starting time of job J_j. If $d_{\text{opt}} \leq C_j$ and $C_i \geq C_{j+1}$, then the completion times of jobs J_j and J_{j+1} are, respectively,

$$C_j = t_j + (p_j + \lambda t_j) j^a = (1+\lambda j^a)t_j + p_j j^a$$

and

$$\begin{aligned} C_{j+1} &= C_j + \Delta + \left[p_{j+1} + \lambda(C_j + \Delta)\right](j+1)^a \\ &= [1+\lambda(j+1)^a]C_j + [1+\lambda(j+1)^a]\Delta + p_{j+1}(j+1)^a. \end{aligned}$$

If we shift job J_{j+1} to the left such that the starting time of job J_{j+1} coincides with the completion time of job J_j (i.e., C_j), then the completion time of job J_{j+1} becomes

$$\overline{C}_{j+1} = C_j + (p_{j+1} + \lambda C_j)(j+1)^a = [1+\lambda(j+1)^a]C_j + p_{j+1}(j+1)^a.$$

Thus, the difference between C_{j+1} and $\overline{C}_{j+1}$ is $C_{j+1} - \overline{C}_{j+1} = [1+\lambda(j+1)^a]\Delta$. Since $\lambda > 0$ and $\Delta > 0$, $(C_{j+1} - \overline{C}_{j+1}) > 0$. That is, if we shift job J_{j+1} to the left, then the tardiness cost of job J_{j+1} decreases. Hence, the total cost decreases.

Case (iii): $d_{\text{opt}} \leq C_j$ and $C_i = C_j$

Let t_j be the starting time of job J_j. Denote by $\Delta = \Delta_1 + (\mu + \sigma C_j)$. Note that $\Delta_1 > 0$. If $d_{\text{opt}} \leq C_j$ and $C_i = C_j$, then the completion times of jobs J_j and J_{j+1} are, respectively,

$$C_j = t_j + (p_j + \lambda t_j) j^a = (1 + \lambda j^a) t_j + p_j j^a$$

and

$$\begin{aligned} C_{j+1} &= C_j + \mu + \sigma C_j + \Delta_1 + p_{j+1}(j+1)^a \\ &= \mu + \Delta_1 + (1+\sigma) C_j + p_{j+1}(j+1)^a, \end{aligned}$$

and the maintenance duration is given by

$$f(t) = \mu + \sigma C_j = \mu + \sigma \left[(1 + \lambda j^a) t_j + p_j j^a \right].$$

If we shift job J_{j+1} to the left such that the starting time of job J_{j+1} coincides with time $(C_j + \mu + \sigma C_j)$, then the completion time of job J_{j+1} becomes

$$\begin{aligned} \overline{C}_{j+1} &= (C_j + \mu + \sigma C_j) + p_{j+1}(j+1)^a \\ &= \mu + (1+\sigma) C_j + p_{j+1}(j+1)^a. \end{aligned}$$

Thus, the difference between C_{j+1} and $\overline{C}_{j+1}$ is $C_{j+1} - \overline{C}_{j+1} = \Delta_1$. Since $\Delta_1 > 0$, $(C_{j+1} - \overline{C}_{j+1}) > 0$. That is, if we shift job J_{j+1} to the left, then the tardiness cost of job J_{j+1} decreases. Hence, the total cost decreases.

Case (iv): $d_{\text{opt}} \geq (C_j + \Delta)$ and $C_i < C_j$

Let t_i be the completion time of the maintenance and t_j be the starting time of job J_j. If $d_{\text{opt}} \geq (C_j + \Delta)$ and $C_i < C_j$, then the completion time of job J_j is

$$C_j = t_j + [p_j + \lambda(t_j - t_i)]\, j^a = (1 + \lambda j^a) t_j + p_j j^a - \lambda t_i j^a.$$

If we shift job J_j to the right such that the finishing time of job J_j coincides with time $(C_j + \Delta)$, then the completion time of job J_j becomes

$$\overline{C}_j = C_j + \Delta = (1 + \lambda j^a) t_j + p_j j^a - \lambda t_i j^a + \Delta.$$

Thus, the difference between C_j and $\overline{C}_j$ is $C_j - \overline{C}_j = -\Delta$. Since $\Delta > 0$, $(C_j - \overline{C}_j) < 0$. That is, if we shift job J_j to the right, then the earliness cost of job J_j decreases. Hence, the total cost decreases.

Case (v): $d_{\text{opt}} \geq (C_j + \Delta)$ and $C_i \geq C_{j+1}$

Let t_i be the starting time of job J_j. If $d_{\text{opt}} \geq (C_j + \Delta)$ and $C_i \geq C_{j+1}$, then the completion time of job J_j is

$$C_j = t_j + (p_j + \lambda t_j) j^a = (1 + \lambda j^a) t_j + p_j j^a.$$

If we shift job J_j to the right such that the finishing time of job J_j coincides with time $(C_j + \Delta)$, then the completion time of job J_j becomes

$$\overline{C}_j = C_j + \Delta = (1 + \lambda j^a)t_j + p_j j^a + \Delta.$$

Thus, the difference between C_j and $\overline{C}_j$ is $C_j - \overline{C}_j = -\Delta$. Since $\Delta > 0$, $(C_j - \overline{C}_j) < 0$. That is, if we shift job J_j to the right, then the earliness cost of job J_j decreases. Hence, the total cost decreases.

Case (vi): $d_{\text{opt}} \geq (C_j + \Delta)$ and $C_i = C_j$

Let t_j be the starting time of job J_j. Denote by $\Delta = \Delta_1 + (\mu + \sigma C_j)$. Note that $\Delta_1 > 0$. If $d_{\text{opt}} \geq (C_j + \Delta)$ and $C_i = C_j$, then the completion time of job J_j and the maintenance duration are, respectively,

$$C_j = t_j + (p_j + \lambda t_j) j^a = (1 + \lambda j^a)t_j + p_j j^a$$

and

$$f(t) = \mu + \sigma C_j = \mu + \sigma \left[(1 + \lambda j^a)t_j + p_j j^a\right].$$

If we shift job J_j and the maintenance to the right such that the finishing time of the maintenance coincides with time $(C_j + \Delta)$, then the completion time of job J_j and the maintenance duration, respectively, becomes

$$\overline{C}_j = (t_j + \overline{\Delta}) + \left[p_j + \lambda(t_j + \overline{\Delta})\right] j^a = (1 + \lambda j^a)t_j + p_j j^a + (1 + \lambda j^a)\overline{\Delta}$$

and

$$\overline{f}(t) = \mu + \sigma \overline{C}_j = \mu + \sigma \left[(1 + \lambda j^a)t_j + p_j j^a + (1 + \lambda j^a)\overline{\Delta}\right],$$

where $\overline{\Delta} > 0$ is the difference between the starting time of job J_j and t_j after shifting job J_j and the maintenance to the right such that the maintenance finishes at time $(C_j + \Delta)$. Obviously, $\overline{C}_j + \overline{f}(t) = C_j + \Delta$. We have that $\overline{\Delta} = (\Delta_1 / (1 + \lambda j^a)(1 + \sigma))$. Thus, the difference between C_j and $\overline{C}_j$ is $C_j - \overline{C}_j = -\frac{\Delta_1}{(1+\sigma)}$. Since $\sigma \geq 0$ and $\Delta_1 > 0$, $(C_j - \overline{C}_j) < 0$. That is, if we shift job J_j to the right, then the earliness cost of job J_j decreases. Hence, the total cost decreases.

Case (vii): $C_j < d_{\text{opt}} < (C_j + \Delta)$ and $C_i < C_j$

If $C_j < d_{\text{opt}} < (C_j + \Delta)$ and $C_i < C_j$, then we shift job J_j to the right and job J_{j+1} to the left such that the finishing time of job J_j coincides with time d_{opt} and the starting time of job J_{j+1} coincides with time d_{opt}. Based on the results of Case (i) and Case (iv), we obtain that the total cost decreases.

Case (viii): $C_j < d_{\text{opt}} < (C_j + \Delta)$ and $C_i \geq C_{j+1}$

If $C_j < d_{\text{opt}} < (C_j + \Delta)$ and $C_i \geq C_{j+1}$, then we shift job J_j to the right and job J_{j+1} to the left such that the finishing time of job J_j coincides with time d_{opt} and the starting time of job J_{j+1} coincides with time d_{opt}. Based on the results of Case (ii) and Case (v), we obtain that the total cost decreases.

Case (ix): $C_j < d_{\text{opt}} < (C_j + \Delta)$, $C_i = C_j$, and $d_{\text{opt}} > (\mu + (1+\sigma)C_i)$

If $C_j < d_{\text{opt}} < (C_j + \Delta)$, $C_i = C_j$, and $d_{\text{opt}} > (\mu + (1+\sigma)C_i)$, we first shift job J_j and the maintenance to the right such that the finishing time of the maintenance coincides with time d_{opt}, and then shift job J_{j+1} to the left such that the starting time of job J_{j+1}, coincides with time d_{opt}. Based on the results of Case (i) and Case (vi), we obtain that the total cost decreases.

Case (x): $C_j < d_{\text{opt}} < (C_j + \Delta)$, $C_i = C_j$, and $d_{\text{opt}} \leq (\mu + (1+\sigma)C_i)$

If $C_j < d_{\text{opt}} < (C_j + \Delta)$, $C_i = C_j$, and $d_{\text{opt}} \leq (\mu + (1+\sigma)C_i)$, then we shift job J_{j+1} to the left such that the starting time of job J_{j+1} coincides with the maintenance finishing time. Based on the result of Case (iii), we see that the total cost decreases.

After at most $(n-1)$ such steps, we get an optimal schedule without idle time between consecutive jobs. □

In Lemma 3.2, we show that there exists an optimal common due-window in which the due-window starting time and the due-window finishing time coincide with some jobs' completion times.

Lemma 3.2 ([58]). *For any given sequence π, there exists an optimal common due-window in which the due-window starting time d and the due-window finishing time w coincide with some jobs' completion times.*

Proof. The proof is similar to that of Lemma 1 of Yang et al. [58]. □

In Lemma 3.3, we show that an optimal schedule exists in which the first job in the schedule starts at time zero.

Lemma 3.3. *An optimal schedule exists in which the schedule starts at time zero.*

Proof. We prove this lemma by contradiction. Suppose that there exists an optimal schedule π in which the first job does not start at time zero. From Lemma 3.1, we have that an optimal schedule exists with no idle time between consecutive jobs. From Lemma 3.2, we have that an optimal schedule exists in which the due-window starting time d and the due-window finishing time w coincide with some jobs' completion times. If we shift all the jobs to the left until the first job starts at time zero for processing, then the actual processing times of all jobs and the maintenance duration become smaller. Meanwhile, the lengths of time from the jobs scheduled before the due-window to the due-window starting time d and those from the jobs scheduled after the due-window to the due-window finishing time w become smaller. Consequently, the total earliness, tardiness, and due-window related costs reduce. This contradicts the optimality of π and Lemma 3.3 follows. □

In what follows, we denote by k and $(k+m)$ the indices of the jobs completed at times d and w, respectively, where m denotes the number of jobs fully contained in the due-window. In Lemma 3.4, we show that the positions k and $(k+m)$ are functions of the cost parameters and are independent of the locations of the maintenance and the maintenance duration. That is, $C_{[k]} = d$ and $C_{[k+m]} = w$. We also show that the deteriorating and learning effects do not affect the values of k and $(k+m)$ obtained by Mosheiov and Sarig [40].

Lemma 3.4 ([40]). *For any given sequence π, there exists an optimal common due-window in which the starting time d equals $C_{[k]}$, where $k = \lceil n(\delta-\gamma)/\alpha \rceil$, and the finishing time w equals $C_{[k+m]}$, where $(k+m) = \lceil n(\beta-\delta)/\beta \rceil$.*

Proof. Consider an optimal schedule and an optimal due-window such that $C_{[k]} = d$ and $C_{[k+m]} = w$. Using the small perturbation technique introduced by Panwalkar et al. [43], we investigate the change in the total cost when the due-window starting time d and finishing time w are shifted.

When the due-window starting time d is moved Δ_1 units of time to the right, the change in the total cost is given by

$$\Delta Z_1 = \alpha k \Delta_1 + \gamma n \Delta_1 - \delta n \Delta_1. \tag{3.2}$$

On the other hand, when the due-window starting time d is moved Δ_1 units of time to the left, the change in the total cost is given by

$$\Delta Z_2 = -\alpha(k-1)\Delta_1 - n\gamma\Delta_1 + n\delta\Delta_1. \tag{3.3}$$

Both (3.2) and (3.3) are non-negative due to the optimality of the original solution. If $\Delta Z_1 = \alpha k \Delta_1 + \gamma n \Delta_1 - \delta n \Delta_1 \geq 0$, we have $k \geq n(\delta-\gamma)/\alpha$, whereas we obtain $k \leq (n(\delta-\gamma)/\alpha + 1)$ from $\Delta Z_2 \geq 0$. Since k is a non-negative integer, it follows that $k = \lceil n(\delta-\gamma)/\alpha \rceil$, where $k = \lceil n(\delta-\gamma)/\alpha \rceil$ denotes the smallest integer larger than or equal to $n(\delta-\gamma)/\alpha$.

Similarly, when the due-window finishing time w is moved Δ_2 units of time to the right, the change in the total cost is given by

$$\Delta Z_1 = -\beta(n-(k+m))\Delta_2 + n\delta\Delta_2. \tag{3.4}$$

On the other hand, when the due-window finishing time w is moved Δ_2 units of time to the left, the change in the total cost is given by

$$\Delta Z_2 = \beta(n-(k+m-1))\Delta_2 - n\delta\Delta_2. \tag{3.5}$$

Again, both (3.4) and (3.5) are non-negative due to the optimality of the original solution. If $\Delta Z_1 = -\beta(n-(k+m))\Delta_2 + n\delta\Delta_2 \geq 0$, we have $(k+m) \geq n(\beta-\delta)/\beta$, whereas we obtain $(k+m) \leq (n(\beta-\delta)/\beta + 1)$ from $\Delta Z_2 \geq 0$. Since $(k+m)$ is a non-negative integer, it follows that $(k+m) = \lceil n(\beta-\delta)/\beta \rceil$. □

3.3.2 Optimal Solutions

Let $J_{[j]}$ denote the job scheduled in the jth position in a job sequence. All the jobs are available for processing at time zero. Then, the actual processing time and the completion time of jobs scheduled before the maintenance can be expressed as follows:

$$\begin{aligned}
p_{[1]1} &= (p_{[1]} + \lambda \cdot 0)1^a = p_{[1]}1^a,\\
C_{[1]} &= t_{[1]} + p_{[1]1} = 0 + p_{[1]}1^a = p_{[1]}1^a,\\
p_{[2]2} &= (p_{[2]} + \lambda C_{[1]})2^a = \lambda p_{[1]}1^a 2^a + p_{[2]}2^a,\\
C_{[2]} &= C_{[1]} + p_{[2]2} = p_{[1]}1^a + \lambda p_{[1]}1^a 2^a + p_{[2]}2^a = p_{[1]}1^a(1+\lambda 2^a) + p_{[2]}2^a,\\
p_{[3]3} &= (p_{[3]} + \lambda C_{[2]})3^a = \lambda p_{[1]}1^a(1+\lambda 2^a)3^a + \lambda p_{[2]}2^a 3^a + p_{[3]}3^a,\\
C_{[3]} &= C_{[2]} + p_{[3]3} = p_{[1]}1^a(1+\lambda 2^a) + p_{[2]}2^a + \lambda p_{[1]}1^a(1+\lambda 2^a)3^a + \lambda p_{[2]}2^a 3^a\\
&\quad + p_{[3]}3^a = p_{[1]}1^a(1+\lambda 2^a)(1+\lambda 3^a) + p_{[2]}2^a(1+\lambda 3^a) + p_{[3]}3^a.
\end{aligned}$$

Hence, by induction, the actual processing time and the completion time of a job scheduled before the maintenance can be, respectively, expressed as follows:

$$\begin{aligned}
p_{[j]j} &= \lambda p_{[1]}1^a(1+\lambda 2^a)\cdots(1+\lambda(j-1)^a)\,j^a\\
&\quad + \lambda p_{[2]}2^a(1+\lambda 3^a)\cdots(1+\lambda(j-1)^a)\,j^a\\
&\quad + \cdots + \lambda p_{[j-2]}(j-2)^a(1+\lambda(j-1)^a)\,j^a + \lambda p_{[j-1]}(j-1)^a j^a + p_{[j]}j^a\\
&= \sum_{u=1}^{j} p_{[u]}v_{ju},
\end{aligned}$$

where $v_{ju} = \lambda u^a j^a \prod_{h=u+1}^{j-1}(1+\lambda h^a)$, $j = 1,2,\ldots,i$, $u = 1,2,\ldots,j-2$, $v_{j,j-1} = \lambda(j-1)^a j^a$, and $v_{jj} = j^a$, and

$$\begin{aligned}
C_{[j]} &= p_{[1]}1^a(1+\lambda 2^a)\cdots(1+\lambda j^a)\\
&\quad + p_{[2]}2^a(1+\lambda 3^a)\cdots(1+\lambda j^a)\\
&\quad + \cdots + p_{[j-1]}(j-1)^a(1+\lambda j^a) + p_{[j]}(j)^a\\
&= \sum_{u=1}^{j} p_{[u]}z_{ju},
\end{aligned}$$

where $z_{ju} = u^a \prod_{h=u+1}^{j}(1+\lambda h^a)$, $j = 1,2,\ldots,i$, $u = 1,2,\ldots,j$, and $\prod_{h=u+1}^{u}(1+\lambda h^a) = 1$.

As mentioned above, after the maintenance, the machine deterioration will reset first and then starts anew. Thus, the actual processing time and the completion time of jobs scheduled after the maintenance can be expressed as follows:

$$\begin{aligned}
p_{[i+1],i+1} &= (p_{[i+1]} + \lambda \cdot 0)(i+1)^a = p_{[i+1]}(i+1)^a,\\
C_{[i+1]} &= C_{[i]} + \mu + \sigma C_{[i]} + p_{[i+1],i+1} = \mu + (1+\sigma)C_{[i]} + p_{[i+1]}(i+1)^a,
\end{aligned}$$

$$\begin{aligned}
p_{[i+2],i+2} &= \left(p_{[i+2]} + \lambda p_{[i+1]}(i+1)^a\right)(i+2)^a \\
&= \lambda p_{[i+1]}(i+1)^a(i+2)^a + p_{[i+2]}(i+2)^a, \\
C_{[i+2]} &= C_{[i+1]} + p_{[i+2],i+2} \\
&= \mu + (i+\sigma)C_{[i]} + p_{[i+1]}(i+1)^a\left(1+\lambda(i+2)^a\right) + p_{[i+2]}(i+2)^a, \\
p_{[i+3],i+3} &= \left(p_{[i+3]} + \lambda\left(p_{[i+1],i+1} + p_{[i+2],i+2}\right)\right)(i+3)^a \\
&= \lambda p_{[i+1]}(i+1)^a\left(1+\lambda(i+2)^a\right)(i+3)^a + \lambda p_{[i+2]}(i+2)^a(i+3)^a \\
&\quad + p_{[i+3]}(i+3)^a, \\
C_{[i+3]} &= C_{[i+2]} + p_{[i+3],i+3} \\
&= \mu + (1+\sigma)C_{[i]} + p_{[i+1]}(i+1)^a\left(1+\lambda(i+2)^a\right)\left(1+\lambda(i+3)^a\right) \\
&\quad + p_{[i+2]}(i+2)^a\left(1+\lambda(i+3)^a\right) + p_{[i+3]}(i+3)^a.
\end{aligned}$$

Hence, by induction, the actual processing time and the completion time of a job scheduled after the maintenance can be, respectively, expressed as follows:

$$\begin{aligned}
p_{[j]j} &= \lambda p_{[i+1]}(i+1)^a\left(1+\lambda(i+2)^a\right)\cdots\left(1+\lambda(j-1)^a\right)j^a \\
&\quad + \lambda p_{[i+2]}(i+2)^a\left(1+\lambda(i+3)^a\right)\cdots\left[1+\lambda\left(j-(i+1)\right)^a\right]j^a \\
&\quad + \cdots + \lambda p_{[j-2]}(j-2)^a\left(1+\lambda(j-1)^a\right)j^a + \lambda p_{[j-1]}(j-1)^a j^a + p_{[j]}(j)^a \\
&= \sum_{u=i+1}^{j} p_{[u]}v_{ju},
\end{aligned}$$

where $v_{ju} = \lambda u^a j^a \prod_{h=u+1}^{j-1}(1+\lambda h^a)$, $j = i+1, i+2, \ldots, n$, $u = i+1, i+2, \ldots, j-2$, $v_{j,j-1} = \lambda(j-1)^a j^a$, and $v_{jj} = j^a$, and

$$\begin{aligned}
C_{[j]} &= \mu + (1+\sigma)C_{[i]} + p_{[i+1]}(i+1)^a\left(1+\lambda(i+2)^a\right)\cdots\left(1+\lambda j^a\right) \\
&\quad + p_{[i+2]}(i+2)^a\left(1+\lambda(i+3)^a\right)\cdots\left(1+\lambda j^a\right) \\
&\quad + \cdots + p_{[j-1]}(j-1)^a\left(1+\lambda j^a\right) + p_{[j]}j^a \\
&= \mu + (1+\sigma)\sum_{u=1}^{i} p_{[u]}z_{iu} + \sum_{u=i+1}^{j} p_{[u]}z_{ju}, \qquad (3.6)
\end{aligned}$$

where $z_{iu} = u^a \prod_{h=u+1}^{i}(1+\lambda h^a)$, $u = 1, 2, \ldots, j$, $z_{ju} = u^a \prod_{h=u+1}^{j}(1+\lambda h^a)$, $j = i+1, i+2, \ldots, n$, $u = i+1, i+2, \ldots, j$, and $\prod_{h=u+1}^{u}(1+\lambda h^a) = 1$.

By Lemma 3.4, we can determine the optimal locations of the due-window starting time d and the due-window finishing time w. If the maintenance is performed before the due-window (i.e., $i < k$), then the total cost is given by

$$\begin{aligned}
f(d, D, \pi) = Z &= \sum_{j=1}^{n}\left(\alpha E_{[j]} + \beta T_{[j]} + \gamma d + \delta D\right) \\
&= \alpha\left[\sum_{j=1}^{i}(d - C_{[j]}) + \sum_{j=i+1}^{k}(d - C_{[j]})\right] + \beta \sum_{j=k+m+1}^{n}(C_{[j]} - w)
\end{aligned}$$

$$
\begin{aligned}
&+n\gamma\left[\sum_{j=1}^{i} p_{[j]j}+\sum_{j=i+1}^{k} p_{[j]j}+\left(\mu+\sigma\sum_{j=1}^{i} p_{[j]j}\right)\right]+n\delta\sum_{j=k+1}^{k+m} p_{[j]j}\\
&=\alpha\left[\sum_{j=1}^{i}(j-1)p_{[j]j}+\sum_{j=i+1}^{k}(j-1)p_{[j]j}+i\left(\mu+\sigma\sum_{j=1}^{i} p_{[j]j}\right)\right]\\
&+\beta\sum_{j=k+m+1}^{n}(n-j+1)p_{[j]j}\\
&+n\gamma\left[\sum_{j=1}^{i} p_{[j]j}+\sum_{j=i+1}^{k} p_{[j]j}+\left(\mu+\sigma\sum_{j=1}^{i} p_{[j]j}\right)\right]+n\delta\sum_{j=k+1}^{k+m} p_{[j]j}\\
&=\mu(n\gamma+i\alpha)+\sum_{j=1}^{i}\sum_{u=1}^{j}[n\gamma(1+\sigma)+\alpha(j-1+\sigma i)]p_{[u]}v_{ju}\\
&+\sum_{j=i+1}^{k}\sum_{u=i+1}^{j}[n\gamma+\alpha(j-1)]p_{[u]}v_{ju}\\
&+\sum_{j=k+1}^{k+m}\sum_{u=i+1}^{j} n\delta p_{[u]}v_{ju}+\sum_{j=k+m+1}^{n}\sum_{u=i+1}^{j}\beta(n-j+1)p_{[u]}v_{ju}\\
&=M+\sum_{j=1}^{n} W_j p_{[j]},
\end{aligned}
\tag{3.7}
$$

where $M=\mu(n\gamma+i\alpha)$, and

$$
\begin{aligned}
W_1 &= (n\gamma(1+\sigma)+\alpha\sigma i)v_{11}+(n\gamma(1+\sigma)+\alpha(1+\sigma i))v_{21}\\
&\quad+(n\gamma(1+\sigma)+\alpha(2+\sigma i))v_{31}+\ldots+(n\gamma(1+\sigma)+\alpha(i-1+\sigma i))v_{i1},\\
W_2 &= (n\gamma(1+\sigma)+\alpha(1+\sigma i))v_{22}+(n\gamma(1+\sigma)+\alpha(2+\sigma i))v_{32}\\
&\quad+\ldots+(n\gamma(1+\sigma)+\alpha(i-1+\sigma i))v_{i2},\\
&\vdots\\
W_i &= (n\gamma(1+\sigma)+\alpha(i-1+\sigma i))v_{ii},\\
W_{i+1} &= (n\gamma+\alpha i)v_{i+1,i+1}+(n\gamma+\alpha(i+1))v_{i+2,i+1}+(n\gamma+\alpha(i+2))v_{i+3,i+1}\\
&\quad+(n\gamma+\alpha(k-1))v_{k,i+1}+n\delta(v_{k+1,i+1}+v_{k+2,i+1}+v_{k+3,i+1}\\
&\quad+\ldots+v_{k+m,i+1})+\beta(n-k-m)v_{k+m+1,i+1}\\
&\quad+\beta(n-k-m-1)v_{k+m+2,i+1}+\beta(n-k-m-2)v_{k+m+3,i+1}\\
&\quad+\ldots+\beta v_{n,i+1},\\
W_{i+2} &= (n\gamma+\alpha(i+1))v_{i+2,i+2}+(n\gamma+\alpha(i+2))v_{i+3,i+2}\\
&\quad+\ldots+(n\gamma+\alpha(k-1))v_{k,i+2}\\
&\quad+n\delta(v_{k+1,i+2}+v_{k+2,i+2}+v_{k+3,i+2}+\ldots+v_{k+m,i+2})\\
&\quad+\beta(n-k-m)v_{k+m+1,i+2}+\beta(n-k-m-1)v_{k+m+2,i+2}\\
&\quad+\beta(n-k-m-2)v_{k+m+3,i+2}+\ldots+\beta v_{n,i+2},
\end{aligned}
$$

$$
\begin{aligned}
&\vdots\\
W_k &= (n\gamma+\alpha(i+1))v_{k,k}+n\delta(v_{k+1,k}+v_{k+2,k}+\ldots+v_{k+m,k})\\
&\quad+\beta(n-k-m)v_{k+m+1,k}+\beta(n-k-m-1)v_{k+m+2,k}\\
&\quad+\beta(n-k-m-2)v_{k+m+3,k}+\ldots+\beta v_{nk},\\
W_{k+1} &= n\delta(v_{k+1,k+1}+v_{k+2,k+1}+\ldots+v_{k+m,k+1})\\
&\quad+\beta(n-k-m)v_{k+m+1,k+1}+\beta(n-k-m-1)v_{k+m+2,k+1}\\
&\quad+\beta(n-k-m-2)v_{k+m+3,k+1}+\ldots+\beta v_{n,k+1},\\
W_{k+2} &= n\delta(v_{k+2,k+2}+v_{k+3,k+2}+\ldots+v_{k+m,k+2})\\
&\quad+\beta(n-k-m)v_{k+m+1,k+2}+\beta(n-k-m-1)v_{k+m+2,k+2}\\
&\quad+\beta(n-k-m-2)v_{k+m+3,k+2}+\ldots+\beta v_{n,k+2},\\
&\vdots\\
W_{k+m} &= n\delta v_{k+m,k+m}+\beta(n-k-m)v_{k+m+1,k+m}+\beta(n-k-m-1)v_{k+m+2,k+m}\\
&\quad+\beta(n-k-m-2)v_{k+m+3,k+m}+\ldots+\beta v_{n,k+m},\\
W_{k+m+1} &= \beta(n-k-m)v_{k+m+1,k+m+1}+\beta(n-k-m-1)v_{k+m+2,k+m+1}\\
&\quad+\beta(n-k-m-2)v_{k+m+3,k+m+1}+\ldots+\beta v_{n,k+m+1},\\
W_{k+m+2} &= \beta(n-k-m-1)v_{k+m+2,k+m+2}+\beta(n-k-m-2)v_{k+m+3,k+m+2}\\
&\quad+\ldots+\beta v_{n,k+m+2},\\
&\vdots\\
W_n &= \beta v_{nn}.
\end{aligned}
$$

Moreover, applying the above analysis in a similar manner to the case where the maintenance is performed within the due-window (i.e., $k \le i \le (k+m)$), we can formulate the problem as follows:

$$
\begin{aligned}
f(d,D,\pi) = Z &= \sum_{j=1}^{n}(\alpha E_{[j]}+\beta T_{[j]}+\gamma d+\delta D)\\
&= \alpha\sum_{j=1}^{k}(d-C_{[j]})+\beta\sum_{j=k+m+1}^{n}(C_{[j]}-w)\\
&\quad+n\gamma\sum_{j=1}^{k}p_{[j]j}+n\delta\left[\sum_{j=k+1}^{i}p_{[j]j}+\sum_{j=i+1}^{k+m}p_{[j]j}+\left(\mu+\sigma\sum_{j=1}^{i}p_{[j]j}\right)\right]\\
&= \alpha\sum_{j=1}^{k}(j-1)p_{[j]j}+\beta\sum_{j=k+m+1}^{n}(n-j-1)p_{[j]j}+n\gamma\sum_{j=1}^{k}p_{[j]j}\\
&\quad+n\delta\left[\sum_{j=k+1}^{i}p_{[j]j}+\sum_{j=i+1}^{k+m}p_{[j]j}+\left(\mu+\sigma\sum_{j=1}^{i}p_{[j]j}\right)\right]
\end{aligned}
$$

$$
\begin{aligned}
&= n\delta\mu + \sum_{j=1}^{k}\sum_{u=1}^{j}[n\gamma + \alpha(j-1) + n\delta\sigma]p_{[u]}v_{ju} \\
&\quad + \sum_{j=k+1}^{i}\sum_{u=1}^{j} n\delta(1+\sigma)p_{[u]}v_{ju} \\
&\quad + \sum_{j=i+1}^{k+m}\sum_{u=i+1}^{j} n\delta p_{[u]}v_{ju} + \sum_{j=k+m+1}^{n}\sum_{u=i+1}^{j}\beta(n-j+1)p_{[u]}v_{ju} \\
&= M + \sum_{j=1}^{n} W_j p_{[j]}.
\end{aligned} \tag{3.8}
$$

Similarly, if the maintenance is performed after the due-window (i.e., $i \geq (k+m)$), then the problem can formulate as follows:

$$
\begin{aligned}
f(d,D,\pi) = Z &= \sum_{j=1}^{n}(\alpha E_{[j]} + \beta T_{[j]} + \gamma d + \delta D) \\
&= \alpha\sum_{j=1}^{k}(d - C_{[j]}) + \beta\left[\sum_{j=k+m+1}^{i}(C_{[j]} - w) + \sum_{j=i+1}^{n}(C_{[j]} - w)\right] \\
&\quad + n\gamma\sum_{j=1}^{k} p_{[j]j} + n\delta\sum_{j=k+1}^{k+m} p_{[j]j} \\
&= \alpha\sum_{j=1}^{k}(j-1)p_{[j]j} + \beta\left[\sum_{j=k+m+1}^{i}(n-j-1)p_{[j]j} + \sum_{j=i+1}^{n}(n-j-1)p_{[j]j}\right] \\
&\quad + \beta(n-i)\left(\mu + \sigma\sum_{j=1}^{i} p_{[j]j}\right) + n\gamma\sum_{j=1}^{k} p_{[j]j} + n\delta\sum_{j=k+1}^{k+m} p_{[j]j} \\
&= \beta\mu(n-i) + \sum_{j=1}^{k}\sum_{u=1}^{j}[n\gamma + \alpha(j-1) + \beta\sigma(n-i)]p_{[u]}v_{ju} \\
&\quad + \sum_{j=k+1}^{k+m}\sum_{u=1}^{j}[n\delta + \beta\sigma(n-i)]p_{[u]}v_{ju} \\
&\quad + \sum_{j=k+m+1}^{i}\sum_{u=1}^{j}[\beta(n-j+1) + \beta\sigma(n-i)]p_{[u]}v_{ju} \\
&\quad + \sum_{j=i+1}^{n}\sum_{u=i+1}^{j}\beta(n-j+1)p_{[u]}v_{ju} \\
&= M + \sum_{j=1}^{n} W_j p_{[j]}.
\end{aligned} \tag{3.9}
$$

Note that if $i = n$, then it means that no maintenance is scheduled in the job sequence.

For solving (3.7)–(3.9), we introduce a simple matching algorithm to find the minimum of the sum of the products of two sequences of numbers.

Lemma 3.5 ([18]). *Let there be two sequences of numbers x_i and y_i. The sum of the products of the corresponding elements $\sum_{i=1}^{n} x_i y_i$ is the least if the sequences are monotonic in the opposite sense.*

By Lemma 3.4, we can determine the optimal positions of the due-window starting time $k = \lceil n(\delta - \gamma)/\alpha \rceil$ and the due-window finishing time $(k+m) = \lceil n(\beta - \delta)/\beta \rceil$. In the following, we develop a polynomial time algorithm to solve the $1/p_{jr} = (p_j + \lambda t)r^a,\ ma/\sum_{j=1}^{n}(\alpha E_j + \beta T_j + \gamma d + \delta D)$ problem.

Algorithm

Step 1. Set $i = 1$.
Step 2. For $j = 1, 2, \ldots, n$, calculate the positional weights W_j.
Step 3. Renumber the jobs in non-decreasing order of their normal processing times p_j. By Lemma 3.5, the job with the largest normal processing time is assigned to the position with the smallest value of the positional weight W_j, the job with the next largest normal processing time is assigned to the position with the next smallest value of the positional weight W_j, and so on. Then, obtain a local optimal schedule and the corresponding total cost.
Step 4. $i = i + 1$. If $i \leq n$, then go to Step 2. Otherwise, go to Step 5.
Step 5. The optimal schedule is the one with the minimum total cost.

The time complexity of Step 2 is $O(1)$ and the time complexity of Step 3 is $O(n\log n)$. Since the maintenance can be scheduled immediately after any one of the jobs, n different maintenance positions must be evaluated to obtain the global optimal schedule in Step 4. Thus, we conclude that the following theorem holds.

Theorem 3.1. *The $1/p_{jr} = (p_j + \lambda t)r^a,\ ma/\sum_{j=1}^{n}(\alpha E_j + \beta T_j + \gamma d + \delta D)$ problem can be optimally solved in $O(n^2 \log n)$ time.*

Clearly, if no maintenance is scheduled in the job sequence (i.e., $i = n$), then the problem reduces to the $1/p_{jr} = (p_j + \lambda t)r^a/\sum_{j=1}^{n}(\alpha E_j + \beta T_j + \gamma d + \delta D)$ problem and can be formulated as follows:

$$
\begin{aligned}
f(d, D, \pi) &= Z = \sum_{j=1}^{n} (\alpha E_{[j]} + \beta T_{[j]} + \gamma d + \delta D) \\
&= \alpha \sum_{j=1}^{k} (d - C_{[j]}) + \beta \sum_{j=k+m+1}^{n} (C_{[j]} - w) + n\gamma \sum_{j=1}^{k} p_{[j]j} + n\delta \sum_{j=k+1}^{k+m} p_{[j]j} \\
&= \sum_{j=1}^{k} [n\gamma + \alpha(j-1)] p_{[j]j} + \sum_{j=k+1}^{k+m} n\delta p_{[j]j} + \sum_{j=k+m+1}^{n} \beta(n-j+1) p_{[j]j} \\
&= \sum_{j=1}^{k} \sum_{u=1}^{j} [n\gamma + \alpha(j-1)] p_{[u]} v_{ju}
\end{aligned}
$$

$$+\sum_{j=k+1}^{k+m}\sum_{u=1}^{j} n\delta p_{[u]}v_{ju} + \sum_{j=k+m+1}^{n}\sum_{u=1}^{j}\beta(n-j+1)p_{[u]}v_{ju}$$
$$=\sum_{j=1}^{n} W_j p_{[j]}. \tag{3.10}$$

where $v_{ju} = \lambda u^a j^a \prod_{h=u+1}^{j-1}(1+\lambda h^a)$, $j = 1,2,\dots,n$, $u = 1,2,\dots,j-2$, $v_{j,j-1} = \lambda(j-1)^a j^a$, and $v_{jj} = j^a$. Thus, the following corollary holds.

Corollary 3.1. *The* $1/p_{jr} = (p_j+\lambda t)r^a/\sum_{j=1}^{n}(\alpha E_j+\beta T_j+\gamma d+\delta D)$ *problem can be optimally solved in* $O(n\log n)$ *time.*

3.4 Problem $1/p_{jr} = (p_j+\lambda t)r^a$, $ma/C_{\max}$

Based on the analysis mentioned above, we know that an optimal schedule exists with no idle time between consecutive jobs and the optimal schedule starts at time zero. From (3.6), the completion time of the last job in a job sequence can be expressed as

$$\begin{aligned}C_{[n]} &= \mu+(1+\sigma)C_{[i]}+p_{[i+1]}(i+1)^a(1+\lambda(i+2)^a)\cdots(1+\lambda n^a)\\&\quad +p_{[i+2]}(i+2)^a(1+\lambda(i+3)^a)\cdots(1+\lambda n^a)\\&\quad +\cdots+p_{[n-1]}(n-1)^a(1+\lambda n^a)+p_{[n]}n^a\\&=\mu+(1+\sigma)\sum_{u=1}^{i}p_{[u]}z_{iu}+\sum_{u=i+1}^{n}p_{[u]}z_{nu},\end{aligned}$$

where $z_{iu} = u^a\prod_{h=u+1}^{i}(1+\lambda h^a)$, $u = 1,2,\dots,i$, $z_{nu} = u^a\prod_{h=u+1}^{n}(1+\lambda h^a)$, $u = i+1,i+2,\dots,n$, and $\prod_{h=u+1}^{u}(1+\lambda h^a) = 1$. Then, the makespan of all jobs is given by

$$\begin{aligned}C_{\max} &= \mu+(1+\sigma)\sum_{u=1}^{i}p_{[u]}z_{iu}+\sum_{u=i+1}^{n}p_{[u]}z_{nu}\\&\quad \mu+\sum_{u=1}^{i}p_{[u]}(1+\sigma)z_{iu}+\sum_{u=i+1}^{n}p_{[u]}z_{nu}\\&=\mu+\sum_{u=1}^{n}p_{[u]}w_u,\end{aligned} \tag{3.11}$$

where

$$w_u=\begin{cases}(1+\sigma)u^a\prod_{h=u+1}^{i}(1+\lambda h^a) & u=1,2,\dots,i\\ u^a\prod_{h=u+1}^{n}(1+\lambda h^a) & u=i+1,i+2,\dots,n\end{cases}.$$

Observe that (3.11) can be viewed as the scalar product of the w_u and $p_{[u]}$ vectors, $u = 1, 2, \ldots, n$. Therefore, by Lemma 3.5, all the jobs are sorted in non-decreasing order of their normal processing times (p_u) first, and then the job with the largest normal processing time is assigned to the position with the smallest value of the positional weight w_u, the job with the next largest normal processing time is assigned to the position with the next smallest value of the positional weight w_u, and so on. The time complexity of a sorting algorithm is $O(n \log n)$. Since the maintenance can be scheduled immediately after any one of the jobs, n different maintenance positions must be evaluated to obtain the global optimal schedule. Therefore, the time complexity for the $1/p_{jr} = (p_j + \lambda t)r^a$, $ma/C_{\max}$ problem is $O(n^2 \log n)$.

Theorem 3.2. *The* $1/p_{jr} = (p_j + \lambda t)r^a$, $ma/C_{\max}$ *problem can be optimally solved in* $O(n^2 \log n)$ *time.*

Moreover, if no maintenance is scheduled in the job sequence, then the problem reduces to the $1/p_{jr} = (p_j + \lambda t)r^a/C_{\max}$ problem. As a result, (3.11) can be reformulated as

$$C_{\max} = \sum_{u=1}^{n} p_{[u]} w_u, \tag{3.12}$$

where $w_u = u^a \prod_{h=u+1}^{n}(1 + \lambda h^a)$, $u = 1, 2, \ldots, n$, and $\prod_{h=n+1}^{n}(1 + \lambda h^a) = 1$. Then (3.12) can be viewed as the scalar product of the w_u and $p_{[u]}$ vectors, $u = 1, 2, \ldots, n$. Since $\lambda > 0$ and $a < 0$, the positional weights w_u are already sorted in a non-increasing order, $u = 1, 2, \ldots, n$, and therefore, by Lemma 3.5, the optimal job sequence for the $1/p_{jr} = (p_j + \lambda t)r^a/C_{\max}$ problem is well-known the shortest processing time (SPT) sequence.

Lemma 3.6 ([48, 56]). *For the* $1/p_{jr} = (p_j + \lambda t)r^a/C_{\max}$ *problem, the optimal schedule is obtained by sequencing jobs in the SPT rule.*

3.5 Problem $1/p_{jr} = (p_j + \lambda t)r^a$, ma/TC

Similarly, the $1/p_{jr} = (p_j + \lambda t)r^a$, ma/TC problem can be formulated as follows:

$$\begin{aligned} TC &= \sum_{j=1}^{n} C_{[j]} \\ &= (n-i)\mu + \sum_{j=1}^{i}\sum_{u=1}^{j} [(n-j+1) + (n-i)\sigma] p_{[u]} v_{ju} \\ &\quad + \sum_{j=i+1}^{n}\sum_{u=i+1}^{j} (n-j+1) p_{[u]} v_{ju}. \end{aligned} \tag{3.13}$$

Hence, following the analysis of the $1/p_{jr} = (p_j + \lambda t)r^a/\sum_{j=1}^{n}(\alpha E_j + \beta T_j + \gamma d + \delta D)$ problem, we conclude that the following theorem holds.

Theorem 3.3. *The $1/p_{jr} = (p_j + \lambda t)r^a$, ma/TC problem can be optimally solved in $O(n^2 \log n)$ time.*

In addition, if no maintenance is arranged in the job sequence, then the problem reduces to the $1/p_{jr} = (p_j + \lambda t)r^a/TC$ problem. Consequently, (3.13) can be rewritten as

$$TC = \sum_{u=1}^{n} p_{[u]} w_u, \tag{3.14}$$

where $w_u = u^a \sum_{k=u}^{n} \prod_{h=u+1}^{k} (1 + \lambda h^a)$, $u = 1,2,\ldots,n$, and $\prod_{h=u+1}^{u} (1 + \lambda h^a) = 1$. Then (3.15) can be viewed as the scalar product of the w_u and $p_{[u]}$ vectors, $u = 1,2,\ldots,n$. Since $\lambda > 0$ and $a < 0$, the positional weights w_u are already sorted in a non-increasing order, $u = 1,2,\ldots,n$, and therefore, by Lemma 3.5, the optimal job sequence for the $1/p_{jr} = (p_j + \lambda t)r^a/TC$ problem is still the well-known SPT sequence.

Lemma 3.7 ([48,56]). *For the $1/p_{jr} = (p_j + \lambda t)r^a/TC$ problem, the optimal schedule is obtained by sequencing jobs in the SPT rule.*

3.6 Problem $1/p_{jr} = (p_j + \lambda t)r^a$, $ma/TADC$

In this section, we consider a scheduling problem with the objective of minimizing the variation of jobs in flow time. This scheduling measure was first considered by Kanet [25]. This type of problem has applications in any service or manufacturing setting whenever it is deemed desirable to provide jobs (or customers) the same treatment, that is, each job (or customer) spends approximately the same time in the system or waits for service the same time as every other job (or customer). The $1/p_{jr} = (p_j + \lambda t)r^a$, $ma/TADC$ problem can be formulated as follows:

$$\begin{aligned}
TADC &= \sum_{j=1}^{n} \sum_{k=j}^{n} |C_j - C_k| \\
&= \sum_{j=1}^{n} (2j - n - 1) C_{[j]} \\
&= \sum_{j=1}^{i} \left[(2j - n - 1) \sum_{u=1}^{j} p_{[u]} z_{ju} \right] \\
&\quad + \sum_{j=i+1}^{n} \left[(2j - n - 1) \left(\mu + (1+\sigma) \sum_{u=1}^{i} p_{[u]} z_{iu} + \sum_{u=i+1}^{j} p_{[u]} z_{ju} \right) \right] \\
&= \sum_{j=1}^{i} \sum_{u=1}^{j} (2j - n - 1) u^a \prod_{h=u+1}^{j} (1 + \lambda h^a) p_{[u]}
\end{aligned}$$

$$+\sum_{j=i+1}^{n}\left[(2j-n-1)\left(\mu+\sum_{u=1}^{i}(1+\sigma)u^a\prod_{h=u+1}^{j}(1+\lambda h^a)p_{[u]}\right)\right.$$
$$\left.+\sum_{u=i+1}^{j}(2j-n-1)u^a\prod_{h=u+1}^{j}(1+\lambda h^a)p_{[u]}\right], \tag{3.15}$$

where $\prod_{h=u+1}^{u}(1+\lambda h^a)=1$. Similar to the analysis of the $1/p_{jr}=(p_j+\lambda t)r^a/C_{\max}$ problem, we conclude that the following theorem holds.

Theorem 3.4. *The $1/p_{jr}=(p_j+\lambda t)r^a$, $ma/TADC$ problem can be optimally solved in $O(n^2\log n)$ time.*

Moreover, if no maintenance is scheduled in the job sequence, then the problem reduces to the $1/p_{jr}=(p_j+\lambda t)r^a/TADC$. Consequently, (3.15) can be rewritten as

$$\begin{aligned}
TADC &= \sum_{j=1}^{n}\sum_{k=j}^{n}|C_j-C_k|\\
&= \sum_{j=1}^{n}(2j-n-1)C_{[j]}\\
&= \sum_{j=1}^{n}\sum_{u=1}^{j}(2j-n-1)p_{[u]}z_{ju}\\
&= \sum_{j=1}^{n}\sum_{u=j}^{n}(2u-n-1)z_{uj}p_{[j]}\\
&= \sum_{j=1}^{n}\left[\sum_{u=j}^{n}\left((2u-n-1)j^a\prod_{h=j+1}^{u}(1+\lambda h^a)\right)p_{[j]}\right].
\end{aligned} \tag{3.16}$$

Let $w_j=\sum_{u=j}^{n}(2u-n-1)j^a\prod_{h=j+1}^{u}(1+\lambda h^a)$, $j=1,\ldots,n$, where $\prod_{h=j+1}^{j}(1+\lambda h^a)=1$. Then (3.16) can be viewed as the scalar product of the w_j and $p_{[j]}$ vectors, $j=1,\ldots,n$. Based on the above analysis and Lemma 3.5, we conclude that the following lemma holds.

Lemma 3.8 ([56]). *The $1/p_{jr}=(p_j+\lambda t)r^a/TADC$ problem can be optimally solved in $O(n\log n)$ time.*

3.7 Conclusions

This paper considered single-machine scheduling problems simultaneous with deteriorating and learning effects under a deteriorating maintenance consideration. The objectives were to minimize the sum of earliness, tardiness, and due-window related costs, the makespan, the total completion time and TADC problems. We showed that all the versions of the problem under study are polynomially solvable. Table 3.1

Table 3.1: The time complexity results for various cases of the problem

Problem	Complexity	Reference
$1/p_{jr} = (p_j + \lambda t)r^a,\ ma/\sum_{j=1}^{n}(\alpha E_j + \beta T_j + \gamma d + \delta D)$	$O(n^2 \log n)$	Theorem 3.1
$1/p_{jr} = (p_j + \lambda t)r^a/\sum_{j=1}^{n}(\alpha E_j + \beta T_j + \gamma d + \delta D)$	$O(n \log n)$	Corollary 1
$1/p_{jr} = (p_j + \lambda t)r^a,\ ma/C_{\max}$	$O(n^2 \log n)$	Theorem 3.2
$1/p_{jr} = (p_j + \lambda t)r^a/C_{\max}$	SPT rule	Lemma 3.6 [48, 56]
$1/p_{jr} = (p_j + \lambda t)r^a,\ ma/TC$	$O(n^2 \log n)$	Theorem 3.3
$1/p_{jr} = (p_j + \lambda t)r^a/TC$	SPT rule	Lemma 3.7 [48, 56]
$1/p_{jr} = (p_j + \lambda t)r^a,\ ma/TADC$	$O(n^2 \log n)$	Theorem 3.4
$1/p_{jr} = (p_j + \lambda t)r^a/TADC$	$O(n \log n)$	Lemma 3.8 [56]

lists the time complexity results for variants of the problem and its special cases. It is worthwhile for future research to study scheduling problems simultaneous with other models of deterioration and learning, in multi-machine settings, and optimizing other performance measures.

Acknowledgments We thank the editors and the anonymous reviewers for their helpful comments and suggestions on an earlier version of the paper. This research was supported by the National Science Council of Taiwan, Republic of China, under grant number NSC 100-2221-E-252-002-MY2.

References

1. Alidaee, B., Womer, N.: Scheduling with time dependent processing times: Review and extensions. Journal of the Operational Research Society **50**, 711–729 (1999)
2. Bachman, A., Janiak, A.: Scheduling jobs with position-dependent processing times. Journal of the Operational Research Society **55**, 257–264 (2004)
3. Biskup, D.: Single-machine scheduling with learning considerations. European Journal of Operational Research **115**, 173–178 (1999)
4. Biskup, D.: A state-of-the-art review on scheduling with learning effects. European Journal of Operational Research **188**, 315–329 (2008)
5. Browne, S., Yechiali, U.: Scheduling deteriorating jobs on a single processor. Operations Research **38**, 495–498 (1990)
6. Cheng, T.C.E.: Optimal common due-date with limited completion time deviation. Computers & Operations Research **15**, 91–96 (1988)
7. Cheng, T.C.E., Ding, Q., Lin, B.: A concise survey of scheduling with time-dependent processing times. European Journal of Operational Research **152**, 1–13 (2004)
8. Cheng, T.C.E., Kang, L., Ng, C.: Due-date assignment and single machine scheduling with deteriorating jobs. Journal of the Operational Research Society **55**, 198–203 (2004)
9. Cheng, T.C.E., Kovalyov, M.: Scheduling with learning effects on job processing times. Working Paper 06/94, Faculty of Business and Information Systems, The Hong Kong Polytechnic University (1994)
10. Cheng, T.C.E., Wang, G.: Single machine scheduling with learning effect considerations. Annals of Operations Research **98**, 273–290 (2000)

11. Cheng, T.C.E., Yang, S.J., Yang, D.L.: Common due-window assignment and scheduling of linear time-dependent deteriorating jobs and a deteriorating maintenance activity. International Journal of Production Economics doi:10.1016/j.ijpe.2010.10.005
12. Gawiejnowicz, S.: A note on scheduling on a single processor with speed dependent on a number of executed jobs. Information Processing Letters **56**, 297–300 (1996)
13. Gawiejnowicz, S.: Scheduling deteriorating jobs subject to job or machine availability constraints. European Journal of Operational Research **180**, 472–478 (2007)
14. Gawiejnowicz, S.: Time-dependent Scheduling. Springer-Verlag, New York (2008)
15. Gawiejnowicz, S., Kononov, A.: Complexity and approximability of scheduling resumable proportionally deteriorating jobs. European Journal of Operational Research **200**, 305–308 (2010)
16. Graham, R., Lawler, E., J.K., L., Rinnooy Kan, A.: Optimization and approximation in deterministic sequencing and scheduling: A survey. Annals of Discrete Mathematics **5**, 287–326 (1979)
17. Gupta, J., Gupta, S.: Single facility scheduling with nonlinear processing times. Computers and Industrial Engineering **14**, 387–393 (1988)
18. Hardy, G., Littlewood, J., Polya, G.: Inequalities. Cambridge University Press, London (1967)
19. Hsu, C.J., Low, C., Su, C.T.: Single-machine scheduling problem with an availability constraint under simple linear deterioration. Journal of the Chinese Institute of Industrial Engineers **27**, 189–198 (2010)
20. Huang, X., Wang, J.B., Wang, L.Y., Gao, W.J., Wang, X.R.: Single machine scheduling with time-dependent deterioration and exponential learning effect. Computers and Industrial Engineering **58**, 58–63 (2010)
21. Janiak, A., Kovalyov, M.Y.: Scheduling deteriorating jobs. In: A. Janiak (ed.) Scheduling in Computer and Manufacturing Systems, pp. 12–25. WKL, Warszawa, Poland (2006)
22. Janiak, A., Rudek, R.: Scheduling problems with position dependent job processing times. In: A. Janiak (ed.) Scheduling in Computer and Manufacturing Systems, pp. 26–32. WKL, Warszawa, Poland (2006)
23. Janiak, A., Rudek, R.: Experience based approach to scheduling problems with the learning effect. IEEE Transactions on Systems, Man and Cybernetics, Part A: Systems and Humans **39**, 344–357 (2009)
24. Ji, M., He, Y., Cheng, T.C.E.: Scheduling linear deteriorating jobs with an availability constraint on a single machine. Theoretical Computer Science **362**, 115–126 (2006)
25. Kanet, J.: Minimizing variation of flow time in single machine systems. Management Science **27**, 1453–1459 (1981)
26. Kubzin, M., Strusevich, V.: Two-machine flow shop no-wait scheduling with machine maintenance. 4OR: A Quarterly Journal of Operations Research **3**, 303–313 (2005)
27. Kubzin, M., Strusevich, V.: Planning machine maintenance in two-machine shop scheduling. Operations Research **54**, 789–800 (2006)
28. Kunnathur, A., Gupta, S.: Minimizing the makespan with late start penalties added to processing times in a single facility scheduling problem. European Journal of Operational Research **47**, 56–64 (1990)
29. Kuo, W.H., Yang, D.L.: Minimizing the total completion time in a single-machine scheduling problem with a time-dependent learning effect. European Journal of Operational Research **174**, 1184–1190 (2006)
30. Lee, I.S.: Single machine scheduling with controllable processing times: A parametric study. International Journal of Production Economics **22**, 105–110 (1991)
31. Lee, W.C.: A note on deteriorating jobs and learning in single-machine scheduling problems. International Journal of Business and Economics **3**, 83–89 (2004)
32. Liman, S., Ramaswamy, S.: Earliness-tardiness scheduling problems with a common delivery window. Operations Research Letters **15**, 195–203 (1994)
33. Liman, S., Panwalkar, S.S., Thongmee, S.: Determination of common due window location in a single machine scheduling problem. European Journal of Operational Research **93**, 68–74 (1996)

34. Liman, S., Panwalkar, S.S., Thongmee, S.: Common due window size and location determination in a single machine scheduling problem. Journal of the Operational Research Society **49**, 1007–1010 (1998)
35. Lodree Jr., E., Geiger, C.: A note on the optimal sequence position for a rate-modifying activity under simple linear deterioration. European Journal of Operational Research **201**, 644–648 (2010)
36. Low, C., Hsu, C.J., Su, C.T.: Minimizing the makespan with an availability constraint on a single machine under simple linear deterioration. Computers & Mathematics with Applications **56**, 257–265 (2008)
37. Ma, Y., Chu, C., Zuo, C.: A survey of scheduling with deterministic machine availability constraints. Computers and Industrial Engineering **58**, 199–211 (2010)
38. Mosheiov, G.: Scheduling jobs with step-deterioration: Minimizing makespan on a single- and multi-machine. Computers and Industrial Engineering **28**, 869–879 (1995)
39. Mosheiov, G., Sarig, A.: A due-window assignment problem with position-dependent processing times. Journal of the Operational Research Society **59**, 997–1003 (2008)
40. Mosheiov, G., Sarig, A.: Scheduling a maintenance activity and due-window assignment on a single machine. Computers & Operations Research **36**, 2541–2545 (2009)
41. Mosheiov, G., Sidney, J.: Scheduling with general job-dependent learning curves. European Journal of Operational Research **147**, 665–670 (2003)
42. Mosheiov, G., Sidney, J.: Scheduling a deteriorating maintenance activity on a single machine. Journal of the Operational Research Society **61**, 882–887 (2010)
43. Panwalkar, S., Smith, M., Seidmann, A.: Common due date assignment to minimize total penalty for the one machine scheduling problem. Operations Research **30**, 391–399 (1982)
44. Sanlaville, E., Schmidt, G.: Machine scheduling with availability constraints. Acta Informatica **9**, 795–811 (1998)
45. Schmidt, G.: Scheduling with limited machine availability. European Journal of Operational Research **121**, 1–15 (2000)
46. Sun, L.: Single-machine scheduling problems with deteriorating jobs and learning effects. Computers and Industrial Engineering **57**, 843–846 (2009)
47. Toksar, M., Oron, D., Güner, E.: Single machine scheduling problems under the effects of nonlinear deterioration and time-dependent learning. Computers & Mathematics with Applications **50**, 401–406 (2009)
48. Wang, J.B.: A note on scheduling problems with learning effect and deteriorating jobs. International Journal of Systems Science **37**, 827–833 (2006)
49. Wang, J.B.: Single-machine scheduling problems with the effects of learning and deterioration. Omega **35**, 397–402 (2007)
50. Wang, J.B.: Single machine scheduling with learning effect and deteriorating jobs. Computers and Industrial Engineering **57**, 1452–1456 (2009)
51. Wang, J.B.: Single machine scheduling with time-dependent learning effect and deteriorating jobs. Journal of the Operational Research Society **60**, 583–586 (2009)
52. Wang, J.B., Cheng, T.: Scheduling problems with the effects of deterioration and learning. Asia-pacific Journal of Operational Research **24**, 245–261 (2007)
53. Wang, J.B., Huang, X., Wang, X.Y., Yin, N., Wang, L.Y.: Learning effect and deteriorating jobs in the single machine scheduling problems. Applied Mathematical Modelling **33**, 3848–3853 (2009)
54. Wang, X., Cheng, T.C.E.: Single-machine scheduling with deteriorating jobs and learning effects to minimize the makespan. European Journal of Operational Research **178**, 57–70 (2007)
55. Wu, C.C., Lee, W.C.: Scheduling linear deteriorating jobs to minimize makespan with an availability constraint on a single machine. Information Processing Letters **87**, 89–93 (2003)
56. Yang, D.L., Kuo, W.H.: Single-machine scheduling with both deterioration and learning effects. Annals of Operations Research **172**, 315–327 (2009)
57. Yang, D.L., Kuo, W.H.: Some scheduling problems with deteriorating jobs and learning effects. Computers and Industrial Engineering **58**, 25–28 (2010)

58. Yang, S.J., Yang, D.L., Cheng, T.C.E.: Single-machine due-window assignment and scheduling with job-dependent aging effects and deteriorating maintenance. Computers & Operations Research **37**, 1510–1514 (2010)
59. Yin, Y., Xu, D., Sun, K., Li, H.: Some scheduling problems with general position-dependent and time-dependent learning effects. Information Sciences **179**, 2416–2425 (2009)

Chapter 4
Scheduling Unrelated Parallel Machines with Sequence Dependent Setup Times and Weighted Earliness–Tardiness Minimization

Eva Vallada and Rubén Ruiz

Abstract This work deals with the unrelated parallel machine scheduling problem with machine and job-sequence dependent setup times. The studied objective is the minimization of the total weighted earliness and tardiness. We study existing Mixed Integer Programming (MIP) mathematical formulations. A genetic algorithm is proposed, which includes a procedure for inserting idle times in the production sequence in order to improve the objective value. We also present a benchmark of small and large instances to carry out the computational experiments. After an exhaustive computational and statistical analysis, the conclusion is that the proposed method shows a good performance.

4.1 Introduction

Parallel machines shops are particularly interesting settings as they model real situations in which a bottleneck stage has been replicated with several machines in order to increase capacity. Busy stages in multi-stage hybrid flowshops can be seen as parallel machines settings. More formally, parallel machines problems aim to schedule a set N of n jobs on a set M of m machines that are disposed in parallel. Each job has to visit exactly one machine and it is assumed that all machines are capable of processing all jobs. Each machine cannot process more than one job at the same time and once started, jobs must be processed through completion. This is known as no preemption allowed. The most specific case of parallel machines problems is when all the m available machines are identical. This means that there is no difference between processing one job in one machine or another. In these cases, the

Eva Vallada and Rubén Ruiz
Grupo de Sistemas de Optimización Aplicada, Instituto Tecnológico de Informática (ITI), Ciudad Politécnica de la Innovación, Edificio 8G, Acceso B, Universidad Politécnica de Valencia, Camino de Vera s/n, 46022 Valencia, Spain, e-mail: {evallada,rruiz}@eio.upv.es

R.Z. Ríos-Mercado and Y.A. Ríos-Solís (eds.), *Just-in-Time Systems*, Springer Optimization and Its Applications 60, DOI 10.1007/978-1-4614-1123-9_4,

input data is a vector of processing times for the jobs $j = 1, \ldots, n$ in the machines that is denoted by p_j. These processing times are non-negative, fixed and known in advance. Furthermore, processing times are usually assumed to be integer. A more general case is when machines are said to be uniform and the processing time of a job j on machine i follows the relationship $p_{ij} = p_j / s_i$, where s_i represents a different speed for machine i when processing the jobs, that is, according to s_i, some machines are "faster" or "slower" for all the jobs. The most general case, which includes the two previous ones as special cases, is the unrelated parallel machines. In this scenario, the processing time for each job depends on the machine to which it is assigned to. Therefore, the input data is a processing time matrix p_{ij}. This paper deals with this last and most general case.

In the literature, the most commonly studied criterion is the minimization of the maximum completion time or makespan ($C_{\max}$). According to the $\alpha/\beta/\gamma$ classification notation of [14], the unrelated parallel machines problem with the makespan criterion is denoted as $R//C_{\max}$. It is interesting to note that this scheduling setting, with this objective, is just a type of assignment problem. Let us explain this last fact in more detail. We denote by J_i the set of jobs assigned to machine i. Machine i will be occupied processing all of its assigned jobs in J_i during $C_i = \sum_{\forall k \in J_i} p_{ik}$ time units. C_i represents the completion time of machine i. Given all completion times C_i for all machines $i \in M$, the makespan is simply calculated as $C_{\max} = \max_{i \in M}\{C_i\}$. It is straightforward to see that the different C_i values are not affected by the order in which the jobs in J_i are processed by the machines. As a result, the total number of possible solutions in the $R//C_{\max}$ problem is m^n and the sequencing problem is irrelevant, only the assignment problem is to be considered. This problem is $\mathcal{NP}$-Hard in the strong sense, as [12] showed that the special case with identical machines (referred to as $P//C_{\max}$), is also $\mathcal{NP}$-Hard. Moreover, [25] demonstrated that even the two machines version, denoted as $P2//C_{\max}$ is already $\mathcal{NP}$-Hard.

Even though makespan is the most widely studied optimization criterion in the literature, it has been criticized countless times due to the fact that client satisfaction is neglected. Jobs typically model, in one way or another, orders placed by clients that usually have to delivered by a specific due date. Due dates are, as other given data, non-negative, fixed and known in advance and are denoted as d_j. Ideally, and given the time C_j at which one job $j \in N$ is finished, one seeks to finish jobs prior to their due dates, that is, $C_j < d_j$. In this context, an optimization criterion closely tied with client satisfaction is the minimization of the total tardiness, or $\sum_{j=1}^{n} T_j$, where T_j the measure of tardiness of job j, defined as $T_j = \max\{C_j - d_j, 0\}$. In some circumstances, it is not economical to finish jobs in advance, that is, before their due dates. In these cases, jobs (products) have to be stored and no revenue is obtained from them until the delivery date is due. We denote by E_j the earliness of a job j, calculated as $E_j = \max\{d_j - C_j, 0\}$. Just-in-Time (JIT) scheduling seeks precisely to reduce both the amount of tardiness as well as earliness, and to complete jobs exactly by their due dates or as close as possible to their due dates. This is referred to as total earliness–tardiness minimization or $\sum_{j=1}^{n}(E_j + T_j)$. Note that a job is either late, or early, or exactly on time. This paper deals with a generalized weighted version of earliness–tardiness minimization. Note that not all jobs are equally important and priorities are often used in practice. Each unit of time that a job is early (tardy) is multiplied by a early (tardy) priority which is different for each job. As a result, the objective considered in this work is $\sum_{j=1}^{n}(w'_jE_j + w_jT_j)$ where w'_j is the early weight or priority and w_j is the tardiness weight or priority, not necessarily equal to w'_j. Earliness–tardiness or JIT scheduling is an active field of research, with some well known reviews available as the one of Baker and Scudder [5]. Also Zhu, and Meredith [46] highlight key aspects of JIT implementation in industries, being JIT scheduling one of them.

In order to make the problem studied in this paper resemble real production shops, we additionally consider setup times. Setup times are non-productive periods needed at machines in order to make cleanings, configurations or preparations for the production in between jobs. Existence of setup times at machines in between different jobs or lots is extremely common in the industry. Setup times have been widely studied and reviewed by Allahverdi et al. [1] and Yang and Liao [42] or a bit more recently, by Allahverdi et al. [2]. The importance of considering the setup times in production scheduling cannot be underestimated as pointed out by Allahverdi and Soroush [3]. Furthermore, in this paper we consider the most difficult variant of setup times, namely separable or anticipatory, sequence and machine dependent setup times. We denote by S_{ijk} the fixed, non-negative and known amount of setup time that will be needed at machine i, $\forall i \in M$ after having processed job j when job k is the next job in the sequence, $\forall j,k, j \neq k, \in N$. In addition, we consider the following special triangular inequality: $S_{ijk} \leq S_{ijl} + p_{il} + S_{ilk}$, $\forall i \in M$, $\forall j,k,l \in N$, $j \neq k, j \neq l, k \neq l$. In other words, in a given machine i, the setup time between jobs j and k is either equal or lower than the setup between job j and any other job l, the processing time of job l and the setup between jobs l and k. This does not seem a strong condition to satisfy. Furthermore, setup times are asymmetric, that is, setup time between jobs j and k on machine i might be different from setup time between jobs k and j on the same machine.

With the total weighted earliness–tardiness minimization objective (wET from now on) and/or with the machine and sequence dependent setup times (SDST in short), the order or sequence of the jobs assigned to each machine in the unrelated parallel machine problem is now of uttermost importance. Depending on when a job j is scheduled inside a machine, its completion time C_j will change, and therefore, its earliness and tardiness values. Furthermore, we have already commented that the setups are sequence dependent, so the amount of setup depends on the sequence at each machine. With all this in mind, the problem considered in this paper is denoted as $R/S_{ijk}/\sum_{j=1}^{n}(w'_jE_j + w_jT_j)$. Obviously, this problem is also $\mathcal{NP}$-Hard as most special cases of it are already so as has been commented. As we will see in later sections, a vast portion of the literature does not consider the insertion of idle time, that is, the scheduling methods proposed do not leave machines idle when these are capable of processing jobs. This is known as non-delay schedule generation and jobs are never left idle if there is a machine available. In this paper we consider a special procedure for building general schedules with inserted idle time, which is obviously beneficial for the criterion considered, specially for the earliness measures. The reader is referred to Kanet and Sridharan [19] for a review on inserted idle time scheduling. We propose a simple and effective genetic algorithm for the considered problem. Several variations of the algorithm are tested, allowing or not insertion of idle times.

The remainder of this paper is organized as follows: Sect. 4.2 provides a review of the literature on this problem. In Sect. 4.3, a Mixed Integer Programming (MIP) model formulation is presented. In Sect. 4.4, we describe in detail the proposed genetic algorithm. Computational results and statistical evaluations are reported in Sect. 4.5 and finally, conclusions are given in Sect. 4.6.

4.2 Literature Review

Parallel machine problems have been studied for quite a long time already, as some of the early works attest (McNaughton [27]). General reviews about parallel machines scheduling are available from Cheng and Sin [9] and Mokotoff [28]. Quite interestingly, the review of Liam and Xing [24] enumerates recent and important trends in parallel machine machines problems, citing setup times and JIT criteria as the most important. Note that the literature on parallel machines scheduling is extensive, specially if we drop the setup times. Therefore, in the following, we concentrate more on the recent work as related as possible to the $R/S_{ijk}/\sum_{j=1}^{n}(w'_jE_j + w_jT_j)$ problem considered in this paper.

Heady and Zhu [17] propose several heuristic and dispatching rules methods for a SDST identical parallel machines problems with unweighted earliness–tardiness or $P/S_{jk}/\sum_{j=1}^{n}(E_j + T_j)$. Later, Balakrishnan et al. [6] studied a closely related problem. The authors presented a Mixed Integer Linear Programming (MILP) model for a uniform parallel machines with setup times and wET objective or $Q/r_j,S_{ijk}/\sum_{j=1}^{n}(w'_jE_j + w_jT_j)$, note that the authors also consider release dates. The authors were able to solve problems of up to ten jobs and eight machines (10×8) back at the time. Zhu and Heady [45] studied the very same problem as Balakrishnan et al. [6] except for the release dates, which were not considered. A similar MILP model is presented, although it is not compared to the previous model of Balakrishnan et al. [6] nor against a complete benchmark. Actually, a close study of both models yields to the conclusion that the model presented in [45] is identical to that of [6]. Later, Omar and Teo [30] studied also a similar problem but with some small differences. Omar and Teo [30] proved that the model presented in Zhu and Heady [45] was indeed incomplete as one important constraint was missing. Omar and Teo [30] presented another model which outperformed that of [45] and also the corrected version. Again, the authors did not check the presented model with the one of [6].

Şerifoğlu and Ulusoy [10] present two genetic algorithms for a problem very similar to the one dealt with in [6]. The authors consider two types of machines, type I and type II. Machines inside the same type are identical but machines of different types are uniform. The presented genetic algorithms contain several interesting characteristics, as for example the use of scheduling rules that allow inserted idle time. New crossover operators are also included in the study. The authors tested problems of up to 60 jobs and 4 machines. Later, Radhakrishnan and Ventura [34] proposed a simulated annealing for an unweighted earliness–tardiness problem where all machines are identical, both as regards processing times and also as regards setups. The authors tested their proposed approach on instances of up to 80 jobs and 15 machines. Guang and Lau [15] decomposed the timetabling (i.e., insertion of idle time) problem from the sequencing problem and proposed a squeaky wheel metaheuristic approach for the same problem with identical machines. Recently, Behnamian et al. [7] have studied also the same problem with identical machines but with a due date window, inside which no earliness or tardiness penalties occur. The authors proposed several metaheuristic methods. As we can see, the specific problem studied in this paper, considering distinct due dates, different weights for earliness and tardiness, unrelated parallel machines regarding both setup times and processing times, has been seldom dealt with in the scientific literature.

Note that the list of similar problems studied in the literature is large and not all papers can be cited here due to reasons of space. Some similar problems are for example the minimization of the total maximum tardiness in an unrelated parallel machine problem with setups in Kim et al. [21] or a similar problem with total weighted tardiness in Kim et al. [22]. Weighted common due dates earliness–tardiness minimization in identical parallel machines with batch-sequence dependent setups is studied in Yi and Wang [43] by means of a fuzzy logic embedded genetic algorithm. A tabu search is presented in Bilge et al. [8] for a uniform parallel machines problem with sequence dependent setups, release dates and unweighted total tardiness. The same problem was later studied by Anghinolfi and Paolucci [4]. Rabadi et al. [33] studied the $R/S_{ijk}/C_{\max}$ problem. Similarly, Logendran et al. [26] proposed several tabu search methods for the $R/S_{ijk}/\sum_{j=1}^{n} w_j T_j$ problem. Strong lower bounds are presented in Kedad-Sidhoum et al. [20] for the identical parallel machines problem without setup times. Similar problem, but this time with common due dates is approached in Rios-Solis and Sourd [36]. Note that this list is far from exhaustive as literally hundreds of papers have been published that consider either earliness-tardiness criteria or sequence-dependent setup times in parallel machines settings.

4.3 Exact Mixed Integer Linear Programming Model

In this section, we provide a MIP mathematical model for the unrelated parallel machine scheduling problem with SDST. Note that this model is an adapted version of that proposed by Guinet [16].

The model involves the following decision variables:

$$
\begin{aligned}
X_{ijk} &= \begin{cases} 1, & \text{if job } j \text{ precedes job } k \text{ on machine } i \\ 0, & \text{otherwise} \end{cases}, \\
C_j &= \text{Completion time of job } j, \\
E_j &= \text{Earliness of job } j, \\
T_j &= \text{Tardiness of job } j.
\end{aligned}
$$

The objective function is

$$\min \sum_{j \in N} (w'_j E_j + w_j T_j).$$

And the constraints are

$$\sum_{i \in M} \sum_{\substack{j \in \{0\} \cup \{N\} \\ j \neq k}} X_{ijk} = 1, \qquad \forall k \in N, \tag{4.1}$$

$$\sum_{i \in M} \sum_{\substack{k \in N \\ j \neq k}} X_{ijk} \leq 1, \qquad \forall j \in N, \tag{4.2}$$

$$\sum_{k \in N} X_{i0k} \leq 1, \qquad \forall i \in M, \tag{4.3}$$

$$\sum_{\substack{h \in \{0\} \cup \{N\} \\ h \neq k, h \neq j}} X_{ihj} \geq X_{ijk}, \qquad \forall j,k \in N, j \neq k, \forall i \in M, \tag{4.4}$$

$$C_k + V_i(1 - X_{ijk}) \geq C_j + S_{ijk} + p_{ik}, \quad \forall j \in \{0\} \cup \{N\}, \forall k \in N, j \neq k, \forall i \in M, \tag{4.5}$$

$$C_j + E_j - T_j = d_j, \qquad \forall j \in N, \tag{4.6}$$

$$C_j \geq 0, \qquad \forall j \in N, \tag{4.7}$$

$$E_j \geq 0, \qquad \forall j \in N, \tag{4.8}$$

$$T_j \geq 0, \qquad \forall j \in N, \tag{4.9}$$

$$X_{ijk} \in \{0,1\}, \forall j \in \{0\} \cup \{N\}, \forall k \in N, j \neq k, \forall i \in M. \tag{4.10}$$

Where $C_0 = S_{i0k} = 0, \forall i \in M, \forall k \in N$. The objective is to minimize the maximum completion time or makespan. Constraint set (4.1) ensures that every job is assigned to exactly one machine and has exactly one predecessor. Notice the usage of dummy jobs 0 as X_{i0k}, $\forall i \in M, k \in N$. With constraints set (4.2) we set the maximum number of successors of every job to one. Set (4.3) limits the number of successors of the dummy jobs to a maximum of one on each machine. With set (4.4) we ensure that jobs are properly linked in the machines, that is, if a given job j is processed on a given machine i, a valid predecessor h must exist on the same machine. Constraint set (4.5) is to control the completion times of the jobs at the machines. Basically, if a job k is assigned to machine i after job j (i.e., $X_{ijk} = 1$), its completion time

C_k must be greater than the completion time of j, C_j, plus the setup time between j and k and the processing time of k. If $X_{ijk} = 0$. Notice a value V_i which is a big constant. However, this big constant can be very much tightened to a value that represents an upper bound on the occupation of a given machine i. Therefore, $V_i = \sum_{j=1}^{n}\left(p_{ij} + \max_{k=1}^{n} S_{ijk}\right)$. In other words, a given machine i cannot be busy during more time that the sum of the processing times of all the jobs on that machine, plus the sum of the maximum setups between all jobs in that machine and all possible succeeding jobs. Set (4.6) is central for the definition of the earliness and tardiness. Note that one job is either early or tardy. Sets (4.7)–(4.10) simply define the nature of all decision variables. Therefore, the model contains n^2m binary variables, $3n$ continuous variables, and $2n^2m - nm + 6n + m$ constraints.

The previous model is straightforward and easy to understand. However, it needs a very large number of binary variables. Instead, the model of [6] contains much less variables. As a result, we also test the model of [6] that, for the sake of completeness, is defined below:

$$
\begin{aligned}
X_{jk} &= \begin{cases} 1, & \text{if job } j \text{ precedes job } k \text{ on the same machine} \\ 0, & \text{otherwise} \end{cases}, \\
Y_{ij} &= \begin{cases} 1, & \text{if job } j \text{ is processed on machine} i \\ 0, & \text{otherwise} \end{cases}, \\
C_j &= \text{Completion time of job } j, \\
E_j &= \text{Earliness of job } j, \\
T_j &= \text{Tardiness of job } j.
\end{aligned}
$$

X_{jk} variables are only defined for $j = 1, \ldots, n-1$ and $k = j+1, j+2, \ldots, n$, which means that there are only $\frac{n(n-1)}{2}$ binary X_{jk} variables. Note that if $X_{jk} = 0$ it means that job k precedes job j. The variables Y_{ij} are defined $\forall j \in N$ and $\forall i \in M$ The objective function is:

$$\min \sum_{j \in N} (w'_j E_j + w_j T_j).$$

The constraints of this model are:

$$\sum_{i \in M} Y_{ij} = 1, \qquad \forall j \in N, \tag{4.11}$$

$$Y_{ij} + \sum_{\substack{l \in M \\ l \neq i}} Y_{lk} + X_{jk} \leq 2, j = \{1, \ldots, n-1\}, k = \{j+1, \ldots, n\}, \quad \forall i \in M, \tag{4.12}$$

$$C_k - C_j + V_i(3 - X_{jk} - Y_{ij} - Y_{ik}) \geq p_{ik} + S_{ijk}, j = \{1, \ldots, n-1\}, k = \{j+1, \ldots, n\}, \quad \forall i \in M, \tag{4.13}$$

$$C_j - C_k + V_i(2 + X_{jk} - Y_{ij} - Y_{ik}) \geq p_{ij} + S_{ikj}, j = \{1, \ldots, n-1\}, k = \{j+1, \ldots, n\}, \quad \forall i \in M, \tag{4.14}$$

$$C_j + E_j - T_j = d_j, \quad \forall j \in N,$$

$$C_j \geq (p_{ij})(Y_{ij}), \quad \forall j \in N, \forall i \in M,$$

$$C_j \geq 0, \quad \forall j \in N,$$

$$E_j \geq 0, \quad \forall j \in N,$$

$$T_j \geq 0, \quad \forall j \in N,$$

$$X_{jk} \in \{0,1\}, \quad j = \{1,\ldots,n-1\}, k = \{j+1,\ldots,n\},$$

$$Y_{ij} \in \{0,1\}, \quad \forall j \in N, \forall i \in M.$$

Constraints in the set (4.11) force all jobs to be assigned to exactly one machine. Set (4.12) ensures that the precedence between jobs is consistent with machine assignments, that is, job j can precede job k only if they are assigned to the same machine. Constraints (4.13) and (4.14) control the processing times of jobs that follow other jobs. Note the clever use of the minimum number of X_{jk} variables. All other constraints are very similar to those of the previous model. Compared with the previous model, which has n^2m binary variables, this one, has $(n(n-1))/2 + nm$ binary variables. To put this into perspective, for a problem with 15 jobs and 5 machines, the first model needs 1,125 binary variables whereas this second model needs just 180 variables. Similarly, the number of constraints in this second model is $(3n(n-1)m)/3 + nm + 5n$. Following the example, the previous model would have 2,270 constraints and this one just 1,725. As we can see, this second model is much smaller. We will later see how this model compares with the initial one.

4.4 Proposed Methods

In this section, a genetic algorithm is proposed for the $R/S_{ijk}/\sum_{j=1}^{n}(w'_jE_j + w_jT_j)$ problem studied in this paper. Genetic algorithms (GAs) are optimization methods commonly used to solve scheduling problems (Goldberg [13], Reeves [35], Etiler et al. [11], Onwubolu and Mutingi [31], Ruiz et al. [38], Zhang and Lai [44], Vallada and Ruiz [40], Vallada and Ruiz [39], among others). They are based on nature, and they are usually classified into what is now referred to as bio-inspired methods (Holland [18]). In general, a GA works with a set of solutions. This set is often called "population" and each member of the population is referred to as an individual. Each individual is evaluated and assigned a fitness value. The key is that better solutions to the problem are correlated with high fitness values. After all individuals have been assigned a fitness value, another operator kicks-in. This operator is known as selection and should be biased towards fitter individuals from the population, that is, better solutions have a greater chance of being selected. This procedure mimics the biological "natural selection" and "survival of the fittest" in the Darwin's Theory of Evolution.

Selected individuals are simply called "parents" and mate into a procedure that is commonly known as "crossover". In crossover, new solutions are generated and the intention is to identify good traits in parents in the hope of generating better solutions or offspring. Once a given number of selections and crossover operations have been carried out, a new generation of solutions is created that often replaces the old generation. Moreover, a mutation scheme can be applied in order to introduce diversification into the population.

As we mentioned, GAs have been applied with great success to several scheduling problems in the last years. In order to reach state-of-the-art performance, GAs have to be carefully instantiated for the studied problem. Otherwise, subpar performance is obtained. This process of instantiation is not straightforward as it requires a good deal of problem-specific knowledge and lots of experimentation. In our case, the proposed GA is based on that developed by Vallada and Ruiz [39] for the same problem with the objective to minimize the makespan. The main features of the new algorithm presented in this chapter are the local search procedure and the insertion of idle times in order to improve the objective function. Basically, the representation of solutions, initialization of the population and selection and mutation operators are similar to those used in Vallada and Ruiz [39]. In the following subsections details about the features of the algorithm are given.

4.4.1 Representation of Solutions, Initialization of the Population, and Selection Procedure

The representation of solutions, initialization of the population, selection and mutation operators are similar to those used in Vallada and Ruiz [39]. Nevertheless, a summary of these characteristics is given in the following.

The solution representation consists of an array of jobs for each machine that represents the processing order of the jobs assigned to that machine, as usual in parallel machine scheduling problems. Then, the GA is formed by a population of P_{size} individuals, where each individual has m arrays of jobs.

Regarding the initialization of the population, a good individual is obtained by means of the Multiple Insertion (MI) heuristic proposed by Kurz and Askin [23]. The rest of the individuals are randomly generated and, in order to obtain a good initial population, the MI heuristic is applied. The MI heuristic is simple and, for each random individual, inserts each job in every position of every machine and finally places the job in the position that results in the lowest weighted earliness-tardiness value.

Finally, the selection mechanism is based on that proposed in Ruiz and Allahverdi [37], which is referred to as n-tournament selection. A given percentage of the population is randomly selected according to a parameter called "pressure". The individual with the lowest weighted earliness–tardiness value among the selected individuals is chosen as one of the parents. The same procedure is repeated to obtain the other parent but removing the first selected parent to avoid repetition.

4.4.2 Crossover Operator

The crossover operator is applied with the objective to generate two good individuals, known as offspring, from two good individuals selected trough the selection operator (parents). One of the most commonly used crossover operators is the One Point Order Crossover (OP). Not only is this operator well regarded to scheduling problems, but it is also simple and easy to code. OP basically starts by drawing a random cut point at a position of both parents. The jobs before the cut point are inherited from parent 1 to children 1 and from parent 2 to children 2. The jobs after the cut point are inherited from parent 1 to children 2 in the relative order in which they appear in parent 1. A similar process is carried out to inherit from parent 2 to children 1. The parallel machine adaptation consists of the application of the OP to each machine, that is, for each machine i, one point p_i is randomly selected from the parent 1 and jobs from the first position to the p_i position are copied to children 1. Jobs from the point $p_i + 1$ position to the end are copied to the second offspring. At this point, we have to decide how to insert the missing jobs from parent 2, that is, the rest of the jobs from parent 2 which are not inserted yet. In [39] these jobs are inserted into all possible positions and finally the job is placed in the position that results in the best value of the objective function (makespan in that paper). Then, a small local search procedure is included in the crossover operator to further improve the generated children. However, the weighted earliness-tardiness objective needs significantly more computational effort to be computed, so this small local search procedure can not be applied in the crossover operator. In this case, missing jobs from parent 2 are inserted at the end of the partial sequence, for each machine and no multiple insertions are carried out.

4.4.3 Local Search

Hybridizing genetic algorithms with local search procedures is a very common occurrence in the scheduling literature. Actually, some authors refer to the resulting methods as "memetic algorithms". The reason for the local search inclusion is the vast improvements obtained as regards the quality of the solutions. Local search based on the insertion neighborhood is probably the most common due to its good results. The concept of insertion neighborhood has to be carefully dealt with in the case of parallel machines as we do not really have common permutations as it is the case of single machine or flowshop scheduling.

We test the inter-machine insertion neighborhood (IMI), like in [39], such that for all the machines, all the jobs are inserted in every position of all the machines, that is, a job from a given machine i is extracted and inserted into all positions of all other machines, except i. This is repeated for all jobs. In order to reduce the computational effort needed by the local search, it is possible to introduce a simple and very efficient speed up procedure: when a job is inserted, it is not necessary to evaluate the complete sequence for obtaining the new weighted earliness-tardiness value of the machine. Jobs placed before the position where the job is inserted are not affected, so the weighted earliness–tardiness value until that position, can be pre-computed and then, it is not necessary to evaluate the whole sequence. More details about the speed up procedure can be found in [40].

4.4.4 Idle Time Insertion

One of the most promising features of the proposed algorithm is the insertion of idle times in machine sequences. Weighted earliness–tardiness objective can be improved when an early job is delayed, that is, when the weight of being early is greater than the weight of being tardy, a delay could improve the objective. Obviously, when a job is delayed, it affects to all the subsequent jobs. The procedure to check if the insertion of idle times could improve the objective is shown in Fig. 4.1. The first step is to find the first early job in every machine (sequence of jobs assigned to each machine). Then the total sum of weights is computed, both for earliness and tardiness. When the sum of weights for earliness is greater than for tardiness, the insertion of idle times improves the total weighted earliness–tardiness. Idle times are inserted just before the first early job and after the insertion at least one job will be JIT. The process is repeated for each machine until the total weights of early jobs is lower than the total weights of tardy jobs (stopping criterion). The procedure can be applied in two different ways: in the first one, it is used like a post-process, that is, when the algorithm is finished, the procedure is applied to the best individual found. The second one is to insert the procedure inside the algorithm. After the initialization of the individuals and every time an individual is evaluated, the procedure is also applied. In this way, the algorithm is always working with a population of individuals which can include idle times in the sequence.

Let us picture an example with six jobs and two machines. In Figs. 4.2 and 4.3, we can see the solution before and after idle times insertion, respectively. The objective value is also showed, an important reduction of the total weighted earliness-tardiness is obtained when idle times are allowed. In Table 4.1, we can see the processing times for each job in both machines (columns P_{1j} and P_{2j}), the due date for each job (DD_j), the weights for earliness and tardiness per job (columns WE_j and WT_j, respectively), the completion time for each job with no insertion of idle times (C_j), the completion time for each job when insertion of idle times is allowed (C'_j), the weighted earliness–tardiness value for each job when idle times are not inserted

```
procedure Idle Time Insertion
    for each machine m do
        while nof false do
            1. Find the first early job in m
            2. Compute the sum of weights of the early subsequent jobs (TotalWE)
            3. Compute the sum of weights of the tardy subsequent jobs (TotalWT)
            3. Compute the minimum earliness among all the early jobs (MinEarly)
            if TotalWE > TotalWT then
                    4. Insert MinEarly Idle times just before the first early job
                    5. Compute the new weighted earliness-tardiness value
            endif
            else
                    6. Break
        endwhile
    endfor
end
```

Fig. 4.1: Pseudocode for idle time insertion procedure.

(ET_j) and the weighted earliness–tardiness value for each job when idle times are inserted (ET'_j). Total weighted earliness–tardiness with and without idle times are also showed.

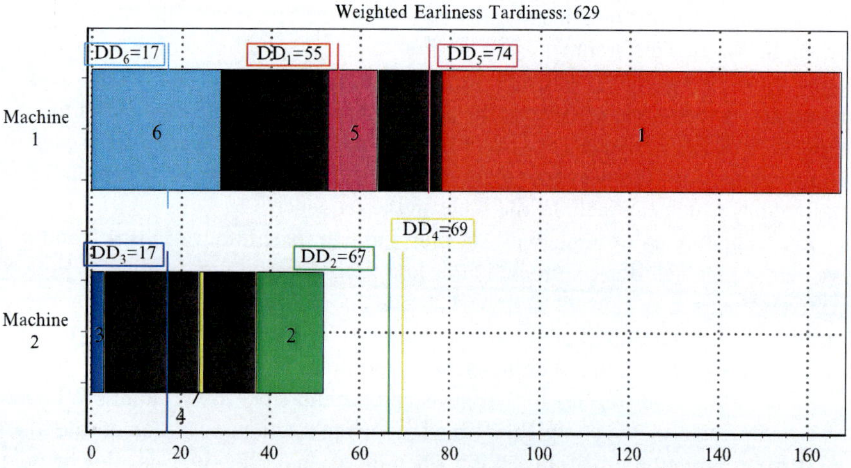

Fig. 4.2: Example for six jobs and two machines without idle times insertion.

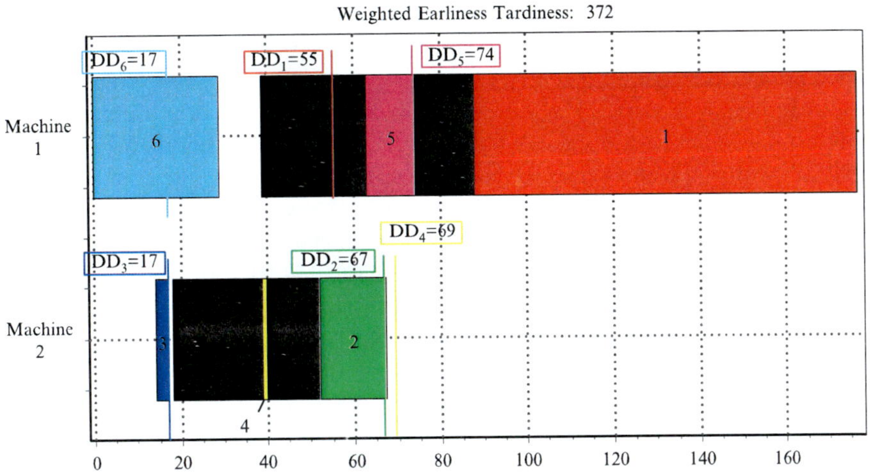

Fig. 4.3: Example for six jobs and two machines after idle times insertion.

Table 4.1: Example for six jobs and two machines.

Jobs (j)	P_{1j}	P_{2j}	DD_j	WE_j	WT_j	C_j	C'_j	ET_j	ET'_j
1	89	82	55	6	2	167	177	224	244
2	6	15	67	9	8	52	67	135	**0**
3	86	3	17	3	1	3	17	42	**0**
4	94	1	69	4	4	25	40	176	116
5	11	99	74	4	8	64	74	40	**0**
6	29	96	17	9	1	29	29	12	12
							Total ET	629	372

Jobs just-in-time after idle insertions are in bold
P_{1j}, (P_{2j}) processing time of job j in machine 1 (machine 2), DD_j due date for job j, WE_j weight of the earliness for job j, WT_j weight of the tardiness for job j, C_j completion time for job j without insertion of idle times, C'_j completion time for job j after insertion of idle times, ET_j weighted earliness-Tardiness of job j without insertion of idle times, ET'_j weighted earliness-Tardiness of job j after insertion of idle times

4.5 Computational and Statistical Evaluation

In this section, we will first present the benchmark of instances employed in this chapter. Then, we will evaluate the two previous mathematical models with a commercial solver in order to gauge their performance on a smaller set of instances. Later, the presented methods will be analyzed. Comprehensive computational and statistical analyses are presented.

We generate a standard set of instances where the processing times are uniformly distributed in the range $[1,99]$ as it is common in scheduling literature. We generate three groups of instances. The first two groups are the test instances, which are divided into small and large. The last group is the calibration instances, different from the test ones and used in the calibration of the methods. Separation between test and calibration instances ensures that no over-fitting or bias in the computational experiments can be attributed to calibration methodologies. Let us detail the test instances first. In the set of small instances we test all combinations of $n = 6,8,10,12$ and $m = 2,3,4,5$. The set of large instances employs $n = 50,100,150,200,250$, and $m = 10,15,20,25,30$. For each $n \times m$ combination, we consider two different distributions of setup times. In the first, setup times are uniformly distributed in the range $[1,49]$, whereas in the second, the range used is $[1,124]$. All generated setup times follow the previously mentioned triangular inequality. These two distributions model small versus large setup times as regards the processing times. When generating the due dates of the jobs, we employ the method presented in Potts and Van Wassenhove [32], also employed by Vallada et al. [41] which uses a uniform distribution for the due dates as $[P(1-T-R/2), P(1-T+R/2)]$. T and R are two parameters called *Tardiness Factor* and *Due Date Range*, respectively. P is commonly a lower bound on the makespan. In this case, P is a good solution obtained by the algorithms proposed by Vallada and Ruiz [39] for the same problem with the objective of minimizing the makespan. In this chapter we use $T = \{0.4, 0.6\}$, $R = \{0.2, 0.6\}$. Moreover, weights for both, earliness and tardiness, are also generated by means of a uniform distribution in the range $[1,9]$. For the test instances set, we generate five replicates for each combination of n, m, S_{ijk}, T and R. Therefore, there are $4 \cdot 4 \cdot 2 \cdot 2 \cdot 2 \cdot 5 = 640$ small instances and $5 \cdot 5 \cdot 2 \cdot 2 \cdot 2 \cdot 5 = 1{,}000$ large instances in the benchmark.

Calibration instances are generated in the same way as test instances but in this case with only one replicate for each large combination of n, m, S_{ijk}, T and R. Therefore, there are $4 \cdot 4 \cdot 2 \cdot 2 \cdot 2 \cdot 1 = 200$ calibration instances.

4.5.1 Testing the Mathematical Models

For testing the two models presented in Sect. 4.3, referred simply as model 1 (based on Guinet [16]) and model 2 (Balakrishnan et al. [6]), we employ the small set of 640 instances with maximum size of 12×5. For each instance, we generate a LP model file. We use IBM-ILOG CPLEX 12.1 in the tests. All tests are run on a cluster of 30 blade servers each one with two Intel XEON E5420 processors running at 2.5 GHz and with 16 GB of RAM memory. Each processor has four cores. The experiments are carried out in virtualised Windows XP machines, each one with two virtualised processors and 2 GB of RAM memory. Starting from version 11.0, IBM-ILOG CPLEX (CPLEX in short) can apply parallel branch and cut methods in multi-core CPUs. Therefore, we run all models once with a single thread and another time with two threads (CPLEX SET THREADS 2 option). Finally, we test each model with two different stopping time criteria: 3,600 and 7,200 s, respectively. This is carried out in order to check how additional CPU time affects results. As a result, each one of the 640 instances is run four times (serial 3,600, serial 7,200, parallel 3,600 and parallel 7,200) for a total number of results of 2,560 for each tested instance (5,120 results in total). The total running time for solving all these 5,120 instances was 31.33 CPU days.

For each result we record a categorical variable type of solution with two possible values: 0 and 1. The value of 0 means that an optimal solution was reached. The optimum wET value and the time needed for obtaining it are stored. The value of 1 means that the CPU time limit was reached and a feasible integer solution was found, albeit not proved to be optimal. The gap is recorded in such a case. We managed to find the optimum solutions in all 620 instances. Basically, model 2 always found the optimum solution, whereas model 1 failed to do so in the largest instances. Table 4.2 presents, for each model, and for all runs, the average gap in percentage (GAP%) and the average time needed in all cases.

Table 4.2: Results of the two tested models under IBM-ILOG CPLEX 12.1 for the small instances.

		Model 1								Model 2							
		Time 3,600				Time 7,200				Time 3,600				Time 7,200			
		1 Thread		2 Threads		1 Thread		2 Threads		1 Thread		2 Threads		1 Thread		2 Threads	
n	m	GAP%	Time	GAP%	Time	GAP%	Time	GAP%	Time	GAP%	Time	GAP%	Time	GAP%	Time	GAP%	Time
6	2	0.00	0.65	0.00	0.42	0.00	0.58	0.00	0.40	0.00	0.08	0.00	0.07	0.00	0.06	0.00	0.05
6	3	0.00	0.53	0.00	0.41	0.00	0.46	0.00	0.37	0.00	0.13	0.00	0.09	0.00	0.08	0.00	0.09
6	4	0.00	0.26	0.00	0,24	0.00	0.21	0.00	0.22	0.00	0.10	0.00	0.07	0.00	0.06	0.00	0.07
6	5	0.00	0.19	0.00	0.20	0.00	0.17	0.00	0.20	0.00	0.09	0.00	0.06	0.00	0.06	0.00	0.06
6	$\overline{m}$	0.00	0.41	0.00	0.32	0.00	0.36	0.00	0.30	0.00	0.10	0.00	0.07	0.00	0.06	0.00	0.07
8	2	0.00	20.49	0.00	11.92	0.00	20.18	0.00	11.34	0.00	0.73	0.00	0.44	0.00	0.56	0.00	0.42
8	3	0.00	11.94	0.00	5.85	0.00	11.48	0.00	5.86	0.00	1.47	0.00	0.86	0.00	1.14	0.00	0.83
8	4	0.00	6.48	0.00	3.66	0.00	6.16	0.00	3.67	0.00	1.70	0.00	1.03	0.00	1.31	0.00	1.17
8	5	0.00	1.96	0.00	1.28	0.00	1.74	0.00	1.28	0.00	0.82	0.00	0.51	0.00	0.61	0.00	0.54
8	$\overline{m}$	0.00	10.22	0.00	5.68	0.00	9.89	0.00	5.54	0.00	1.18	0.00	0.71	0.00	0.90	0.00	0.74
10	2	3.02	1,195.61	1.27	788.03	1.38	1,512.61	0.39	966.02	0.00	6.93	0.00	5.54	0.00	6.10	0.00	4.64
10	3	0.00	449.23	0.00	266.60	0.00	496.41	0.00	269.39	0.00	11.61	0.00	5.62	0.00	10.25	0.00	6.37
10	4	0.00	273.45	0.00	147.99	0.00	302.23	0.00	140.96	0.00	21.25	0.00	12.59	0.00	17.44	0.00	14.97
10	5	0.00	56.24	0.00	28.75	0.00	63.16	0.00	29.77	0.00	21.44	0.00	12.96	0.00	17.18	0.00	12.40
10	$\overline{m}$	0.75	493.63	0.32	307.84	0.34	593.60	0.10	351.53	0.00	15.31	0.00	9.18	0.00	12.74	0.00	9.59
12	2	56.20	3,485.18	52.01	3,396.60	51.05	6,751.44	47.10	6,622.09	0.00	138.78	0.00	85.29	0.00	119.62	0.00	91.47
12	3	57.72	3,552.49	51.43	3,464.87	49.72	6,881.03	41.06	6,558.80	0.00	400.79	0.00	199.49	0.00	331.38	0.00	279.75
12	4	25.06	2,899.35	12.70	2,327.60	13.59	4,892.60	7.14	3,708.26	0.00	228.33	0.00	123.24	0.00	186.81	0.00	154.29
12	5	1.47	906.55	1.22	489.51	1.09	972.38	0.92	611.81	0.00	162.93	0.00	94.93	0.00	128.53	0.00	105.23
12	$\overline{m}$	35.11	2,710.89	29.34	2,419.65	28.86	4,874.37	24.06	4,375.24	0.00	232.71	0.00	125.74	0.00	191.59	0.00	157.69
Average		8.97	803.79	7.41	683.37	7.30	1,369.55	6.04	1,183.15	0.00	62.32	0.00	33.93	0.00	51.33	0.00	42.02

As we can see, Model 1 produces large gaps for instances of ten jobs. It is easy to see how running CPLEX with two cores (2 threads) improves results significantly. As a matter of fact, using two threads is almost as effective as doubling the allowed CPU time. Note how thc largest gaps are not observed for the largest number of machines but rather for instances with just two or three machines. The explanation is that in these instances, many jobs have to be sequenced at each machine and therefore, setup times are more important. It is clear that Model 2 is vastly superior to Model 1. First, all instances are solved to optimality and all gaps are zero. Second, the CPU times needed are much shorter. It is interesting to see how results vary with 2 h (runs were independent to those of 1 h). This is due to the heuristic nature of some of the initializations and cutting rules of CPLEX that while not changing the optimality values, do change the CPU times. As for all the models, and although not shown due to reasons of space, T values of 0.4 resulted in worse solutions. Similarly, R values of 0.2 resulted in larger gaps for Model 1 and larger CPU times for Model 2. Lastly, shorter setup times of $[1,49]$ produced slightly worse results. As a final note, we attempted to solve instances of more jobs with Model 2 and quickly found a ceiling at about 16 jobs and 2–3 machines.

4.5.2 Computational Evaluation for Small Instances

Before testing the proposed algorithms with the set of instances, a small calibration experiment by means of a Design of Experiments (Montgomery [29]) is carried out. In Table 4.3, the parameters considered are shown: population size (P_{size}), crossover probability (P_c), mutation probability (P_m), local search probability (P_{ls}) Moreover, in order to reduce the computational effort, the parameter pressure of the selection operator (*Pressure*) is set to 10%, as in [39]. We can see in Table 4.3, the values tested for the calibration experiment (details about the statistical analysis are not shown due to space restrictions). The best combination of values is in bold. It is interesting to notice that the probability for all the operators is set to 1, that is, the operators are always applied, which is quite a departure from the standard values used normally.

Table 4.3: Values tested for the parameters in the calibration experiment (best combination in bold).

Parameter	Calibration of GAIdle version
Population size (P_{size})	60;**80**
Probability of crossover (P_c)	0.5;**1**
Probability of mutation (P_m)	0.5;**1**
Probability of local search (P_{ls})	0.5;**1**
Pressure of the selection operator (*Pressure*)	**10**

After running the mathematical models, the optimal solution is obtained for all the instances. In this section, a computational evaluation for the proposed method is carried out for the small instances. Four variants of the genetic algorithm are tested: without insertion of idle times, with insertion of idle times in a post-process procedure and finally, considering insertion of idle times inside the algorithm, with standard parameter values and calibrated parameter values, respectively. Regarding the response variable for the experiments, the Average Relative Percentage

Deviation (RPD) is computed for each instance according to the following expression:

$$\text{Relative Percentage Deviation } (RPD) = \frac{Method_{\text{sol}} - Optimal_{\text{sol}}}{Optimal_{\text{sol}}} \times 100.$$

Where $Optimal_{\text{sol}}$ is the optimal solution obtained after running the mathematical models, and $Method_{\text{sol}}$ is the solution obtained with a given version of the algorithm. We run five replicates of each algorithm. The stopping criterion is set to a maximum elapsed CPU time of $n \times (m/2) \times t$ ms, where t is set to different values, specifically 60 and 120. In this way, we can study the behavior of the methods when the amount of time is decreased or increased. For small instances, the maximum CPU time varies from 0.36 s for six jobs and two machines to 1.8 s for 12 jobs and 5 machines, when the value of t is set to 60. When t value is 120, maximum CPU time varies from 0.72 to 3.6 s.

Table 4.4 contains the results. Obviously, the RPD for the mathematical model is 0, as the optimal solution is always found. We can observe that results are much better when the insertion of idle times is allowed. Specifically, the algorithm with the insertion of idle times in a post-process procedure (GAIdlePost) improves by up to 47% the results obtained by the same algorithm without insertion of idle times (GANoIdle). If we focus our attention on the calibrated algorithm which includes the insertion of idle times (GAIdleCal), results are up to 198% and 106% better than GANoIdle and GAIdlePost, respectively. Finally, we can observe that for small instances, the calibrated version (GAIdleCal) does not improve significantly the standard version (GAIdleSt). Parameter values for the standard version are: 80 individuals for the population size, 10% for the pressure of the selection operator and 0.5 for all the probability rates (crossover, mutation, and local search). In the next subsection, we will see that the picture is completely different for large instances. The best version of the algorithm (GAIdleCal) is 7.16% off the optimal solution provided by the mathematical models. However, we have to remember that the CPU times for the proposed algorithms are much smaller than the times needed by the mathematical models.

In order to obtain more conclusive results, a statistical analysis by means of an analysis of variance (ANOVA) is carried out ([29]). We can see in Fig. 4.4 the means plot for all t values, on average, with HSD Tukey intervals ($\alpha = 0.05$). We can see that differences between GAIdleSt and GAIdleCal are not statistically different (confidence intervals are overlapped), that is, on average, the behavior of both versions is the same for small instances.

4.5.3 Computational Evaluation for Large Instances

In this section, we evaluate the behavior of the proposed algorithm under the set of large instances. The computational evaluation is carried out in the same conditions that for the small instances. Regarding the response variable for the experiments,

Table 4.4: Average relative percentage deviation (RPD) for the proposed algorithms (small instances), setting of t to 60;120 in the stopping criterion.

Instance	GANoIdle	GAIdlePost	GAIdleSt	GAIdleCal
6×2	7.48 ; 7.48	4.46 ; 4.46	0.01 ; 0.01	0.01 ; 0.01
6×3	26.11 ; 26.17	14.85 ; 13.42	3.44 ; 3.44	3.51 ; 3.51
6×4	39.99 ; 40.21	30.48 ; 32.56	17.44 ; 17.07	17.46 ; 17.46
6×5	71.56 ; 71.18	64.72 ; 65.34	51.86 ; 51.03	51.24 ; 51.24
8×2	7.47 ; 7.47	5.32 ; 5.36	1.28 ; 1.01	1.05 ; 1.00
8×3	12.63 ; 12.90	7.91 ; 8.00	1.73 ; 1.77	1.95 ; 2.46
8×4	15.54 ; 15.61	9.84 ; 9.54	3.96 ; 4.28	4.00 ; 3.53
8×5	41.82 ; 41.89	28.27 ; 28.15	15.59 ; 15.61	16.22 ; 16.64
10×2	7.78 ; 7.65	4.08 ; 4.22	0.86 ; 0.74	0.74 ; 0.74
10×3	10.18 ; 10.02	4.69 ; 5.72	0.32 ; 0.22	0.37 ; 0.35
10×4	12.49 ; 12.40	6.95 ; 7.35	2.36 ; 2.83	2.50 ; 1.79
10×5	30.71 ; 27.46	17.79 ; 18.60	6.39 ; 5.95	5.94 ; 6.39
12×2	3.20 ; 3.04	1.71 ; 1.50	0.40 ; 0.30	1.15 ; 0.89
12×3	7.69 ; 7.45	4.96 ; 3.59	1.09 ; 1.68	1.22 ; 1.14
12×4	13.56 ; 11.47	7.02 ; 6.39	1.46 ; 1.14	0.93 ; 1.13
12×5	35.60 ; 34.22	20.28 ; 21.93	10.96 ; 6.67	6.83 ; 6.22
Average	21.49 ; 21.04	14.58 ; 14.76	7.45 ; 7.11	7.19 ; 7.16

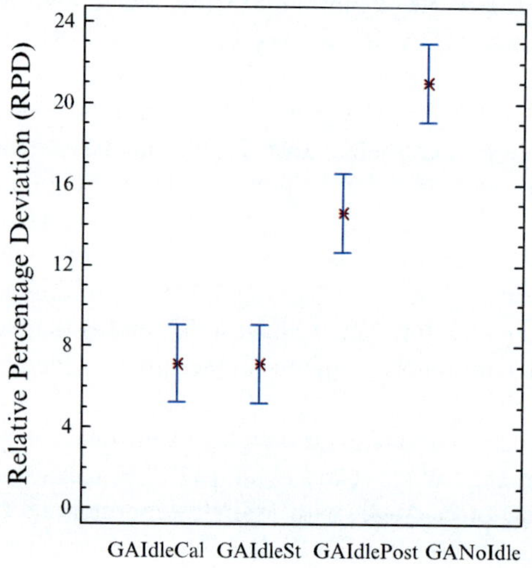

Fig. 4.4: Means plot and tukey HSD intervals at the 95% confidence level for the versions of the genetic algorithm proposed (small instances).

the Average RPD is computed for each instance according to the following expression:

$$\text{Relative Percentage Deviation } (RPD) = \frac{Method_{\text{sol}} - Best_{\text{sol}}}{Best_{\text{sol}}} \times 100.$$

In this case, optimal solution is not known, so the RPD value is computed using the best known solution ($Best_{sol}$) obtained among all the versions. As previously, the stopping criterion is set to a maximum elapsed CPU time of $n \times (m/2) \times t$ ms, where t is set to 60 and 120.

In Table 4.5, we can see results for large instances. Results for both t values are separated by a semicolon ($t = 60$; $t = 120$), as in the previous case. The first interesting outcome is the good performance of the calibrated version (GAIdleCal) which clearly overcomes the remaining versions. However, if we focus our attention on the different groups, we can see differences are much greater for the smallest instances (50, 100 and 150 jobs) when $t = 60$. When t value is set to 120, GAIdleCal obtains the best results for all groups of instances. This result makes sense since two important aspects have an important influence on the computational effort needed: the first one is the fact that all the operators (including local search) are always applied in the calibrated version. The other one is that the insertion of idle times is inside the algorithm. Therefore, as we will see later, t value is really important for the calibrated version (GAIdleCal). In any case, on average, results obtained by GAIdleCal are up to 66% better than with the standard version (GAIdleSt). Remember that in this case, the algorithm is exactly the same only the parameter values are changed. Regarding the other two versions, differences are even much larger, up to 294% respect to the version inserting idle times in a post-process (GAIdlePost) and up to 311% respect to the version without insertion of idle times (GANoIdle).

As in the previous case, a statistical analysis by means of an ANOVA is carried out to check if the observed differences are statistically significant. In Fig. 4.5, we can see the means plot for all t values, on average, with HSD Tukey intervals ($\alpha = 0.05$). In this case, intervals are not overlapped and we can conclude differences are statistically significant, then, calibrated version of the algorithm (GAIdleCal) obtains the best results of the comparison. We can also observe that in spite of differences between GANoIdle and GAIdlePost seemed not very important according to the Table 4.5, these differences are statistically significant, so we can conclude GAIdlePost shows a better performance than GANoIdle.

Finally, a last statistical analysis is carried out to check the importance of t factor in the performance of the algorithms. Results in Table 4.5 showed that performance of the calibrated version of the algorithm (GAIdleCal) could be affected by the t factor. In Fig. 4.6, we can observe the interaction plot between the versions of the algorithm and t factor. We can see that for GAIdleCal version, the *RPD* value is much better as the t parameter increases. For the other three versions of the algorithm, value of t does not have an influence on the performance (intervals are overlapped for both t values).

4.6 Conclusions and Future Research

In this chapter a genetic algorithm for the parallel machine scheduling problem with the objective of minimizing the total weighted earliness–tardiness is proposed. A MIP model is also formulated for the same problem. The algorithm includes a

Table 4.5: Average relative percentage deviation (RPD) for the proposed algorithms (large instances), setting of t to 60;120 in the stopping criterion.

Instance	GANoIdle	GAIdlePost	GAIdleSt	GAIdleCal
50×10	140.71 ; 130.86	126.87 ; 121.47	49.41 ; 44.32	43.32 ; 38.39
50×15	299.72 ; 298.84	281.94 ; 299.61	70.36 ; 72.16	52.09 ; 42.64
50×20	369.52 ; 378.12	339.24 ; 348.80	76.26 ; 77.01	50.36 ; 40.00
50×25	415.41 ; 421.03	398.03 ; 389.88	122.90 ; 101.90	75.56 ; 48.15
50×30	391.86 ; 385.35	368.50 ; 368.66	108.49 ; 104.34	101.12 ; 75.99
100×10	16.66 ; 15.45	15.20 ; 15.40	21.61 ; 21.89	14.13 ; 10.83
100×15	27.97 ; 25.98	26.97 ; 26.84	32.02 ; 30.21	16.00 ; 14.20
100×20	44.32 ; 41.29	44.97 ; 43.07	36.40 ; 33.67	17.59 ; 14.75
100×25	58.82 ; 58.75	59.22 ; 55.78	41.91 ; 40.38	20.85 ; 17.17
100×30	60.07 ; 59.24	59.21 ; 56.27	40.61 ; 38.05	26.35 ; 18.18
150×10	8.53 ; 6.58	8.85 ; 7.03	7.30 ; 6.04	4.77 ; 3.44
150×15	12.60 ; 11.82	12.87 ; 11.96	10.94 ; 11.26	7.10 ; 4.02
150×20	15.56 ; 14.42	17.33 ; 16.39	15.64 ; 14.30	10.09 ; 5.52
150×25	22.43 ; 20.54	23.13 ; 20.47	21.31 ; 19.04	21.96 ; 11.06
150×30	16.42 ; 15.69	17.11 ; 16.63	17.46 ; 17.03	36.54 ; 26.83
200×10	8.65 ; 6.62	8.95 ; 6.59	12.99 ; 10.32	18.47 ; 9.41
200×15	11.58 ; 10.36	12.72 ; 10.45	16.62 ; 13.48	16.04 ; 8.91
200×20	15.79 ; 15.10	16.38 ; 15.25	22.10 ; 19.25	17.91 ; 9.57
200×25	19.28 ; 19.08	21.05 ; 19.32	29.52 ; 24.75	20.25 ; 9.51
200×30	27.55 ; 24.84	27.87 ; 24.21	27.10 ; 27.37	26.64 ; 10.18
250×10	8.77 ; 5.27	8.33 ; 5.84	16.95 ; 13.86	35.66 ; 17.93
250×15	10.10 ; 7.96	10.65 ; 8.10	17.39 ; 12.15	24.59 ; 13.39
250×20	12.70 ; 10.78	13.12 ; 10.67	20.61 ; 14.94	25.34 ; 13.41
250×25	15.19 ; 13.70	16.42 ; 13.08	27.04 ; 21.28	26.63 ; 13.75
250×30	18.82 ; 15.70	19.30 ; 16.45	29.04 ; 23.44	26.18 ; 12.25
Average	81.96 ; 80.53	78.17 ; 77.13	35.68 ; 32.50	29.42 ; 19.58

procedure for inserting idle times in order to improve the objective function. Four versions of the algorithm are proposed according to the application of the idle times insertion procedure.

We have carried out an extensive evaluation of the versions of the algorithm, including a small calibration experiment to set some of the parameter values. Results show that the calibrated version of the algorithm which includes the idle times insertion inside, obtains the best performance by a large margin.

Future research stems from the improvement of the idle time insertion routines, as these are of paramount importance for reducing the total weighted earliness-tardiness values.

Acknowledgments

This work is partially funded by the Spanish Ministry of Science and Innovation, under the project "SMPA – Advanced Parallel Multiobjective Sequencing: Practical and Theoretical Advances" with reference DPI2008-03511/DPI. The authors should

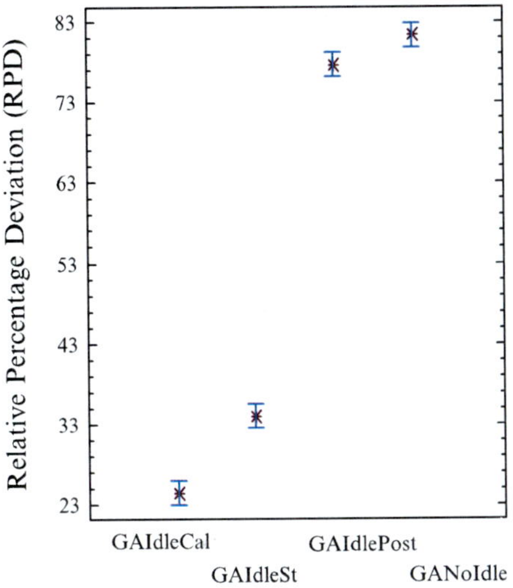

Fig. 4.5: Means plot and tukey HSD intervals at the 95% confidence level for the versions of the genetic algorithm proposed (large instances).

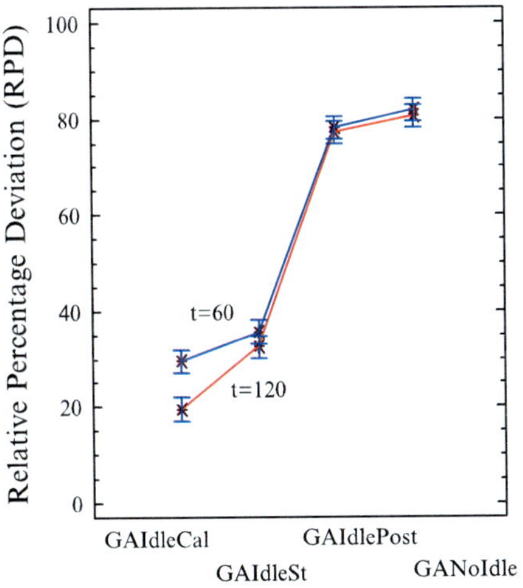

Fig. 4.6: Means plot and tukey HSD intervals at the 95% confidence level for the interaction between time (t) and the proposed algorithm (large instances).

also thank the IMPIVA – Institute for the Small and Medium Valencian Enterprise, for the project OSC with references IMIDIC/2008/137, IMIDIC/2009/198, and IMIDIC/2010/175 and the Polytechnic University of Valencia, for the project PPAR with reference 3147.

References

1. Allahverdi, A., Gupta, J.N.D., Aldowaisan, T.: A review of scheduling research involving setup considerations. Omega, The International Journal of Management Science **27**(2), 219–239 (1999)
2. Allahverdi, A., Ng, C.T., Cheng, T.C.E., Kovalyov, M.Y.: A survey of scheduling problems with setup times or costs. European Journal of Operational Research **187**(3), 985–1032 (2008)
3. Allahverdi, A., Soroush, H.M.: The significance of reducing setup times/setup costs. European Journal of Operational Research **187**(3), 978–984 (2008)
4. Anghinolfi, D., Paolucci, M.: Parallel machine total tardiness scheduling with a new hybrid metaheuristic approach. Computers & Operations Research **34**(11), 3471–3490 (2007)
5. Baker, K.R., Scudder, G.D.: Sequencing with earliness and tardiness penalties. a review. Operations Research **38**(1), 22–36 (2000)
6. Balakrishnan, N., Kanet, J.J., Sridharan, V.: Early/tardy scheduling with sequence dependent setups on uniform parallel machines. Computers & Operations Research **26**(2), 127–141 (1999)
7. Behnamian, J., Zandieh, M., Ghomi, S.M.T.F.: Due window scheduling with sequence-dependent setup on parallel machines using three hybrid metaheuristic algorithms. International Journal of Advanced Manufacturing Technology **44**(7-8), 795–808 (2009)
8. Bilge, U., Kıraç, F., Kurtulan, M., Pekgün, P.: A tabu search algorithm for parallel machine total tardiness problem. Computers & Operations Research **31**(3), 397–414 (2004)
9. Cheng, T.C.E., Sin, C.C.S.: A state-of-the-art review of parallel-machine scheduling research. European Journal of Operational Research **47**(3), 271–292 (1990)
10. Şerifoğlu, F.S., Ulusoy, G.: Parallel machine scheduling with earliness and tardiness penalties. Computers & Operations Research **26**(8), 773–787 (1999)
11. Etiler, O., Toklu, B., Atak, M., Wilson, J.: A genetic algorithm for flow shop scheduling problems. Journal of the Operational Research Society **55**(8), 830–835 (2004)
12. Garey, M.R., Johnson, D.S.: Computers and Intractability: A Guide to the Theory of NP-Completeness. Freeman, San Francisco (1979)
13. Goldberg, D.E.: Genetic Algorithms in Search, Optimization and Machine Learning. Addison-Wesley, Reading (1989)
14. Graham, R.L., Lawler, E., Lenstra, J., Kan, A.R.: Optimization and approximation in deterministic sequencing and scheduling: a survey. Annals of Discrete Mathematics **5**, 287–326 (1979)
15. Guang, F., Lau, H.C.: Efficient algorithms for machine scheduling problems with earliness and tardiness penalties. Annals of Operations Research **159**(1), 93–95 (2008)
16. Guinet, A.: Scheduling sequence-dependent jobs on identical parallel machines to minimize completion time criteria. International Journal of Production Research **31**(7), 1579–1594 (1993)
17. Heady, R.B., Zhu, Z.: Minimizing the sum of job earliness and tardiness in a multimachine system. International Journal of Production Research **36**(6), 1619–1632 (1998)
18. Holland, J.H.: Adaptation in Natural and Artificial Systems. The University of Michigan Press, Ann Arbor (1975)
19. Kanet, J.J., Sridharan, V.: Scheduling with inserted idle time: Problem taxonomy and literature review. Operations Research **48**(1), 99–110 (2000)
20. Kedad-Sidhoum, S., Rios-Solis, Y.A., Sourd, F.: Lower bounds for the earliness-tardiness scheduling problem on parallel machines with distinct due dates. European Journal of Operational Research **189**(3), 1305–1316 (2008)
21. Kim, D.W., Kim, K.H., Jang, W., Chen, F.F.: Unrelated parallel machine scheduling with setup times using simulated annealing. Robotics and Computer Integrated Manufacturing **18**(3-4), 223–231 (2002)

22. Kim, D.W., Na, D.G., Chen, F.F.: Unrelated parallel machine scheduling with setup times and a total weighted tardiness objective. Robotics and Computer Integrated Manufacturing **19**(1-2), 173–181 (2003)
23. Kurz, M., Askin, R.: Heuristic scheduling of parallel machines with sequence-dependent set-up times. International Journal of Production Research **39**(16), 3747–3769 (2001)
24. Lam, K., Xing, W.: New trends in parallel machine scheduling. International Journal of Operations & Production Management **17**(3), 326–338 (1997)
25. Lenstra, J.K., Rinnooy Kan, A.H.G., Brucker, P.: Complexity of machine scheduling problems. Annals of Discrete Mathematics **1**, 343–362 (1977)
26. Logendran, R., McDonell, B., Smucker, B.: Scheduling unrelated parallel machines with sequence-dependent setups. Computers & Operations Research **34**(11), 3420–3438 (2007)
27. McNaughton, R.: Scheduling with deadlines and loss functions. Management Science **6**(1), 1–12 (1959)
28. Mokotoff, E.: Parallel machine scheduling problems: A survey. Asia-Pacific Journal of Operational Research **18**(2), 193–242 (2001)
29. Montgomery, D.: Design and Analysis of Experiments, fifth edn. John Wiley & Sons, New York (2007)
30. Omar, M.K., Teo, S.C.: Minimizing the sum of earliness/tardiness in identical parallel machines schedule with incompatible job families: An improved mip approach. Applied Mathematics and Computation **181**(2), 1008–1017 (2006)
31. Onwubolu, G., Mutingi, M.: Genetic algorithm for minimizing tardiness in flow-shop scheduling. Production Planning & Control **10**(5), 462–471 (1999)
32. Potts, C., Van Wassenhove, L.: A decomposition algorithm for the single machine total tardiness problem. Operations Research Letters **1**(5), 177–181 (1982)
33. Rabadi, G., Moraga, R., Al-Salem, A.: Heuristics for the unrelated parallel machine scheduling problem with setup times. Journal of Intelligent Manufacturing **17**(1), 85–97 (2006)
34. Radhakrishnan, S., Ventura, J.A.: Simulated annealing for parallel machine scheduling with earliness-tardiness penalties and sequence-dependent set-up times. International Journal of Production Research **38**(10), 2233–2252 (2000)
35. Reeves, C.R.: A Genetic Algorithm for Flowshop Sequencing. Computers & Operations Research **22**(1), 5–13 (1995)
36. Rios-Solis, Y.A., Sourd, F.: Exponential neighborhood search for a parallel machine scheduling problem. Computers & Operations Research **35**(5), 1697–1712 (2008)
37. Ruiz, R., Allahverdi, A.: No-wait flowshop with separate setup times to minimize maximum lateness. International Journal of Advanced Manufacturing Technology **35**(5-6), 551–565 (2007)
38. Ruiz, R., Maroto, C., Alcaraz, J.: Two new robust genetic algorithms for the flowshop scheduling problem. OMEGA, The International Journal of Management Science **34**(5), 461–476 (2006)
39. Vallada, E., Ruiz, R.: A genetic algorithm for the unrelated parallel machine scheduling problem with sequence dependent setup times. European Journal of Operations Research **211**(3), 611–622 (2011)
40. Vallada, E., Ruiz, R.: Genetic algorithms with path relinking for the minimum tardiness permutation flowshop problem. Omega, The International Journal of Management Science **38**(1-2), 57–67 (2010)
41. Vallada, E., Ruiz, R., Minella, G.: Minimising total tardiness in the *m*-machine flowshop problem: a review and evaluation of heuristics and metaheuristics. Computers & Operations Research **35**(4), 1350–1373 (2008)
42. Yang, W.H., Liao, C.J.: Survey of scheduling research involving setup times. International Journal of Systems Science **30**(2), 143–155 (1999)
43. Yi, Y., Wang, D.W.: Soft computing for scheduling with batch setup times and earliness-tardiness penalties on parallel machines. Journal of Intelligent Manufacturing **14**(3-4), 311–322 (2003)

44. Zhang, L., Wang, L., Zheng, D.Z.: An adaptive genetic algorithm with multiple operators for flowshop scheduling. International Journal of Advanced Manufacturing Technology **27**(5), 580–587 (2006)
45. Zhu, Z., Heady, R.B.: Minimizing the sum of earliness/tardiness in multi-machine scheduling: a mixed integer programming approach. Computers & Industrial Engineering **38**(2), 297–305 (2000)
46. Zhu, Z., Meredith, P.H.: Defining critical elements in JIT implementation: a survey. Industrial Management & Data Systems **95**(8), 21–28 (1995)

Chapter 5
Mixed-Integer Programming Models for Flowshop Scheduling Problems Minimizing the Total Earliness and Tardiness

Débora P. Ronconi and Ernesto G. Birgin

Abstract Scheduling problems involving both earliness and tardiness costs have received significant attention in recent years. This type of problem became important with the advent of the just-in-time (JIT) concept, where early or tardy deliveries are highly discouraged. In this work we examine the flowshop scheduling problem with no storage constraints and with blocking in-process. In this latter environment, there are no buffers between successive machines; therefore, intermediate queues of jobs waiting in the system for their next operations are not allowed. Performance is measured by the minimization of the sum of earliness and tardiness of the jobs. Mixed-integer models that represent these scheduling flowshop problems are presented. The models are evaluated and compared in several problems using commercial known software.

5.1 Introduction

This paper addresses the flowshop scheduling problem. This environment is characterized by n jobs being processed on m machines always in the same order, that is, the k-th operation of every job must be conducted on machine k. We will consider the case with the same job sequence in all machines, known as permutation schedule. As pointed out in [7], permutation schedules do not always include the optimal schedule for a given problem, but their importance should not be underestimated,

Débora P. Ronconi
Department of Production Engineering, EP-USP, University of São Paulo, Av. Prof. Almeida Prado, 128, Cidade Universitária, 05508-900 São Paulo SP, Brazil,
e-mail: dronconi@usp.br

Ernesto G. Birgin
Department of Computer Science, IME-USP, University of São Paulo, Rua do Matão, 1010, Cidade Universitária, 05508-090 São Paulo SP, Brazil, e-mail: egbirgin@ime.usp.br

R.Z. Ríos-Mercado and Y.A. Ríos-Solís (eds.), *Just-in-Time Systems*, Springer Optimization and Its Applications 60, DOI 10.1007/978-1-4614-1123-9_5,

since only permutation schedules are feasible in most real-life situations. A static and deterministic environment is assumed here, where the processing times and due dates are known and all jobs are available for processing since the beginning. Preemptions are not allowed, that is, when a job starts to be processed on a machine, it cannot be interrupted. Among its several applications, one of the most relevant applications of flowshop can be found in the chemical industry [3, 13]. Since specific equipment for chemical processes is expensive, there is a strong motivation to optimize this production environment.

Two different storage conditions are considered in this work: (a) unlimited buffer and (b) blocking in-process. In the flowshop scheduling problem with blocking there is no buffer storage between machines, and then queues of jobs waiting in the system for their next operation are not allowed. A job completed on one machine blocks it until the next machine is available for processing. As mentioned in [5], blocking can be related to the production process itself. Some examples of blocking can be found in concrete block manufacturing, which does not allow stock in some stages of the manufacturing process [4]. Another example appears in a robotic cell, where a job may block a machine while waiting for the robot to pick it up and move it to the next stage [16].

The considered objective function is the minimization of the total deviation of job completion times in relation to their corresponding due dates. Meeting due dates is a common objective for many manufacturing processes. Tardy jobs may generate contractual penalties and loss of credibility, causing damages to the company's image and loss of clients [15]. Early jobs were discouraged since the advent of Just-in-Time approaches due to the costs generated by those jobs, such as tied-up capital and inventory costs [1].

Mixed-integer linear programming (MILP) models, can be used to generate optimal schedules for moderate size problems. With the evolution of computers, in addition to the development of specific software to solve this kind of problem, research in this field has been greatly leveraged. See, for example, [11,17,18,22]. The main goal of this paper is to evaluate, in terms of computational cost, mixed-integer linear programming formulations for the job scheduling problem in the flowshop environment with unlimited and zero buffer, for the minimization of total earliness and tardiness. The rest of this work is organized as follows. In Sect. 5.2, models for the unlimited buffer environment are presented. Models for the blocking in-process situation are introduced in Sect. 5.3. Section 5.4 is devoted to the numerical evaluation of the different formulations. Concluding remarks are given in Sect. 5.5.

5.2 MILP Models for Flowshop with Unlimited Buffer

The production environment with unlimited buffer is characterized by the existence of intermediate buffers between all machines with no constraint in terms of their capacity. At first, we introduce three formulations to be analyzed for this environment.

According to [11], the flowshop environment with unlimited buffer minimizing the makespan is better represented by Wagner's [20], Wilson's [21], and Manne's [10] formulations. Therefore, these first three introduced models are based on them. Note that, differently from the problem considered in [10,20,21], in the present work, the insertion of idle time in the schedule can be advantageous. See [6] for a literature review about scheduling with inserted idle time.

The problem parameters for all the formulations studied in the rest of this work are (a) n: the number of jobs, (b) m: the number of machines, (c) p_{ik}: the processing time of job i on machine k, and (d) d_i: the due date of job i.

The model for the minimization of total earliness and tardiness based on the Wagner's formulation [20] with some modifications suggested in [17], called MUB1 from now on, is presented below. Variables of the model, with their corresponding bound constraints, are: $x_{ij} \in \{0,1\}, i = 1,\ldots,n, j = 1,\ldots,n$, $E_j \geq 0, j = 1,\ldots,n$, $T_j \geq 0, j = 1,\ldots,n$, $W_{jk} \geq 0, j = 1,\ldots,n, k = 1,\ldots,m-1$, $I_{jk} \geq 0, j = 1,\ldots, n-1, k = 1,\ldots,m$, and $C_{jm}, j = 1,\ldots,n$. The meanings of the variables are: x_{ij} is equal to 1 if job i is in j-th position of the sequence, 0 otherwise; T_j is the tardiness of the j-th job, E_j is the earliness of the j-th job, C_{jm} is the completion time of the j-th job on machine m, I_{jk} is the idle time between the j-th and the $(j+1)$-th jobs on machine k, and W_{jk} is the waiting time of the j-th job in the buffer between machines k and $k+1$.

$$\min \sum_{j=1}^{n} E_j + T_j \tag{5.1}$$

$$\text{s. t.} \quad T_j \geq C_{jm} - \sum_{i=1}^{n} x_{ij} d_i, \qquad j = 1,\ldots,n, \tag{5.2}$$

$$E_j \geq \sum_{i=1}^{n} x_{ij} d_i - C_{jm}, \qquad j = 1,\ldots,n, \tag{5.3}$$

$$C_{1m} = \sum_{k=1}^{n-1} \left(\sum_{i=1}^{n} x_{i1} p_{ik} + W_{1k} \right) + \sum_{i=1}^{n} x_{i1} p_{im}, \tag{5.4}$$

$$C_{jm} = C_{j-1,m} + I_{j-1,m} + \sum_{i=1}^{n} x_{ij} p_{im}, \qquad j = 2,\ldots,m, \tag{5.5}$$

$$I_{jk} + \sum_{i=1}^{n} x_{i,j+1} p_{ik} + W_{j+1,k} = W_{jk} + \sum_{i=1}^{n} x_{ij} p_{i,k+1} + I_{j,k+1}, \quad j = 1,\ldots,n-1, \; k = 1,\ldots,m-1, \tag{5.6}$$

$$\sum_{i=1}^{n} x_{ij} = 1, \qquad j = 1,\ldots,n, \tag{5.7}$$

$$\sum_{j=1}^{n} x_{ij} = 1, \qquad i = 1,\ldots,n. \tag{5.8}$$

Constraint (5.2), in conjunction with the non-negativity constraint $T_j \geq 0$, correspond to the linearization of $T_j = \max\{C_{jm} - \sum_{i=1}^{n} x_{ij}d_i, 0\}$ and provide us with the value of the individual tardiness of each job. Constraint (5.3), in conjunction with the non-negativity constraint $E_j \geq 0$, play the analogous role related to the individual earliness of each job. Constraints (5.4) and (5.5) are involved in obtaining the completion times of each job on machine m. Constraint (5.4) applies only to the job ordered in the first position, whose completion time depends on its own processing times and the waiting time between the machines. On the other hand, constraint (5.5) is general, based on the sum of the completion time of the previous job, the idle time of the last machine between two consecutive jobs and the processing time of the job. Constraints (5.7) and (5.8) ensure that a job can only be allocated to a sequence position and that each sequence position can only be related to one job. Constraint (5.6) expresses the relationship among several times, such as processing, waiting and idle times, among machines and jobs. Figure 5.1, based on Gantt chart, illustrates these relationships.

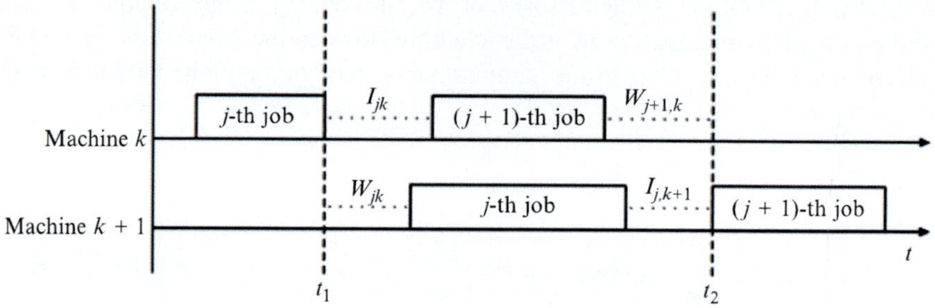

Fig. 5.1: Graphical representation of the MUB1 model constraint (5.6) that expresses the relationship among processing, waiting, and idle times.

The following formulation, based on Wilson's model [21] and called MUB2 from now on, uses neither variables of idle time nor waiting time, but it introduces the variables of processing start time. Variables of the model, with their corresponding bound constraints, are: $x_{ij} \in \{0,1\}, i = 1,\ldots,n, j = 1,\ldots,n$, $E_j \geq 0, j = 1,\ldots,n$, $T_j \geq 0, j = 1,\ldots,n$, and $S_{jk}, j = 1,\ldots,n, k = 1,\ldots,m$. $C_{jm}, j = 1,\ldots,n$, defined in (5.12), representing the completion time of each job on the last machine, are not variables but intermediate values used to simplify the earliness and tardiness expressions in (5.10) and (5.11). Variables S_{jk} represent the starting time of the j-th job on machine k. The other variables have the previously defined meaning.

$$\text{Minimize } \sum_{j=1}^{n} E_j + T_j \tag{5.9}$$

$$\text{subject to } T_j \geq C_{jm} - \sum_{i=1}^{n} x_{ij}d_i, \qquad j = 1,\ldots,n, \tag{5.10}$$

$$E_j \geq \sum_{i=1}^{n} x_{ij} d_i - C_{jm}, \qquad j = 1,\dots,n, \tag{5.11}$$

$$C_{jm} = S_{jm} + \sum_{i=1}^{n} x_{ij} p_{im}, \qquad j = 1,\dots,m, \tag{5.12}$$

$$S_{j+1,k} \geq S_{jk} + \sum_{i=1}^{n} x_{ij} p_{ik}, \qquad j = 1,\dots,n-1, k = 1,\dots,m, \tag{5.13}$$

$$S_{j,k+1} \geq S_{jk} + \sum_{i=1}^{n} x_{ij} p_{ik}, \qquad j = 1,\dots,n, k = 1,\dots,m-1, \tag{5.14}$$

$$S_{11} \geq 0, \tag{5.15}$$

$$\sum_{i=1}^{n} x_{ij} = 1, \qquad j = 1,\dots,n, \tag{5.16}$$

$$\sum_{j=1}^{n} x_{ij} = 1, \qquad i = 1,\dots,n. \tag{5.17}$$

Constraints (5.13)–(5.15) impose the rules for the starting time of each job on each machine; while, as mentioned above, equality (5.12) gives the completion time of each job on the last machine as a function of its start time and its processing time. Constraint (5.13) says that between the starting times of consecutive jobs on a machine, there must be enough time for the first (of the two jobs) to be processed. Constraint (5.14) indicates that between the starting times of a job on two consecutive machines, there must enough time for the job to be processed on the first machine. Constraint (5.15) simply says that the starting time of the first job on the first machine must be non-negative, that is, the insertion of idle time at the beginning of the schedule is allowed. The remaining constraints were already explained before.

The next formulation, based on Manne's model [10] and called MUB3 from now on, uses a different definition of decision binary variables, based on whether a job occurs before others: z_{ij} is equal to 1 if job i precedes job j in the sequence (not necessarily immediately before it), 0 otherwise. In addition, the completion time C_{jk} now refers to job j and not to the j-th job in the sequence as defined earlier. Constant M, used in the formulation, represents a very large positive number. The formulation is presented below. Variables of the model, with their corresponding bound constraints, are: $z_{ij} \in \{0,1\}, i = 1,\dots,n-1, j = i+1,\dots,n$, $E_i \geq 0, i = 1,\dots,n$, $T_i \geq 0, i = 1,\dots,n$, and $C_{ik}, i = 1,\dots,n, k = 1,\dots,m$.

$$\min \sum_{i=1}^{n} E_i + T_i \tag{5.18}$$

$$\text{s. t. } T_i \geq C_{im} - d_i, \qquad i = 1,\dots,n, \tag{5.19}$$

$$E_i \geq d_i - C_{im}, \qquad i = 1,\dots,n, \tag{5.20}$$

$$C_{i1} \geq p_{i1}, \qquad i = 1,\dots,n, \tag{5.21}$$

$$C_{ik} \geq p_{ik} + C_{i,k-1}, \qquad i = 1,\ldots,n, k = 2,\ldots,m, \tag{5.22}$$
$$C_{ik} \geq p_{ik} + C_{jk} - Mz_{ij}, \qquad i = 1,\ldots,n-1, j = i+1,\ldots,n, k = 1,\ldots,m, \tag{5.23}$$
$$C_{jk} \geq p_{jk} + C_{ik} - M(1 - z_{ij}), \quad i = 1,\ldots,n-1, j = i+1,\ldots,n, k = 1,\ldots,m. \tag{5.24}$$

Constraints (5.21) and (5.22) deal with completion times of individual jobs. Constraint (5.21) indicates that the completion time of a job on the first machine must be greater than or equal to the processing time of the job on the first machine. Constraint (5.22) says that between the completion time of a job on two consecutive machines there must be enough time to process the job on the first of the two machines. Constraints (5.23) and (5.24) ensure that only one operation is processed on each machine at any given time and that some order must exist between different jobs on the same machine. The remaining constraints were already explained before.

Finally, we elaborate on an improvement to Manne's model proposed by Liao [9]. Constraints (5.23) and (5.24) can be re-written as

$$C_{ik} - C_{jk} - p_{ik} + Mz_{ij} \geq 0, \tag{5.25}$$
$$-C_{ik} + C_{jk} + p_{ik} - Mz_{ij} \geq p_{jk} + p_{ik} - M, \tag{5.26}$$

respectively. Defining

$$q_{ijk} = C_{ik} - C_{jk} - p_{ik} + Mz_{ij},$$

inequalities (5.25) and (5.26) reduces to

$$0 \leq q_{ijk} \leq M - p_{ik} - p_{jk}. \tag{5.27}$$

Note that inequalities in (5.27) are in fact a lower and upper bound constraint for q_{ijk}. Thus, to solve subproblems of the branch and bound tree, the bounded simplex method can be used and achieving an optimal solution may be easier. It is worth noticing that with these new relationships, there is a reduction of one constraint for each pair of jobs in each machine, but it leads to an increase in variables q_{ijk} in the same amount.

Model MUB4 below incorporates the Liao's suggestion into model MUB3. Variables of the model, with their corresponding bound constraints, are: $z_{ij} \in \{0,1\}, i = 1,\ldots,n-1, j = i+1,\ldots,n$, $E_i \geq 0, i = 1,\ldots,n$, $T_i \geq 0, i = 1,\ldots,n$, $C_{ik}, i = 1,\ldots,n, k = 1,\ldots,m$, and $0 \leq q_{ijk} \leq M - p_{ik} - p_{jk}, i = 1,\ldots,n-1, j = i+1,\ldots,n, k = 1,\ldots,m$.

$$\text{Minimize} \sum_{i=1}^{n} E_i + T_i, \tag{5.28}$$
$$\text{subject to } T_i \geq C_{im} - d_i, \qquad i = 1,\ldots,n, \tag{5.29}$$
$$E_i \geq d_i - C_{im}, \qquad i = 1,\ldots,n, \tag{5.30}$$
$$C_{i1} \geq p_{i1}, \qquad i = 1,\ldots,n, \tag{5.31}$$

$$C_{ik} \geq p_{ik} + C_{i,k-1}, \qquad i = 1,\dots,n, k = 2,\dots,m, \tag{5.32}$$

$$q_{ijk} = C_{ik} - C_{jk} - p_{ik} + Mz_{ij}, \qquad i = 1,\dots,n-1, j = i+1,\dots,n, \quad k = 1,\dots,m. \tag{5.33}$$

Summing up the characteristics of the four models presented in this section for the flowshop scheduling problem with unlimited buffer and minimizing earliness and tardiness, Table 5.1 shows the sizes of each model, expressed by their number of constraints, binary, and continuous variables.

Table 5.1: Number of variables and constraints of the unlimited-buffer formulations.

Model	Binary variables	Continuous variables	Constraints
MUB1	n^2	$nm+2n-m$	$nm+3n-1$
MUB2	n^2	$nm+3n$	$2nm+3n-m$
MUB3	$n(n-1)/2$	$nm+2n$	n^2m+2n
MUB4	$n(n-1)/2$	$n^2m/2-nm/2+2n$	$n^2m/2+nm/2+2n$

5.3 MILP Models for Flowshop with Zero Buffer

The zero buffer environment is characterized by not having intermediary buffers between machines. Therefore, a situation known as blocking can occur, when a machine interrupts its production cycle (and it can also interrupt the production of previous machines), even though it has completed a job, because the next machine is not free. Moreover, when minimizing tardiness *and* earliness, it may be convenient for a job to stay on a machine after being completed even if the next machine is ready to process it.

A model, based on MUB1 (for the unlimited buffer environment) and on Ronconi's model [14] for the zero-buffer situation minimizing the total tardiness, is presented below. This model will be called MZB1 from now on. Variables of the model, with their corresponding bound constraints, are: $x_{ij} \in \{0,1\}, i = 1,\dots,n, j = 1,\dots,n$, $E_j \geq 0, j = 1,\dots,n$, $T_j \geq 0, j = 1,\dots,n$, $B_{jk} \geq 0, j = 1,\dots,n, k = 1,\dots,m$, $I_{jk} \geq 0, j = 1,\dots,n, k = 1,\dots,m$, and $D_{jm}, j = 1,\dots,n$. The new variable B_{jk} stands for the blocking time of the j-th job that, after being completed on machine k, stays on the machine. D_{jm} represents the departure time of the j-th job from the last machine.

$$\min \sum_{j=1}^{n} E_j + T_j, \tag{5.34}$$

$$\text{s. t.} \quad T_j \geq D_{jm} - \sum_{i=1}^{n} x_{ij} d_i, \quad j = 1,\dots,n, \tag{5.35}$$

$$E_j \geq \sum_{i=1}^{n} x_{ij} d_i - D_{jm}, \qquad j = 1, \ldots, n, \tag{5.36}$$

$$D_{1m} = \sum_{k=1}^{m} \left(\sum_{i=1}^{n} x_{i1} p_{ik} + B_{1k} \right), \tag{5.37}$$

$$D_{jm} = D_{j-1,m} + I_{j-1,m} + \sum_{i=1}^{n} x_{ij} p_{im} + B_{jm}, \qquad j = 2, \ldots, n, \tag{5.38}$$

$$I_{jk} + \sum_{i=1}^{n} x_{i,j+1} p_{ik} + B_{j+1,k} = I_{j,k+1} + \sum_{i=1}^{n} x_{ij} p_{i,k+1} + B_{j,k+1}, \qquad j = 1, \ldots, n-1, \quad k = 1, \ldots, m-1, \tag{5.39}$$

$$\sum_{i=1}^{n} x_{ij} = 1, \qquad j = 1, \ldots, n, \tag{5.40}$$

$$\sum_{j=1}^{n} x_{ij} = 1, \qquad i = 1, \ldots, n. \tag{5.41}$$

Constraints (5.35) and (5.36), which are used to identify the tardiness and the earliness of each job, are analogous to those of the unlimited buffer models, replacing the completion time of a job on a machine by its departure time. Constraint (5.37) states that the time when the first job leaves the last machine is equal to the sum of its processing and blocking times in each machine. For the departure time of the other jobs from the last machine, we have constraint (5.38), which also involves the idle times of the last machine. Constraint (5.38) says that the departure time of a job from the last machine is equal to the departure time of the previous job, plus the idle time between both jobs, plus the processing time of the job plus the time the job blocks the machine. Constraint (5.39) establishes the relationships required to keep the consistency between machines idle times and machines blocking times. These relationships are illustrated in Fig. 5.2. Constraints (5.40) and (5.41) are those already presented in previous models, which ensure that each job will only have one position in the sequence and that each position will be taken only by one job.

Model MZB1 presented above uses variables to measure times related to two different states of the machines: empty and busy. When a machine k is empty, its idle time I_{jk} between every pair of consecutive jobs in positions j and $j+1$ is used in the formulation. When machine k is being occupied by the j-th job, the time is divided into the time needed to process the j-th job, given by

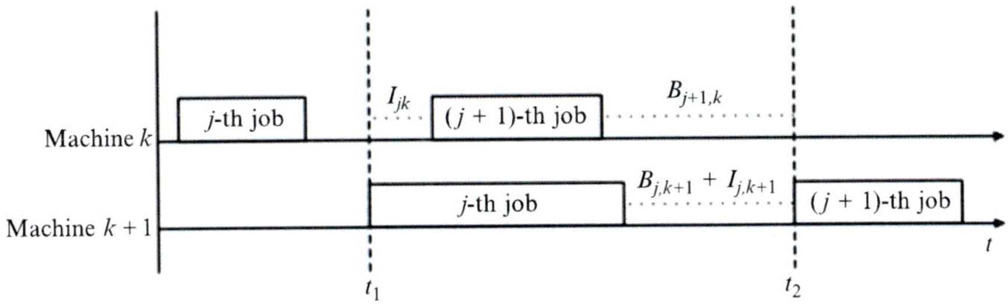

Fig. 5.2: Graphical representation of model MZB1 constraint (5.39) that expresses the relationship among processing, blocking and idle times.

$$\sum_{i=1}^{n} x_{ij} p_{i,k},$$

and the time B_{jk} during which the j-th job blocks machine k after having been processed.

In fact, it is not hard to see that there is no need to divide the time on this way, as the j-th job can be processed on machine k any time between its arrival time and its departure time. In this way, we eliminate the unnecessary imposition of processing the job as soon as it arrives. Therefore, the job can arrive to the machine, block it for a while, then be processed and finally block the machine again before leaving it. The alternative formulation below, called MZB2 from now on, models the problem by imposing very intuitive relations between the departure times of the jobs from the machines.

Variables of the model, with their corresponding bound constraints, are: $x_{ij} \in \{0,1\}, i = 1,\ldots,n, j = 1,\ldots,n$, $E_j \geq 0, j = 1,\ldots,n$, $T_j \geq 0, j = 1,\ldots,n$, and $D_{jk} \geq 0, i = 1,\ldots,n, k = 1,\ldots,m$. D_{jk} stands for departure of the j-th job from machine k. The other variables have the same meanings defined before.

$$\text{Minimize } \sum_{j=1}^{n} E_j + T_j, \tag{5.42}$$

$$\text{subject to } T_j \geq D_{jm} - \sum_{i=1}^{n} x_{ij} d_i, \qquad j = 1,\ldots,n, \tag{5.43}$$

$$E_j \geq \sum_{i=1}^{n} x_{ij} d_i - D_{jm}, \qquad j = 1,\ldots,n, \tag{5.44}$$

$$D_{j1} > \sum_{i=1}^{n} x_{ij} p_{i1}, \qquad j = 1,\ldots,n, \tag{5.45}$$

$$D_{jk} \geq D_{j,k-1} + \sum_{i=1}^{n} x_{ij} p_{ik}, \qquad j = 1,\ldots,n, k = 2,\ldots,m, \tag{5.46}$$

$$D_{jk} \geq D_{j-1,k} + \sum_{i=1}^{n} x_{ij} p_{ik}, \qquad j = 2, \ldots, n, k = 1, \ldots, m, \tag{5.47}$$

$$D_{jk} \geq D_{j-1,k+1}, \qquad j = 2, \ldots, n, k = 1, \ldots, m-1, \tag{5.48}$$

$$\sum_{i=1}^{n} x_{ij} = 1, \qquad j = 1, \ldots, n, \tag{5.49}$$

$$\sum_{j=1}^{n} x_{ij} = 1, \qquad i = 1, \ldots, n. \tag{5.50}$$

Constraints (5.45) and (5.46) indicate that a job must be processed on a machine before leaving it. Constraint (5.45) is a special case for the first machine and says that the departure time of a job from the first machine must be greater than or equal to its processing time on the machine. Constraint (5.46) indicates that the departure time of a job from a machine must be greater than or equal to its arrival time (i.e., departure time from the previous machine) plus its processing time. Constraint (5.47) says that between the departure times of two consecutive jobs from a given machine, there must be enough time for the second job to be processed. In other words, the departure time of a job from a machine minus its processing time (that is something greater than or equal to its arrival time to the machine) must be greater than or equal to the departure time of the previous job from the machine. Constraint (5.48), deeply related to the zero-buffer situation, says that a job can departure from a machine, and arrive to the next one, only when the previous job in the sequence leaved the machine. Note that, for most of the combinations of indices j and k, constraints (5.46)–(5.48) are the linearization of

$$D_{jk} \geq \max \left\{ D_{j,k-1} + \sum_{i=1}^{n} x_{ij} p_{ik},\ D_{j-1,k} + \sum_{i=1}^{n} x_{ij} p_{ik},\ D_{j-1,k+1} \right\}. \tag{5.51}$$

That is, for each D_{jk}, if $D_{j,k-1} + \sum_{i=1}^{n} x_{ij} p_{ik} \neq D_{j-1,k} + \sum_{i=1}^{n} x_{ij} p_{ik} \neq D_{j-1,k+1}$, only one constraint will be active. It is worth noticing that inequality (5.51) resembles a formulation presented by Leisten [8] for the minimization of the makespan in the flowshop problem with blocking.

Summing up the characteristics of the two models presented in this section for the flowshop scheduling problem with zero buffer and minimizing earliness and tardiness, Table 5.2 shows the sizes of each model, expressed by their number of constraints, binary and continuous variables.

Table 5.2: Number of variables and constraints of the zero-buffer formulations.

Model	Binary variables	Continuous variables	Constraints
MZB1	n^2	$2nm + 3n$	$nm + 4n - m + 1$
MZB2	n^2	$nm + 2n$	$3nm + 3n - 2m + 1$

5.4 Numerical Experiments

In this section, we aim to compare the presented formulations based on the performance of the commercial solver CPLEX when applied to a set of instances varying the number of jobs n, the number of machines m, and considering different scenarios for the jobs due dates.

In the numerical experiments, we considered instances with

$$(n,m) \in \{(5,3),(10,3),(10,7),(10,10),(15,3),(15,7),(15,10),(20,3)\}.$$

Processing times were uniformly distributed between 1 and 99 (as suggested in [19]). Due dates were uniformly distributed between $P(1-TF-DR/2)$ and $P(1-TF+DR/2)$ (as suggested in [12]), where TF and DR are the tardiness factor of jobs and dispersion range of due dates, respectively, while P is a lower bound of the makespan on the flowshop with unlimited buffer [19] defined as

$$P = \max\left\{ \max_{1\le v\le m}\left\{ \sum_{k=1}^{n} p_{kv} + \min_{1\le k\le n}\left\{ \sum_{q=1}^{v-1} p_{kq} \right\} + \min_{1\le k\le n}\left\{ \sum_{q=v+1}^{m} p_{kq} \right\}\right\}, \max_{1\le k\le n}\left\{ \sum_{v=1}^{m} p_{kv} \right\}\right\}. \quad (5.52)$$

The scenarios represent different configurations by varying TF and DR, as follows:

Scenario 1: low tardiness factor $TF = 0.2$ and small due date dispersion range $DR = 0.6$,
Scenario 2: low tardiness factor $TF = 0.2$ and wide due date range $DR = 1.2$,
Scenario 3: high tardiness factor $TF = 0.4$ and small due date range $DR = 0.6$,
Scenario 4: high tardiness factor $TF = 0.4$ and wide due date range $DR = 1.2$.

For each combination of n, m and scenario, we considered ten different randomly generated instances. Therefore, as a whole, the test set consists of $8 \times 4 \times 10 = 320$ problems. We set the value of the big M parameter in formulations MUB3 and MUB4 as $M = 100 \sum_{j=1}^{n} \sum_{k=1}^{m} p_{jk}$.

Formulations were written in AMPL (IBM ILOG AMPL 12.1.0) modelling language and solved using CPLEX 12.1. The experiments were performed in a 2.4 GHz Intel Core2 Quad Q6600 with 4.0 GB of RAM memory and Linux Operating System. The computer implementation of the models as well as the data sets used in our experiments and the solutions found are publicly available for benchmarking purposes at `http://www.ime.usp.br/$\sim$egbirgin/`.

Table 5.3 shows the results of solving formulations MUB1, MUB2, MUB3 and MUB4 for all the instances in the test set. In the table, "CPU Time" means computational CPU time in seconds used by CPLEX to solve the problem, "Simplex It" is the total number of simplex iterations, and "B&B nodes" is the total number of explored nodes in the branch and bound tree. Each cell in the table represents 40 instances (four scenarios and ten instances per scenario). The mean was computed excluding two instances from the top and two instances from the bottom tails of the CPU times. From the table, it can be seen that formulations MUB1 and MUB2 are very similar, when referring to the used CPU time. On the other hand, solving instances of formulations MUB3 and MUB4 is very time consuming. As MUB3 and MUB4 have half the number of integer variables of MUB1 and MUB2, the experiment confirms that, at least for the present set of test instances, it is not true that the computational effort needed to solve a MILP problem is proportional to the number of binary variables of the model, as claimed in [11]. When comparing the effort needed to solve MUB3 and MUB4, it is not possible to confirm either the claim that replacing regular constraints by new variables with lower and upper bounds (as in the modification to the Manne's model suggested by Liao [9]) makes the subproblems easier to be solved using CPLEX. This result may be related, as pointed out in [18], to the chosen value for M that greatly affects the solver performance.

A note on solving the large instances using models MUB3 and MUB4 is in order. Table 5.3 shows that those models need one order of magnitude more computational time to be solved when $n = 10$ and $m = 10$. Moreover, in instance with $n = 10$, $m = 10$, scenario 2, 8-th seed of both models, CPLEX solver with its default parameters reported as final point a non-integer solution with objective function value 1,148.99678 (the optimal value of this instance is $1,162$). As pointed out in [18], the performance of the solver when applied to solve those models is strongly related to the choice of the big M parameter in (5.23), (5.24), and (5.33). In [18], the authors mention that better performances were achieved for large values of M. However, our numerical experiments suggested that, due to rounding issues and scaling, large values of M prevent the solver from finding optimal integer solutions in many cases. Therefore, for using models MUB3 and MUB4, a user should face the problem of setting an adequate value for parameter M capable of providing optimal solutions in a reasonable computational time.

Table 5.4 shows the results of solving formulations MZB1 and MZB2 for all the instances in the test set. From the table, it can be seen that these formulations are very similar, MZB1 been a slightly easier to be solved in the smaller cases while the opposite situation occurs in the larger instances. The quotients between the number of branch and bound nodes used to solve models MZB1 and MZB2 are 1.32, 1.10, 0.99, 1.00, 1.10, 0.97, 0.94 and 1.13, for each of the instances dimensions listed in the table, respectively. Their average is 1.07, showing that 7% more nodes need to be explored in the branch and bound tree to solve model MZB1 in comparison with MZB2. The average number of simplex iterations needed to solve each node of the branch and bound tree is 38.91 for model MZB1 and 32.81 for model MZB2. Both models have the same binary variables, while model MZB2 has fewer continuous variables and more constraints than model MZB1. It seems that this situation helps to reduce the number of explored nodes and the number of simplex iterations needed to solve each node.

Table 5.3: Numerical evaluation of formulations MUB1, MUB2, MUB3, and MUB4.

	Model MUB1			Model MUB2		
n m	CPU Time	Simplex It	B&B nodes	CPU Time	Simplex It	B&B nodes
5 3	0.03	131.03	14.17	0.02	111.89	10.33
10 3	0.61	7,517.03	533.08	0.49	7,341.31	497.03
10 7	1.94	25,943.47	1,082.28	1.65	23,295.50	1,093.39
10 10	4.59	66,336.97	2,065.44	3.65	51,774.64	2,037.33
15 3	10.81	194,452.58	8,230.92	9.63	159,715.39	7,343.39
15 7	91.25	1,501,181.78	34,280.94	75.50	1,207,768.81	36,477.86
15 10	281.25	4,506,249.72	86,486.28	227.75	3,566,364.53	91,435.28
20 3	105.90	1,719,383.14	68,800.81	89.59	1,456,790.72	54,698.81

	Model MUB3			Model MUB4		
n m	CPU Time	Simplex It	B&B nodes	CPU Time	Simplex It	B&B nodes
5 3	0.05	182.31	22.17	0.05	331.08	33.17
10 3	2.11	37,891.92	2,956.08	3.28	58,925.89	5,075.83
10 7	27.90	316,057.22	28,688.97	39.45	461,485.92	44,965.47
10 10	40.17	417,254.75	32,041.25	61.87	635,131.11	50,661.58

Table 5.4: Numerical evaluation of formulations MZB1 and MZB2.

	Model MZB1			Model MZB2		
n m	CPU Time	Simplex It	B&B nodes	CPU Time	Simplex It	B&B nodes
5 3	0.02	147.56	14.72	0.04	154.50	11.17
10 3	0.79	12,171.28	753.50	0.94	12,850.72	685.78
10 7	2.67	39,232.25	1,561.97	3.08	37,722.58	1,578.14
10 10	6.06	101,268.08	3,058.47	6.86	90,915.25	3,073.31
15 3	20.33	392,951.31	14,895.53	21.10	331,480.81	13,569.50
15 7	203.06	3,334,785.25	71,643.31	188.79	2,740,222.69	74,074.36
15 10	645.71	9,658,414.00	179,901.81	629.83	8,440,086.14	190,989.08
20 3	558.15	8,680,575.50	299,238.36	458.06	6,320,653.61	263,867.06

5.5 Concluding Remarks

In [2], it is stated that the problem solution time using branch and bound algorithms is unpredictable and that it depends on the numerical values of data. In the tests conducted, it could be noted that several factors interact, in addition to the number of binary variables, so that favorable or adverse conditions can be established to solve the model. That is, the time to solve a branch and bound algorithm does not have a direct relationship with the size of the model, although it is one of the factors involved in this relationship.

Observing the numerical results, it can be seen that the number of binary variables by itself does not indicate precisely the level of difficulty in solving mixed integer problems, contradicting the claims in [11]. This can be confirmed by the analysis of Table 5.3, which make it clear that formulations with a small number of binary variables (MUB3 and MUB4) took longer to achieve an optimal solution in the various scenarios analyzed.

Our numerical study strongly suggests that models based on n^2 positioning binary variables x_{ij}, named MUB1, MUB2, MZB1, and MZB2, are easier to be solved than models based on $n(n-1)/2$ precedence binary variables z_{ij}, named MUB3 and MUB4. The experiments also seem to suggest that, for models with the same number of binary variables, the ones with similar or less number of continuous variables and more constraints (MUB2 versus MUB1 in the unlimited buffer case, and MZB2 versus MZB1 in the zero buffer case) are slightly easier to be solved.

Acknowledgment This work was supported by PRONEX-CNPq/FAPERJ (E-26/171.1510/2006-APQ1), FAPESP (Grants 2006/53768-0, 2006/03496-3 and 2009/10241-0), and CNPq (308000/2009-9 and 304484/2007-5).

References

1. Biskup, D., Feldmann, M.: Benchmarks for scheduling on a single machine against restrictive and unrestrictive common due dates. Computers & Operations Research **28**, 787–801 (2001)
2. Brandimarte, P.: Neighbourhood search-based optimization algorithms for production scheduling: a survey. Running Series - Designing Production Order Scheduling Tools **5**(2) (1992)
3. Dudek, R.A., Panwalkar, S.S., Smith, M.L.: The lessons of flowshop scheduling research. Operations Research **40**, 7–13 (1992)
4. Grabowski, J., Pempera, J.: The permutation flow shop problem with blocking: A Tabu Search approach. Omega **35**, 302–311 (2007)
5. Hall, N.G., Sriskandarajah, C.: A survey of machine scheduling problems with blocking and no-wait in process. Operations Research **44**, 510–525 (1996)
6. Kanet, J.J., Sridharan, V.: Scheduling with inserted idle time: problem taxonomy and literature review. Operations Research **48**, 99–110 (2000)
7. Kim, Y.D.: Minimizing total tardiness in permutation flowshops. European Journal of Operational Research **85**, 541–555 (1995)
8. Leisten, R.: Flowshop sequencing with limited buffer storage. International Journal of Production Research **28**, 2085–2100 (1990)
9. Liao, C.J., You, C.T.: An improved formulation for the job-shop scheduling problem. Journal of the Operational Research Society **43**, 1047–1054 (1992)
10. Manne, A.S.: On the job-shop scheduling problem. Operations Research **8**, 219–223 (1960)
11. Pan, C.H.: A study of integer programming formulations for scheduling problems. International Journal of Systems Science **28**, 33–41 (1997)
12. Potts, C.N., van Wassenhove, L.N.: A decomposition algorithm for the single machine total tardiness problem. Operations Research Letters **1**, 177–181 (1982)
13. Reklaitis, G.V.: Review of scheduling of process operations. AIChE Symposium Series **78**, 119–133 (1982)
14. Ronconi, D.P.: A contribution to the study of the flowshop problem minimizing the total tardiness. Ph.D. Dissertation (in portuguese), UNICAMP, Campinas, Brazil (1997)

15. Sen, T., Gupta, S.K.: A state-of-art survey of static scheduling research involving due dates. Omega **12**, 63–76 (1984)
16. Sethi, S.P., Sriskandarajah, C., Sorger, G., Blazewicz, J., Kubiak, W.: Sequencing of parts and robot moves in a robotic cell. International Journal of Flexible Manufacturing Systems **4**, 331–358 (1992)
17. Stafford, E.F.: On the development of a mixed-integer linear programming model for the flowshop sequencing problem. Journal of the Operational Research Society **39**, 1163–1174 (1988)
18. Stafford, E.F., Tseng, F.T.,J. N., Gupta, J.N.D.: Comparative evaluation of MILP flowshop models. Journal of the Operational Research Society **56**, 88–101 (2005)
19. Taillard, E.: Benchmarks for basic scheduling problems. European Journal of Operational Research **64**, 278–285 (1993)
20. Wagner, H.M.: An integer linear-programming model for machine scheduling. Naval Research Logistic **6**, 131–140 (1959)
21. Wilson, J.M.: Alternative formulations of a flow-shop scheduling problem. Journal of the Operational Research Society **40**, 395–399 (1989)
22. Zhu, Z., Heady, R.B.: Minimizing the sum of earliness/tardiness in multimachine scheduling: a mixed integer programming approach. Computers & Industrial Engineering **38**, 297–305 (2000)

Chapter 6
Just-in-Time Scheduling with Equal-Size Jobs

Ameur Soukhal and Nguyen Huynh Toung

Abstract This chapter deals with common due date scheduling problem on single and parallel machines in which job processing times are identical. The objective is to minimize the total weighted earliness–tardiness. According to the common due date, two cases are considered. In the first case, the due date is given which involves that the common due date is enough early (restrictive version) or not (non-restrictive version) to constraint the optimal solution. In this case, there is no cost related to its value. The second case deals with unknown due date which is a decision variables. It means that the decision maker not only takes sequencing and scheduling decisions to minimize the total weighted earliness–tardiness but also determines the optimal value of due date which is specified as controllable parameter.

6.1 Introduction

The just-in-time (JIT) scheduling problems have been the subject of intensive researches for many years since these problems have industrial applications and involve exciting theoretical problems. It means that a job is completed as closely as possible to its due date. This approach comes from the "JIT" philosophy in management and production theory: an item should be produced only when it is required. In reality, the earliness penalty implies the loss caused by deterioration of goods, storage cost, and insurance cost, while the tardiness penalty implies contract penalties, dissatisfaction of customers, loss of sales, loss of goodwill, etc. (see [4, 14]).

Ameur Soukhal
Laboratoire d'Informatique, Université François Rabelais, 64 Avenue Jean Portalis, 37200 Tours, France, e-mail: ameur.soukhal@univ-tours.fr

Nguyen Huynh Toung
Faculty of Computer Science and Engineering, University of Technology of Ho Chi Minh City, Ho Chi Minh City, Vietnam, e-mail: nguyen.huynh@yahoo.com

R.Z. Ríos-Mercado and Y.A. Ríos-Solís (eds.), *Just-in-Time Systems*, Springer Optimization and Its Applications 60, DOI 10.1007/978-1-4614-1123-9_6,

In this chapter, equal size jobs should be scheduled as closely as possible to a common due date to minimize the total weighted earliness–tardiness on machine scheduling problems. All jobs are of equal size means that they require the same processing time. This is often the case in real-life problems, notably scheduling problems occurring in batch oriented production systems. An example is that of shampoo packing systems which consist in packing into bottles batches of shampoo. In order to smooth the production (make the resource setup times negligible regarding to the processing times), batches are split into jobs requiring the same processing time (for instance, half an hour of packing time).

Scheduling problems with equal-size jobs have been the subject of several works published in the literature, e.g., in [5] the authors present a state-of-the-art with several regular criteria, but to the best of our knowledge few deals with JIT objective functions.

One of the principle features of JIT scheduling relies on the way the due dates are considered: they can be given parameters or decision variables.

For the former case, the main motivation comes from the need to measure the quality in inventory management and production management [4, 14, 26, 41]. According to the literature, two versions of a given common due date are defined: *restrictive* common due date and *non-restrictive* common due date. It involves that the common due date is enough early (restrictive version) or not (non-restrictive version) to constraint the optimal solution. However, Kanet identifies the non-restrictive common due date d as the case where d is greater or equal to the total processing times [29].

For the latter case, the due dates assignment becomes a challenging issue related to the interaction between various participants of the supply chain: Practice shows that during the negotiations between customers and the producers, the agreement on delivery dates depends on both the production policy, which leads to limit the stocks and the capacity to meet due dates, and the needs of the customers (they may have idea of when ideally to receive their orders). Henceforth, decision makers may need an efficient method for quoting due dates and job scheduling (for more details see some motivation surveys in [14, 15, 26, 28]).

Minimizing total weighted earliness–tardiness on a single machine with only a common due date is shown to be NP-Hard [17–19, 44]. However, according to the value of earliness and tardiness penalties, several polynomial and pseudo-polynomial cases are identified [26, 41].

This chapter is organized as follows. The first part deals with l given common due dates. Both cases are considered: restrictive version (Sect. 6.3) and non-restrictive version (Sect. 6.4) of due dates.

Table 6.1 summarizes the main results on JIT scheduling problem with given common due date.

The studied problems that are considered in the first part of this chapter are summarized in Table 6.2.

Table 6.1: JIT with given common due date

Problem	Method and Complexity status	Reference
$1\|p_i = p\|\sum(E_i + T_i)$	Polynomial $O(n \log n)$	[12]
$1\|d_i = d$, *non-restrictive* $\|\sum(E_i + T_i)$	Polynomial $O(n \log n)$	[29]
$1\|d_i = d\|\sum(E_i + T_i)$	$\mathcal{NP}$-hard	[17,19]
	Dynamic programming $O(n \sum p_i)$ et $O(n^2 d)$	[17,19]
	4/3-approximation	[20]
$1\|d_i = d, p_i = w_i\|\sum w_i(E_i + T_i)$	Polynomial $O(n)$	[18]
$1\|d_i = d, p_i = 1\|\sum w_i(E_i + T_i)$	Polynomial $O(n \log n)$	[18]
$1\|d_i = d\|\sum w_i(E_i + T_i)$	Polynomial if $w_{max} = Poly(n)$	[27]
$1\|d_i = d$, *non-restrictive* $\|\sum w_i(E_i + T_i)$	$\mathcal{NP}$-difficile	[18]
	Dynamic programming $O(n \sum p_i)$ and $O(n \sum w_i)$	[18,27]
	FPTAS	[32]
$1\|p_i = 1, d_i\|\sum(\alpha_i E_i + \beta_i T_i)$	Polynomial $O(n^4)$	[31]
$1\|p_i = p, d_i = d$, *non-restrictive* $\|\sum(\alpha_i E_i + \beta_i T_i)$	Polynomial $O(n^3)$	[24,25,35]
$1\|p_i = p, d_i \in D, \|D\| = k\|\sum(\alpha_i E_i + \beta_i T_i)$	Polynomial	[24,25]
$1\|d_i = d$, *non-restrictive* $\|\sum(\alpha_i E_i + \beta_i T_i)$	B&B $n = 1{,}000$	[39]
$1\|d_i = d$, *non-restrictive* $\|\sum(\alpha_i E_i + \beta_i T_i)$	PTAS	[24,25]
$Qm\|d_i = d$, *non-restrictive* $\|\sum(\alpha E_i + \beta T_i)$	Polynomial $O(n \log n)$	[11]
$Rm\|d_i = d$, *non-restrictive* $\|\sum(E_i + T_i)$	Polynomial $O(n^3)$	[2,33]
$Pm\|d_i = d$, *non-restrictive* $\|\sum w_i(E_i + T_i)$	Strongly $\mathcal{NP}$-hard	[42]
$Pm\|d_i = d, p_i = w_i\|\sum w_i(E_i + T_i)$	Ordinary $\mathcal{NP}$-hard	[40]
	Dynamic programming $O(mnd^m P^{m-1})$	[40]

Table 6.2: Studied problems for a given due date (restrictive and non-restrictive cases)

Problem	Reference
$1\|p_j = p, d_j = d\|\sum(\alpha_j E_j + \beta_j T_j)$	Sect. 6.3.1
$1\|p_j = p, d_j \in D, \|D\| \leq l, l$ *fixed* $\|\sum(\alpha_j E_j + \beta_j T_j)$	Sect. 6.3.2
$Qm\|p_j = p, d_j \in D, \|D\| \leq l, l$ *fixed* $\|\sum(\alpha_j E_j + \beta_j T_j)$	Sect. 6.3.3
$Qm\|p_j = p, d_j = d$, *non-restrictive* $\|\sum(\alpha E_j + \beta T_j)$	Sect. 6.4.1
$Qm\|p_j = p, d_j = d$, *non-restrictive* $\|\sum(\alpha_j E_j + \beta_j T_j)$	Sect. 6.4.2

The second part deals with due dates assignment (due date is a decision variable). Table 6.3 summarizes the main known results on JIT scheduling problem with assignment common due date and the studied problems are summarized in Table 6.4.

6.2 Preliminary Results

Without loss of generality, we assume that m identical parallel machines are available from time 0 onwards to process n independent jobs without preemption. Each job J_j, $j = 1, \ldots, n$, is ready at time zero and has the same integer processing time

Table 6.3: JIT with assignment common due date

Problem	Method and Complexity status	Reference
$1\|d_i = d\|\sum(\alpha E_i + \beta T_i + \gamma d)$	Polynomial $O(n \log n)$	[37]
$1\|d_i = d\|\sum(\alpha E_i + \beta T_i)$	Polynomial $O(n \log n)$	[3]
$P\|p_i = p, d_i = d\|\sum(\alpha E_i + \beta T_i + \gamma d)$	Polynomial	[7]
$Pm\|d_i = d\|\sum(\alpha E_i + \beta T_i + \gamma d)$	Ordinary NP-hard	[7,9]
	Dynamic programming $O(n^{2m+1} p_{\max}^{2m})$	[9]
	3-approximation algorithm in $O(n \log n)$	[43]
	FPTAS	[43]
$P\|d_i = d\|\sum(\alpha E_i + \beta T_i + \gamma d)$	Strongly NP-hard	[9]
$P\|d_i = d\|\sum w_i(E_i + T_i)$	Strongly NP-hard	[42]
$Qm\|p_i = p, d_i = d\|\sum(\alpha E_i + \beta T_i + \gamma d)$	Polynomial $O(3^m)$	[7]
$Qm\|d_i = d\|\sum(\alpha E_i + \beta T_i)$	Polynomial $O(n \log n)$	[11]
$Rm\|d_i = d\|\sum(E_i + T_i)$	Polynomial $O(n^3)$	[2,33]

Table 6.4: Studied problems for an unknown due date (restrictive and non-restrictive cases)

Problem	Reference
$1\|p_j = p, d_j = d\|\sum(\alpha_j E_j + \beta_j T_j + \gamma d)$	Sect. 6.5.2
$1\|p_j = p, d_j \in D, \|D\| \leq l, l\ fixed\|\sum(\alpha_j E_j + \beta_j T_j + \gamma d)$	Sect. 6.5.3
$P\|p_j = p, d_j \in D, \|D\| \leq l, l\ fixed\|\sum(\alpha_j E_j + \beta_j T_j + \gamma d)$	Sect. 6.5.4
$Qm\|p_j = p, d_j = d, nmit\|\sum(\alpha E_j + \beta T_j + \gamma d)$	Sect. 6.5.5

and a due date belonging in the set $\{D_1, D_2, \ldots, D_l\}$ of l unknown integer due dates which are to be minimized. Whenever necessary, we will consider the single machine problem by setting $m = 1$ and the common due date configuration by setting $l = 1$ (in that case, the common due date is denoted by d to follow the classical notation). In this chapter, we make use of the following notations:

- $\mathcal{N}_k$: set of jobs that have due date $D_k, k = 1, \ldots, l$.
- $n_k = |\mathcal{N}_k|$: the number of jobs having due date $D_k, k = 1, \ldots, l$.
- $J_{[j]}$ or $[j]$: the job scheduled at the jth position in a given sequence.
- $p_j = p$: processing time of job $J_j, j = 1, 2, \ldots, n$.
- d_j: due date of job $J_j, j = 1, 2, \ldots, n$.
- S_j and C_j: the starting time and the completion time of job $J_j, j = 1, \ldots, n$ where $C_j = S_j + p$.
- E_j: the earliness of job $J_j, j = 1, \ldots, n$, with $E_j = \max(d_j - C_j, 0)$.
- T_j: the tardiness of job $J_j, j = 1, \ldots, n$, with $T_j = \max(C_j - d_j, 0)$.
- α_j, β_j and γ: unit earliness, tardiness, and due date penalty of job J_j.
- $\alpha_{[j]}$ and $\beta_{[j]}$: the earliness and tardiness costs per time unit of $J_{[j]}$.

- $w^{(1)}$: an earliness vector, given by $(\alpha_1\ \alpha_2\ \ldots\ \alpha_n)$.
- $w^{(2)}$: a tardiness vector, given by $(\beta_1\ \beta_2\ \ldots\ \beta_n)$.
- $Z = f(S) = \sum_j(\alpha_j E_j + \beta_j T_j + \gamma d_j)$: objective function cost for a given sequence S.

Henceforth, for each job J_i one of the three following situations can occur:

- J_j is an early job, that is, $C_j \leq d_j$.
- J_j is a splitting job, that is, $S_j \leq d_j < C_j$ (of course this job is the first tardy job).
- J_j is a fully-tardy job, that is, $d_j < C_j$.

Moreover, we say that job J_j is on-time if and only if $C_j = d_j$.

6.2.1 Assignment Problem

Consider a complete bipartite graph, $G = (U \cup V, U \times V)$ with $U = \{u_1, u_2, \ldots, u_n\}$ and $V = \{v_1, v_2, \ldots, v_m\}$. Assume that $n \leq m$. There is a real cost c_{ij} associated with each arc (u_i, v_j). An assignment is given by a one-to-one transformation function $\varphi : U \to V$. The assignment problem is to determine an assignment set such that $\sum_{u \in U} c_{u\varphi(u)}$ is minimized.

We present the assignment problem by a linear model (**LP**) with 0-1 binary variables x_{ij}:

$$
\begin{aligned}
&min \ \textstyle\sum_{i=1}^{n} \sum_{j=1}^{m} c_{ij} x_{ij} \\
&st. \\
&\qquad \textstyle\sum_{j=1}^{m} x_{ij} = 1 \quad (i = \overline{1,n}) \\
&\qquad \textstyle\sum_{i=1}^{n} x_{ij} \leq 1 \quad (j = \overline{1,m}) \\
&\qquad x_{ij} \in \{0,1\} \quad (i = \overline{1,n}; j = \overline{1,m}).
\end{aligned}
$$

Here, $x_{ij} = 1$ if and only if u_i is assigned to v_j. Each $u_i \in U$ is assigned to only one element in V, and each$v_j \in V$ is implicated up to an assignment. Hence, this problem is also defined by a matrix $C_{(m \times n)} = (c_{ij})$ (Fig. 6.1):

The first algorithm is known as Hungarian method introduced by Kuhn [34]. It solves the assignment problem with an $n \times n$ matrix and the complexity is $O(n^3)$. Several improved and generalized algorithms are proposed up to now, e.g., [8, 21, 30]. However, our main contribution is to prove well solvable cases of studied problems. Then, we consider solely that an assignment problem with an $n \times m$ matrix $(m \geq n)$ can be solved with a generalized Hungarian algorithm in $O(n^2 m)$.

$$C = \begin{bmatrix} c_{11} & c_{12} & \cdots & c_{1n} \\ c_{21} & c_{22} & \cdots & c_{2n} \\ \cdots & \cdots & \cdots & \cdots \\ \cdots & \cdots & \cdots & \cdots \\ c_{m1} & c_{m2} & \cdots & c_{mn} \end{bmatrix}$$

Fig. 6.1: Assignment matrix

6.3 Restrictive Common Due Date and Just-in-Time Scheduling

This section deals with minimizing the total weighted earliness–tardiness on $(m \geq 1)$ uniform parallel machines. Each machine has a different speed. n independent jobs J_i $_{(i=1,\ldots,n)}$ with identical processing times should be completed without preemption at their due date d_j ($d_j \in D$ and $|D| = l$, for given l with $l \geq 1$). We assume that all jobs are ready at time zero. We show that minimizing the total weighted earliness–tardiness on $m \geq 1$ uniform parallel machines with identical processing times ($p_i = p$) and given l due dates is polynomially solvable. Indeed, there is a small difference between unit processing time scheduling and equal processing time scheduling even though with one machine. Of course, in the case of unit processing times, there exists always a JIT job. However, with identical processing times, we cannot assure that there exists a JIT job.

In what follows, we base on the observation that, since all tasks have the same processing time, scheduling of these tasks can be accomplished in two stages, where the first stage determines the completion times without specifying what tasks have these completion times and the second stage assigns jobs to the already determined completion times. For determining the first stage of a considered problem, it is possible to construct in polynomial time a family of sets of completion times that always contains a set of completion times of some optimal schedule. Then, after constructing all sets of the above mentioned family, the second stage of scheduling should be processed in such a way that the assignment of jobs to completion times can be obtained as a solution of an assignment problem using, for instance, the standard mathematical programming methods (*LP*).

6.3.1 $1|p_j = p, d_j = d|\sum(\alpha_j E_j + \beta_j T_j)$

First, we present some well-known properties.

Property 6.1. There are no intermediate idle times between two adjacent jobs.

Property 6.2. The optimal schedule is V-shaped around the common due date. It means that the jobs completed before or on the common due date d are scheduled in non-increasing order of the ratio α_j/p_j, and the jobs starting on or after d are scheduled in non-decreasing order of the ratio β_i/p_i.

Property 6.3. If the starting time of the first job is not zero then there exists one job for which its completion time is d.

Let J_{sp} be a splitting job with $S_{\text{sp}} \le d < C_{\text{sp}}$. Hence, there exists a "JIT" job whether $S_{\text{sp}} = d$.

Let k be the maximum number of early jobs. If $d = (k+1)p$, then $k = \left\lfloor \frac{d}{p} \right\rfloor - 1$; otherwise, $k = \left\lfloor \frac{d}{p} \right\rfloor$. Hence, we can write $k = \left\lfloor \frac{d-1}{p} \right\rfloor$.

It is clear that the job scheduled at the $(k+1)$-th position noted $J_{[k+1]}$ is the splitting job J_{sp}.

Let $\delta_1 = d - kp$, $\delta_2 = p - \delta_1$ be the distances between S_{sp}, C_{sp} and d (see Fig. 6.2).

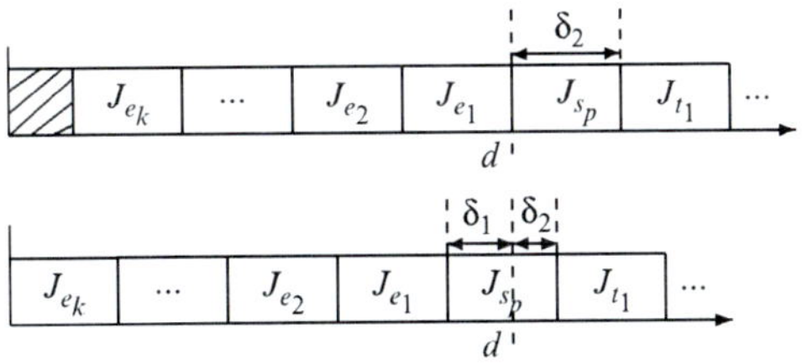

Fig. 6.2: Job positions on single machine with a given common due date

The job positions are defined as following (see Fig. 6.2):

- Concerning the early job set: the last job processed just before J_{sp} is at the first early-position noted J_{e_1}, this job may be JIT; the job processed just before the previous one is at the second early-position noted J_{e_2}; and so on.
- Concerning the tardy job set: the first job processed just after J_{sp} is at the first tardy-position noted J_{t_1}; the job processed just after the previous one is at the second tardy-position noted J_{t_2}; and so on.

Then, the cost of assigning a job to ith early-position is given by: $\alpha_{[k-i+1]}((i-1)p+\delta_1)$ which corresponds to the contribution of this job to the total earliness cost. Similarly, the cost of assigning a job to the ith tardy-position is given by $\beta_{[k+1+i]}(ip+\delta_2)$.

Hence the assignment costs define the matrix $Q_{((k+n)\times n)}$ in which each element is noted by $(q_{i,j})_{1\le i\le 2u+1; 1\le j\le n}$, where jth column corresponds to job J_j; first row corresponds to job J_{sp}; $(2i)$th row corresponds to job J_{e_i} $_{(1\le i\le k)}$; $(2i+1)$th row corresponds to job J_{t_i} $_{(1\le i\le k)}$; and ith row corresponds to job $J_{t_{i-k-1}}$ $_{(2k+2\le i\le n+k)}$ (see Fig. 6.3).

Consequently, element $(q_{i,j})$ is defined as follows:

$$\begin{cases} q_{1,j} = \delta_2\alpha_j & (j=\overline{1,n}) \\ q_{2i,j} = (\delta_1+(i-1)p)\alpha_j & (i=\overline{1,k}; j=\overline{1,n}) \\ q_{2i+1,j} = (\delta_2+ip)\beta_j & (i=\overline{1,k}; j=\overline{1,n}) \\ q_{i,j} = (\delta_2+(i-k-1)p)\beta_j & (i=\overline{2k+2,n+k}; j=\overline{1,n}) \end{cases}.$$

In another point of view more precisely, $Q_{((k+n)\times n)}$ can be determined as follows:

$$\begin{aligned} Q^{\mathrm{T}} = [&\delta_2 w_i^{(2)^{\mathrm{T}}} \quad \delta_1 w_i^{(1)^{\mathrm{T}}} \quad (\delta_2+p)w_i^{(2)^{\mathrm{T}}} \quad (\delta_1+p)w_i^{(1)^{\mathrm{T}}} \\ &(\delta_2+2p)w_i^{(2)^{\mathrm{T}}} \quad \ldots \quad (\delta_1+(k-1)p)w_i^{(1)^{\mathrm{T}}} \\ &(\delta_2+kp)w_i^{(2)^{\mathrm{T}}} \quad (\delta_2+(k+1)p)w_i^{(2)^{\mathrm{T}}} \\ &(\delta_2+(k+2)p)w_i^{(2)^{\mathrm{T}}} \quad \ldots \quad (\delta_2+(u-1)p)w_i^{(2)^{\mathrm{T}}}]. \end{aligned}$$

$$Q = \left[\begin{array}{cccc} \delta_2\beta_1 & \delta_2\beta_2 & \cdots & \delta_2\beta_n \\ \hdashline \delta_1\alpha_1 & \delta_1\alpha_2 & \cdots & \delta_1\alpha_n \\ (\delta_2+p)\beta_1 & (\delta_2+p)\beta_2 & \cdots & (\delta_2+p)\beta_n \\ \hdashline (\delta_1+p)\alpha_1 & (\delta_1+p)\alpha_2 & \cdots & (\delta_1+p)\alpha_n \\ (\delta_2+2p)\beta_1 & (\delta_2+2p)\beta_2 & \cdots & (\delta_2+2p)\beta_n \\ \hdashline \cdots & \cdots & \cdots & \cdots \\ \cdots & \cdots & \cdots & \cdots \\ \hdashline (\delta_1+(k-1)p)\alpha_1 & (\delta_1+(k-1)p)\alpha_2 & \cdots & (\delta_1+(k-1)p)\alpha_n \\ (\delta_2+kp)\beta_1 & (\delta_2+kp)\beta_2 & \cdots & (\delta_2+kp)\beta_n \\ \hdashline (\delta_2+(k+1)p)\beta_1 & (\delta_2+(k+1)p)\beta_2 & \cdots & (\delta_2+(k+1)p)\beta_n \\ \hdashline \cdots & \cdots & \cdots & \cdots \\ \hdashline (\delta_2+(n-1)p)\beta_1 & (\delta_2+(n-1)p)\beta_2 & \cdots & (\delta_2+(n-1)p)\beta_n \end{array}\right]$$

Fig. 6.3: Matrix of job assignment costs

A set $S=(q_{k_1,1}; q_{k_2,2}\ldots; q_{k_n,n})$ of n elements from Q is called a *feasible solution* with $Z=f(S)=\sum_{r=1}^{n} q_{k_r,r}$ if and only if no two elements are from the same row (obviously, two elements in S cannot be from the same column). Therefore, to determine an optimal solution S^* we should find a feasible set of n elements from Q

that minimizes Z. Hence, this problem is equivalent to the well-known assignment problem which is polynomially solvable [34].

Theorem 6.1. *An optimal solution for the scheduling problem* $1|p_j = p, d_j = d|\sum(\alpha_j E_j + \beta_j T_j)$ *with restrictive common due date can be found in* $O(n^3)$ *time.*

Example 6.1. Five jobs will be scheduled on the single machine. The processing times, due dates, and earliness and tardiness penalties are given in the Table 6.5.

Table 6.5: Processing times, due dates and penalty costs

	J_1	J_2	J_3	J_4	J_5
p_i	3	3	3	3	3
d_i	7	7	7	7	7
α_i	7	5	4	3	5
β_i	5	2	7	4	5

We have then two possible values of pair (δ_1, δ_2): $(0,3)$ (Fig. 6.4, Case(a)) and $(1,2)$ (Fig. 6.4, Case(b)). The following figures show two corresponding cases with job positions:

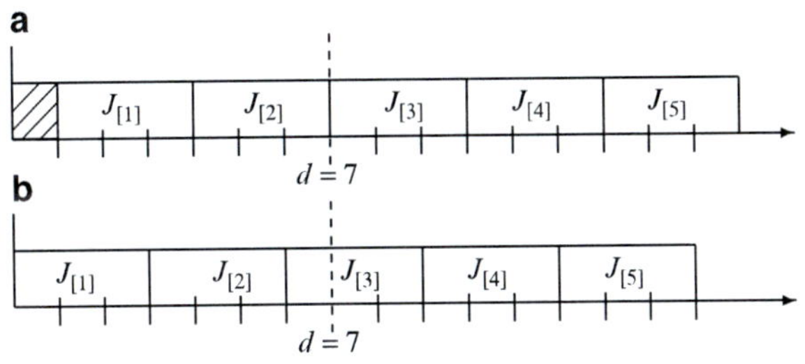

Fig. 6.4: Example with five jobs

In both cases, we can write the matrix Q (see Fig. 6.5).

In the case (a) with $(\delta_1, \delta_2) = (0,3)$, the corresponding matrix and selected elements are as follows (Fig. 6.6).

The corresponding optimal solution is then (3,1,5,4,2) and the optimal value in Case (a) is **69** (see Fig. 6.7).

In the case (b) with $(\delta_1, \delta_2) = (1,2)$, the corresponding matrix and selected elements are as follows (Fig. 6.8).

The corresponding optimal solution is then (4, 3, 5, 1, 2) and the optimal value in Case (b) is **67** (see Fig. 6.9).

$$
Q = \left[\begin{array}{ccccc}
\delta_2\beta_1 & \delta_2\beta_2 & \delta_2\beta_3 & \delta_2\beta_4 & \delta_2\beta_5 \\
\hline
\delta_1\alpha_1 & \delta_1\alpha_2 & \delta_1\alpha_3 & \delta_1\alpha_4 & \delta_1\alpha_5 \\
(\delta_2+p)\beta_1 & (\delta_2+p)\beta_2 & (\delta_2+p)\beta_3 & (\delta_2+p)\beta_4 & (\delta_2+p)\beta_5 \\
\hline
(\delta_1+p)\alpha_1 & (\delta_1+p)\alpha_2 & (\delta_1+p)\alpha_3 & (\delta_1+p)\alpha_4 & (\delta_1+p)\alpha_5 \\
(\delta_2+2p)\beta_1 & (\delta_2+2p)\beta_2 & (\delta_2+2p)\beta_3 & (\delta_2+2p)\beta_4 & (\delta_2+2p)\beta_5 \\
\hline
(\delta_2+3p)\beta_1 & (\delta_2+3p)\beta_2 & (\delta_2+3p)\beta_3 & (\delta_2+3p)\beta_4 & (\delta_2+3p)\beta_5 \\
\hline
(\delta_2+4p)\beta_1 & (\delta_2+4p)\beta_2 & (\delta_2+4p)\beta_3 & (\delta_2+4p)\beta_4 & (\delta_2+4p)\beta_5
\end{array}\right]
$$

Fig. 6.5: Assignment matrix with five jobs

$$
Q = \left[\begin{array}{ccccc}
15 & 6 & 21 & 12 & \underline{\mathbf{15}} \\
\hline
\underline{\mathbf{0}} & 0 & 0 & 0 & 0 \\
30 & 12 & 42 & \underline{\mathbf{24}} & 30 \\
\hline
21 & 15 & \underline{\mathbf{12}} & 9 & 15 \\
45 & \underline{\mathbf{18}} & 63 & 36 & 45 \\
\hline
60 & 24 & 84 & 48 & 60 \\
\hline
75 & 30 & 105 & 60 & 75
\end{array}\right]
$$

Fig. 6.6: Assignment matrix with five jobs and $(\delta_1, \delta_2) = (0,3)$

a

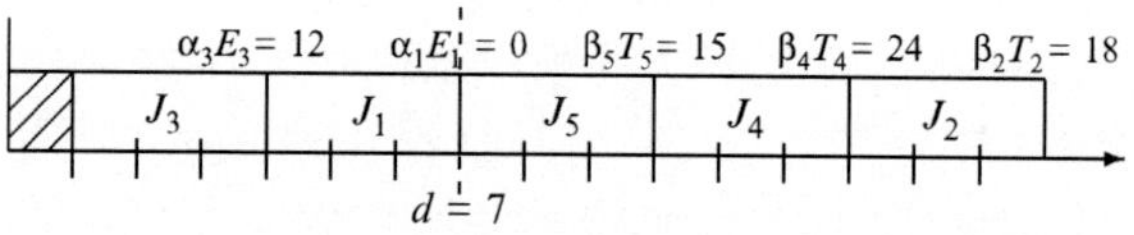

Fig. 6.7: Optimal solution corresponding to Case (a)

$$
Q = \left[\begin{array}{ccccc}
10 & 4 & 14 & 8 & \underline{\mathbf{10}} \\
\hline
7 & 5 & \underline{\mathbf{4}} & 3 & 5 \\
\underline{\mathbf{25}} & 10 & 35 & 20 & 25 \\
\hline
28 & 20 & 16 & \underline{\mathbf{12}} & 20 \\
40 & \underline{\mathbf{16}} & 56 & 32 & 40 \\
\hline
55 & 22 & 77 & 44 & 55 \\
\hline
70 & 28 & 98 & 56 & 70
\end{array}\right]
$$

Fig. 6.8: Assignment matrix with five jobs and $(\delta_1, \delta_2) = (1,2)$

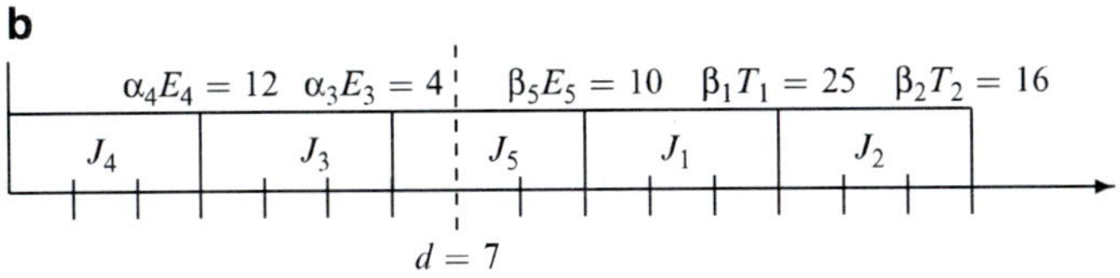

Fig. 6.9: Optimal solution corresponding to Case (b)

Hence, the optimal solution is the best solution from two previous cases. The minimum objective value is then **67** defined by the optimal sequence (4, 3, 5, 1, 2) according to the Case (b).

6.3.2 $1|p_j = p, d_j \in D, |D| = l, l\ fixed|\sum(\alpha_j E_j + \beta_j T_j)$

First, to illustrate our approach we consider the case of two distinct due dates ($l = 2$) noted d_1 and d_2. Without losts of generality, we suppose that $d_1 < d_2$. Let $d_0 = 0$. Then we show how to generalize this approach in the case of l distinct due dates.

6.3.2.1 Two Distinct Due Dates d_1 and d_2

Remark that it is possible to have an optimal solution with an idle time between d_1 and d_2 even if d_1 and d_2 are not quite different.

It is easy to show that the corresponding problem is polynomially solvable when d_1 and d_2 are quite different. In this case, we determine an optimal schedule by solving each sub-problem defined by $d_{l,l=1,2}$ in $0(n^3)$ times (see Sect. 6.2.1).

Let us now consider the general case.

W.l.o.g. we note J_{sp_1} and J_{sp_2} the splitting jobs. By definition, we have $S_{\mathrm{sp}_1} \leq d_1 < C_{\mathrm{sp}_1}$ and $S_{\mathrm{sp}_2} \leq d_2 < C_{\mathrm{sp}_2}$.

In a given sequence, let T_i be the ith idle time and let $|T_i|$ be the duration of this ith idle time.

Property 6.4. There exists an optimal solution that satisfies the following conditions:

1. There are at most an idle time in the time interval $[d_0, d_1[$ and $[d_1, d_2[$, and no idle time after d_2.
2. If there is no idle time in the time interval $[d_1, d_2[$, then there exists a job J_i such that $S_i \in \{0, d_1, d_2\}$.
3. If there is an idle time in the time interval $[d_1, d_2[$, then there exists a job J_{i_1} such that $S_{i_1} \in \{0, d_1\}$ and $S_{i_2} = d_2$.

Proof. To prove property 6.4.1, we consider a solution S with more idle times than those announced (see Fig. 6.10). We can improve this solution S by moving a subsequence of jobs on the direction of their due date. Thus, the moved jobs are executed close to their due dates and the total earliness–tardiness cost is reduced.

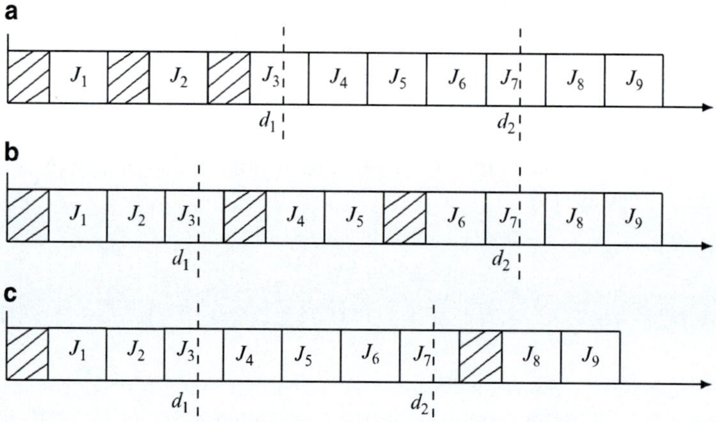

Fig. 6.10: Example with two idle times in $[0,d_1]$ and $]d_1,d_2]$ and an idle time after d_2

To prove property 6.4.2, we suppose that there exists a solution S in which there is no idle time between d_1 and d_2 and there does not exist any job J_i such that $S_i \in \{0,d_1,d_2\}$ (see Fig. 6.11). It means that there is an idle time at the beginning.

Let $\Delta_1 = min(S_{[1]}; d_1 - S_{\mathrm{sp}_1}; d_2 - S_{\mathrm{sp}_2})$ and $\Delta_2 = min(C_{\mathrm{sp}_1} - d_1; C_{\mathrm{sp}_2} - d_2)$; let A_1 (resp. A_2) be the early job set (resp. tardy job set) in S; and let α (resp. β) be the total earliness (resp. tardiness) penalty costs (per unit time) of the early (resp. tardy) jobs in S. Hence, we have $\alpha = \sum_{i\in A_1} \alpha_i$ and $\beta = \sum_{i\in A_2} \beta_i$. Moreover, we remark that

- If sequence S is moved on the left an amount of Δ_1 unit times, the variation of overall cost is defined by $\Delta_1(\sum_{i\in A_1} \alpha_i - \sum_{i\in A_2} \beta_i) = \Delta_1(\alpha - \beta)$.
- If sequence S is moved on the right an amount of Δ_2 unit times, the variation of overall cost is defined by $\Delta_2(\sum_{i\in A_2} \beta_i - \sum_{i\in A_1} \alpha_i) = \Delta_2(\beta - \alpha)$.

Consequently, if $\alpha - \beta \geq 0$ then we can improve this solution by moving the sequence S on the left; otherwise, the sequence should be moved on the right. Hence, there exists a job J_i such that $S_i \in \{0,d_1,d_2\}$.

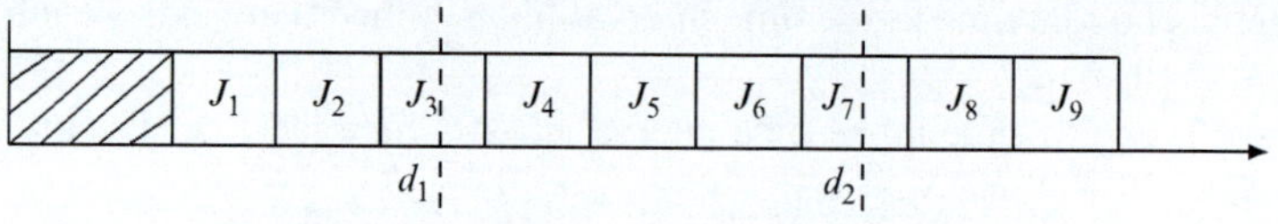

Fig. 6.11: Example with no idle time in $]d_1,d_2]$

To prove property 6.4.3, we consider now a solution S in which there is an idle time between d_1 and d_2. Thus, there exists two sub-sequences of jobs: the first one defined by the jobs scheduled around d_1 and the last one defined by the jobs scheduled around d_2. Note that there is no any idle time between two adjacent jobs in each sub-sequence (see Fig. 6.12).

Let J_{k_1} be the last job belong the first sub-sequence and let J_{k_2} be the first job belong to the second sub-sequence (according to the Fig. 6.12, J_{k_1} =J_5 and J_{k_2} = J_6).

Consider the case (b): there is an idle time at the beginning of S (Fig. 6.12, Case (b)).

Let $\Delta_1 = min(S_{[1]}; d_1 - S_{sp_1})$ and $\Delta_2 = min(C_{sp_1} - d_1; S_{k_2} - C_{k_1})$; let A_1 (resp. A_2) be the early (resp. tardy) job set in the first sub-sequence; and let α (resp. β) be the total earliness (resp. tardiness) penalty costs of the early (resp. tardy) jobs belong the first sub-sequence. Following the same approach as used to prove property 6.4.2, we can show that

- If $\alpha - \beta \geq 0$, then we can improve this solution by moving the first sub-sequence on the left an amount of Δ_1 unit times.
- Otherwise, the first sub-sequence should be moved on the right an amount of Δ_2 unit times.

Consequently, if there is no idle time at the beginning (Fig. 6.12, Case (a)), it means that there exists a job J_{i_1} with $S_{i_1} = 0$.

Consider now the second sub-sequence of jobs scheduled around d_2, we note that $\Delta_1 = min(S_{k_2} - C_{k_1}; d_2 - S_{sp_2})$; $\Delta_2 = min(C_{sp_2} - d_2)$; A_1 (resp. A_2) corresponds to the early (resp. tardy) job set; and α (resp. β) the total earliness (resp. tardiness) penalty costs of the early (resp. tardy) jobs belong the second sub-sequence. We can show that:

- If $\alpha - \beta \geq 0$, then this solution can be improved by moving the second sub-sequence on the left an amount of Δ_1 unit times.
- Otherwise, the second sub-sequence should be moved on the right an amount of Δ_2 unit times.

Hence, for a given sequence S, if there is an idle time in the time interval $[d_1, d_2[$, then there exists a job J_{i_2} with $C_{i_2} = d_2$.

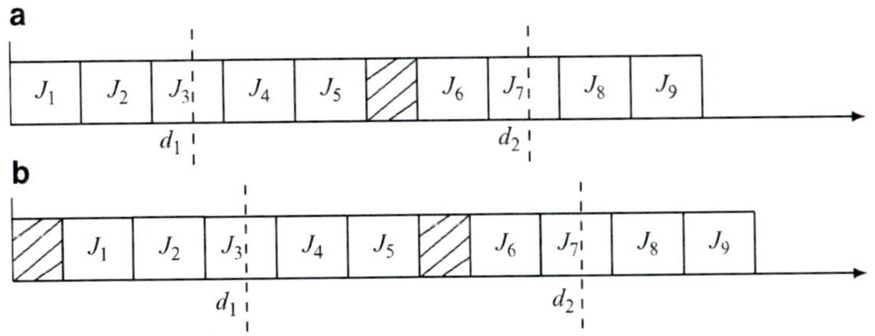

Fig. 6.12: Example with an idle time in $]d_1, d_2]$

Consequently, there exists an optimal schedule satisfied at least one of the following cases (see Fig. 6.13):

- Case (a): $S_{sp_1} = d_1$ and $S_{sp_2} = d_2$.
- Case (b): $S_{sp_1} = d_1$ and only idle time at the beginning (no idle time between d_1 and d_2).
- Case (c): $S_{sp_2} = d_2$ only idle time at the beginning (no idle time between d_1 and d_2).
- Case (d): $S_{sp_2} = d_2$ only idle time between d_1 and d_2 (no idle time at the beginning).
- Case (e): $S_{sp_1} \neq d_1$ and $S_{sp_2} \neq d_2$ and there is neither idle time at the beginning nor between d_1 and d_2.

Note that if there exists idle time between d_1 and d_2 then no job is scheduled from the xth tardy position after d_1 with $0 \leq x \leq (n-1)$. We denote by 0th the tardy position that corresponds to the splitting position. So, we have n cases for the position of starting idle time. The case of no idle time should be also taking into account. In each possible case, we can determine the job positions (or the job starting times) although we don't know which job should be executed on such a position. However, the number of job positions is bounded by $(|D|+1) \times n = 3n$ (we have three intervals $\{[d_0, d_1], [d_1, d_2], [d_2, +\infty)\}$).

Consequently, this problem is equivalent to the assignment problem with the assignment matrix dimension bounded by $(3n \times n)$. Hence, for each case, we can determine the minimum cost in $O(n^3)$ times. Thus, in the case of $l = 2$ an optimal schedule can be found in $O(n^4)$. □

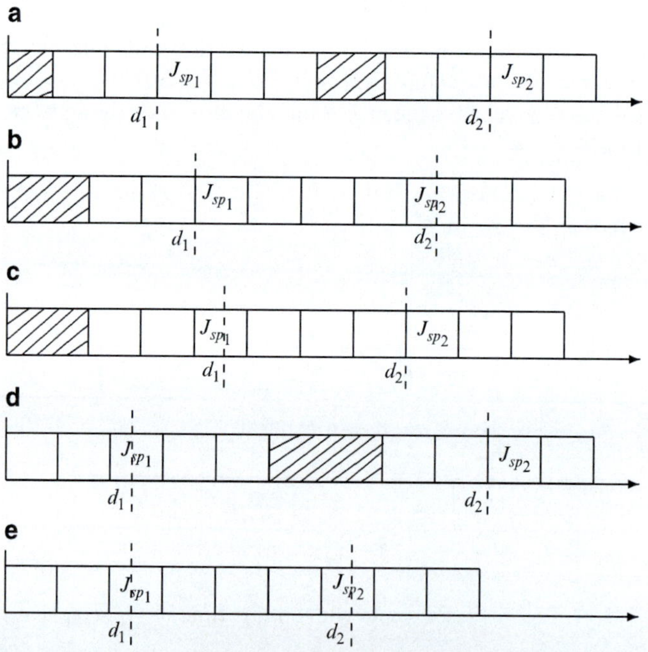

Fig. 6.13: Example with five jobs around d_1 and four jobs around d_2

6.3.2.2 l Distinct Due Dates $d_{j,1\leq j\leq l}$

Consider now l distinct values of due dates. To show that the considered scheduling problem is polynomial, we define several matrices of job contribution assignment costs in which each one associates with a possible case of assigning idle times and of job positions depending on l due dates.

Without lost of generality, we assume that the due dates are arranged according to a non-decreasing order. Let d_j be the jth due date ($d_j \in D, 0 \leq j \leq l$) with $d_0 = 0$ and zero job with due date d_0.

Let X_j be the integer variable represented the idle time between d_{j-1} and d_j: $X_0 \in \{0,1\}$ and $X_{j\ 1<j<l} \in \{0,1,\ldots,n-1\}$. So, the following cases must be studied:

- $X_0 = 0$: there does not exist any idle time at the beginning.
- $X_0 = 1$: there exists an idle time at the beginning.
- $X_j = 1$ and $j > 0$: related to d_{j-1}, there is no tardy job and no splitting job.
- $X_j > 1$ and $j > 0$: related to d_{j-1}, there is no tardy job from the $(X_j - 1)$th position.

For example, in Fig. 6.13, the values of X_j are given as follows:

- Case (a) : $X_1 = 1$ and $X_2 = 4$
- Case (b) : $X_1 = 1$ and $X_2 = 0$
- Case (c) : $X_1 = 1$ and $X_2 = 0$
- Case (d) : $X_1 = 0$ and $X_2 = 4$
- Case (e) : $X_1 = 0$ and $X_2 = 0$

Let X be the research area of all possible values of $X_{j\ j=0..(l-1)}$. Thus, the size of X is bounded by $O(n^{l-1})$.

For each vector of X, we determine a set of sub-sequences of the jobs such that there is no idle time in each sub-sequence. We have at most l sub-sequences.

In each sub-sequence, we can have one or more due dates (clearly, at most l due dates). We show that on each sub-sequence, there exists a job J_i such that $S_i = 0$ (i.e., J_i starts at time zero) or $S_i = d_j$ for some d_j even if $d_i \neq d_j$. Thereafter, with job J_i and d_j, the starting time and completion time of each job position belong the considered sub-sequence is determined. Consequently, there are at most $l+1$ possible cases to define such splitting job J_i.

After positioning of all sub-sequence (at most $(l+1)^l$ possible cases), we determine the minimum cost of assigning a job to a position among those defined. Note that each job can be assigned to an early or tardy position. As proved previously where $l = 2$, the number of job positions is bounded by $O((l+1)n)$. The minimum assigning cost can be given by Kuhn's algorithm in $O((l+1)n^3)$.

Theorem 6.2. *An optimal solution for the single machine scheduling problem* $1|p_j = p, d_j \in D, |D| = l, l\ fixed|\sum(\alpha_j E_j + \beta_j T_j)$ *can be found in* $O(l^{l+1}n^{l+2})$ *times.*

6.3.3 $Qm|p_j = p, d_j \in D, |D| = l, l\ fixed|\sum(\alpha_j E_j + \beta_j T_j)$

Following the same approach as presented in Sect. 6.3.1 we generalize the previous results to the case of m uniform parallel machines. We have n independent jobs that should be scheduled without preemption on m uniform parallel machines. All jobs $J_j, j = 1, \ldots, n$ are ready at the time zero and share a restrictive common due date $d_j \in D$ where $|D| = l$ (l is constant).

Let $X_{(j,l)}$ be the integer variable represents the idle time between d_{j-1} and d_j on the machine M_l: $X_{(0,l)} \in \{0,1\}_{\ (1 \le l \le m)}$ and $X_{(j,l)} \in \{0,1,\ldots,n-1\}$ $(1 < j < l,\ 1 \le l \le m)$. So, the following cases must be taken into account:

- $X_{(0,l)} = 0$: there does not exist an idle time at the beginning on machine M_l, $\forall l \in 1,2,\ldots,m$.
- $X_{(0,l)} = 1$: there exists an idle time at the beginning on machine M_l, $\forall l \in 1,2,\ldots m$.
- $X_{(j,l)} = 1$ and $j > 0$: related to d_{j-1}, there is no tardy job and no splitting job on machine M_l, $\forall l \in 1,2,\ldots,m$.
- $X_{(j,l)} > 1$ and $j > 0$: related to d_{j-1}, there is no tardy job from the $(X_j - 1)$-th position on machine M_l, $\forall l \in 1,2,\ldots,m$.

Now, let X denote the research area of all possible values of $X_{(j,l)\ j=0,\ldots,l-1;}$ $l = 1,\ldots,m$. Thus, the size of X is bounded by $O(2^m n^{m(l-1)})$.

For each vector of X and on each machine, we determine a set of sub-sequence of the jobs such that there is no idle time in each sub-sequence. We have at most ml sub-sequences.

For a given machine, the number of splitting jobs belong to each sub-sequence is bounded by l.

According to Property 6.1 and for each sub-sequence, there exist a due date d_j and a job J_i where $S_i = d_j$, unless $S_i = 0$. With starting time S_i, the job starting times can be determined. Consequently, we have at most $l+1$ possible cases of S_i because of l cases of due date d_j and the case $S_i = 0$.

For each of $(l+1)^{ml}$ possible cases (at most) presented as above to define job starting times of a schedule, we determine the cost of assigning a job to each position belong to the considered sub-sequence. The number of position is bounded by $O(m(l+1)n)$. Hence, the minimum assigning cost can be given by Kuhn's algorithm in $O(m(l+1)n^3)$.

Theorem 6.3. *An optimal solution for the scheduling problem* $Qm|p_j = p, d_j \in D, |D| = l, l\ fixed|\sum(\alpha_j E_j + \beta_j T_j)$ *can be found in* $O(m2^m l^{ml+1} n^{m(l-1)+3})$ *time.*

6.4 Non-Restrictive Common Due Date and Just-in-Time Scheduling

W.l.o.g, with m uniform parallel machines we assume that machines are numbered according to their performances in decreasing order, that is, M_1 is the fastest. Let s_i be the starting execution time on machine M_i and X_i total idle time on M_i. According to Kanet, the non-restrictive common due date d corresponds to the case where d is greater or equal to the total processing times ($d \geq \sum_{i=1,\ldots,n} p_i$) [29]. Hence, in this case we remark that the common due date does not constraint the optimal solution.

For an optimal solution, the following properties can be easily shown

Property 6.5. There is no idle time between adjacent jobs.

Property 6.6. On each machine, there exists either a job starting at time zero or finishing at d (the second case would be happened when $d^{\text{opt}} > 0$).

Property 6.7. There exists at least one job starting at time zero, that is, there exists at least one machine without starting-idle-time.

Property 6.8. If $d^{\text{opt}} > 0$, there exists a machine on which one job starts at time 0 and another job where its completion time is d^{opt}.

Property 6.9. If there is a starting-idle-time X_i on machine M_i, we have: $X_i \leq \lambda_m$ where λ_1 is the performance of M_m, the slowest machine.

Property 6.10. ([37]) If $\gamma \geq \beta/m$, then for an optimal solution we have $d^{\text{opt}} = 0$. Hence, the problem is equivalent to $Qm||\sum C_j$ which is known to be polynomially solvable [6].

Remark 6.1. An optimal solution can be represented by vector $\mathscr{Y} = (y_1^{\text{e}}, y_2^{\text{e}}, \ldots, y_m^{\text{e}}, y_1^{\text{t}}, y_2^{\text{t}}, \ldots, y_m^{\text{t}})$ where y_i^{e} (resp. y_i^{t}) represents the number of early (resp. tardy) jobs scheduled on machine M_i (all jobs have the same processing time) where the starting times of jobs are given according to the previous properties. Thus, the overall cost $f(S)$ can be performed in $O(m)$ time.

6.4.1 $Qm|p_j = p, d_j = d$, *non-restrictive* $|\sum(\alpha E_j + \beta T_j)$

First, let's study the case where the weights (unit earliness and tardiness) do not depend on the jobs, that is, $\forall j = 1, \ldots, n$, $\alpha_j = \alpha$ and $\beta_j = \beta$. The problem is denoted by $Qm|p_j = p, d_j = d$, *non-restrictive* $|\sum(\alpha E_j + \beta T_j)$. W.l.o.g, we can assume that $d = np\lambda_m$. This problem can be solved in $O(mn)$ times by a trivial algorithm where the jobs are scheduled around the value of d (see Fig. 6.14):

Example 6.2. $n = 13$ jobs should be scheduled on $m = 3$ machines. The performances of machines are $\lambda_i = i, \forall i \in \{1, 2, 3\}$. For instance, let consider $\alpha_j = \beta_j = 1$. According to algorithm 6.14, an optimal solution is described by Fig. 6.15.

However, in the following we will show how this algorithm can be better used to determine an optimal solution in $O(m^2)$ time.

Proposition 6.1. *An optimal solution S^* derived by algorithm 6.14 can be determined, however, in constant execution time bounded by $O(m^2)$.*

```
Let L be the set of scheduled jobs
L =∅
For each job J_j, j = 1,..., n
 | For each machine M_i in non-increasing order of λ_i ,i = 1,..., m
 |  | Schedule job J_j at the least penalized position in respect of jobs that have been scheduled already
 |  | (early position is a lower priority than tardy position)
 |  | L = L ∪ {j}
 | End For
End For
```

Fig. 6.14: Algorithm for $Qm|p_j = p, d_j = np\lambda_i|\sum(\alpha E_j + \beta T_j)$

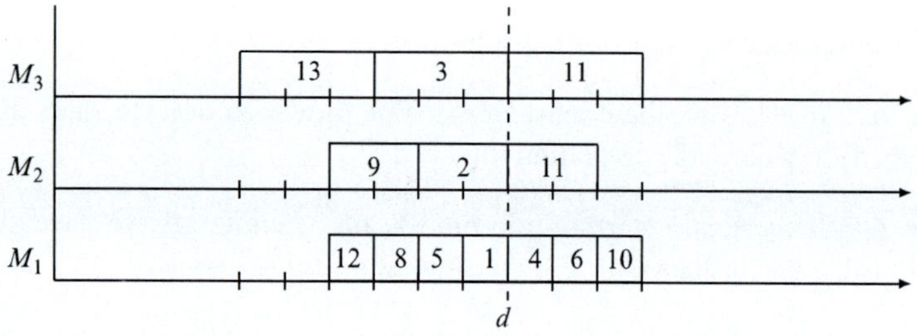

Fig. 6.15: An example with 13 jobs

The basic idea of the proof is based on reducing the input-size of the instance, that is, instead of considering n jobs, we will show that to compute an optimal solution we should solve this problem with a limited number of jobs bounded by $O(m)$. For that, we introduce a mathematical integer linear programming MILP where an obtained solution corresponds to the solution returned by algorithm 6.14. Note that this MILP can be considered as a proof of optimality of algorithm 6.14. So, the proof of proposition 6.1 is also based on the following properties.

6.4.1.1 Some Properties

Let $u_{(i,1)}$ (resp. $u_{(i,2)}$) be the number of early (resp. tardy) jobs executed on machine M_i. Some properties are first observed:

- In single machine case ($m = 1$):
  1. If $\alpha \geq (n-1)\beta$ then $u_{(1,1)} = 0$
  2. If $\beta > (n-1)\alpha$ then $u_{(1,2)} = 0$
- In m identical parallel machine case ($\lambda_i = \lambda, \forall i$):
  1. If $\alpha > \frac{(n-m)}{m}\beta$ then $u_{(i,1)} = 0,\ \forall i = 1,\ldots,m$
  2. If $\beta > \frac{(n-m)}{m}\alpha$ then $u_{(i,2)} = 0,\ \forall i = 1,\ldots,m$

- In m uniform machine case:
  1. If $\alpha > (\frac{(n-m)}{\sum 1/\lambda_i} + \max \lambda_i)\beta$, then $u_{(i,1)} = 0\ \forall i = 1,\ldots,m$
  2. If $\beta > (\frac{(n-m)}{\sum 1/\lambda_i} + \max \lambda_i)\alpha$, then $u_{(i,2)} = 0\ \forall i = 1,\ldots,m$

6.4.1.2 Linear Integer Model

Consider a linear integer model with $2m$ non-negative decision variables $u_{(i,j)}, i = \overline{1,m}, j = \overline{1,2}$ as follows:

$$\sum_{i=1}^{m}\sum_{j=1}^{2} u_{(i,j)} = n - m, \tag{6.1}$$

$$\alpha u_{(i,1)} < \beta(u_{(i,2)} + 1) \qquad \forall i = 1,\ldots,m, \tag{6.2}$$

$$\beta u_{(i,2)} \leq \alpha(u_{(i,1)} + 1) \qquad \forall i = 1,\ldots,m, \tag{6.3}$$

$$\lambda_i u_{(i,1)} \leq \lambda_k(u_{(k,1)} + 1) \qquad \forall i = 1,\ldots,m, \forall k = i+1,\ldots,m, \tag{6.4}$$

$$\lambda_k u_{(k,1)} < \lambda_i(u_{(i,1)} + 1) \qquad \forall i = 1,\ldots,m, \forall k = i+1,\ldots,m, \tag{6.5}$$

$$\lambda_i u_{(i,2)} \leq \lambda_k(u_{(k,2)} + 1) \qquad \forall i = 1,\ldots,m, \forall k = i+1,\ldots,m, \tag{6.6}$$

$$\lambda_k u_{(k,2)} < \lambda_i(u_{(i,2)} + 1) \qquad \forall i = 1,\ldots,m, \forall k = i+1,\ldots,m. \tag{6.7}$$

The number of early and retard jobs is exactly equal to $n - m$ because there are m JIT jobs (constraint (6.1)).

For an optimal solution obtained by algorithm, we can model the relationship between each pair $(u_{(i,1)}, u_{(i,2)})$ by the constraints (6.2) and (6.3):

- The job assigned to the first position on machine M_i cannot be scheduled after the last tardy job scheduled on this machine. It means that the cost of the first early job scheduled on M_i cannot be greater or equal to its cost if it were scheduled tardy behind M_i at the last position (constraint (6.2)).
- The job assigned to the last position on machine M_i cannot be scheduled early at the first position on this machine. It means that the cost of the last tardy job scheduled on M_i cannot be greater or equal to its cost if it were scheduled early behind M_i at the first position (constraint (6.3)).

In the same way, the relationship between each pair $(u_{(i,1)}, u_{(k,1)})$ is modeled by constraints (6.4) and (6.5):

- The completion time of the first job executed on M_i is greater or equal to the starting time of the first job performed on machine M_k for all $i \leq k$ (constraint (6.4)).
- Conversely, the completion time of the first job executed on M_k is strictly greater than the starting time of the first job performed on M_i for all $i \leq k$ (constraint (6.5)).

However, the relationship between each pair $(u_{(i,2)}, u_{(k,2)})$ is modeled by constraints (6.6) and (6.7):

- The starting time of the last job scheduled on M_i is less or equal to the completion time of the last job performed on machine M_k for all $i \leq k$ (constraint (6.6)).
- The starting time of the last job executed on machine M_k is less than the completion time of the last job scheduled on machine M_i for all $i < k$ (constraint (6.7)).

Obviously, the values of $u_{(i,j)}, (i = 1,\ldots,m; j = 1,2)$ satisfying these constraints determine an optimal solution for $Qm|p = p_j, d_j = d, \textit{non-restrictive}\,|\sum(\alpha E_j + \beta T_j)$ with n jobs of which there are m JIT jobs.

First analyze the special cases where $u_{(i,j)} = 0$ for certain (i, j). It is clear that if $u_{(i,j)} = 0$, we have:

1. If $u_{(i,1)} = 0$, we can deduce:
   - $u_{(k,1)} = 0\ (\forall i = 1,\ldots,m; \forall k \geq i)$
   - $u_{(k,2)} \leq \left\lfloor \frac{\lambda_k}{\lambda_i} \right\rfloor\ (\forall i = 1,\ldots,m; \forall k < i)$
   - $u_{(i,2)} \leq \left\lfloor \frac{\beta}{\alpha} \right\rfloor\ (\forall i = 1,\ldots,m)$
2. If $u_{(i,2)} = 0$, we can deduce:
   - $u_{(k,2)} = 0\ (\forall i = 1,\ldots,m; \forall k \geq i)$
   - $u_{(k,1)} \leq \left\lfloor \frac{\lambda_k}{\lambda_i} \right\rfloor\ (\forall i = 1,\ldots,m; \forall k < i)$
   - $u_{(i,1)} < \left\lfloor \frac{\beta}{\alpha} \right\rfloor\ (\forall i = 1,\ldots,m)$

We show that the analysis of specific cases is done in $O(m)$. Indeed, we have a number of indicators that allow us to identify particular cases. Hence, checking these specific cases can be done in $O(m)$. Indeed, some indicators allow us to identify particular cases. These indicators are defined as follows:

- If $\alpha \geq \beta$, then:
  1. $u_{(i,2)} = 0$ if and only if $n \leq m + \Sigma_{k=1}^{i-1}\left(\left\lfloor \frac{\lambda_i}{\lambda_k} \right\rfloor + \left\lfloor \frac{\lambda_i \alpha}{\lambda_k \beta} \right\rfloor\right)$
  2. $u_{(i,1)} = 0$ if and only if $n \leq m + \Sigma_{k=1}^{i-1}\left(\left\lfloor \frac{\lambda_i}{\lambda_k} \right\rfloor + \left\lfloor \frac{\lambda_i \alpha}{\lambda_k \beta} \right\rfloor\right) + \frac{\alpha}{\beta}$
- If $\beta \geq \alpha$, then:
  1. $u_{(i,1)} = 0$ iif $n \leq m + \Sigma_{k=1}^{i-1}\left(\left\lfloor \frac{\lambda_i}{\lambda_k} \right\rfloor + \left\lfloor \frac{\lambda_i \beta}{\lambda_k \alpha} \right\rfloor\right)$
  2. $u_{(i,2)} = 0$ iif $n \leq m + \Sigma_{k=1}^{i-1}\left(\left\lfloor \frac{\lambda_i}{\lambda_k} \right\rfloor + \left\lfloor \frac{\lambda_i \beta}{\lambda_k \alpha} \right\rfloor\right) + \frac{\beta}{\alpha}$

These indicators therefore allow us to identify which variables $u_{(i,j)}$ are null. It remains then to determine the values of variables which are strictly positives $u_{(i,j)} \geq 1$. These values can be deducted from the reduced MILP where the constraints dealing the zero values of some $u_{(i,1)}$ are neglected (since they have already been calculated). Hence, in general case where $u_{(i,j)} \geq 1\ (\forall i = 1,\ldots,m; j = 1,2)$ we proceed as follows.

6.4.1.3 Linear Relaxation

Starting from the MILP presented above, we will deduce a new linear model (PLS) by the linear relaxation of MILP. So, the decision variables in PLS are positive real values, denoted by $u'_{(i,j)}, i = 1,\dots,m; j = 1,2$. The objective here is to obtain a PLS for which a feasible solution corresponds to a feasible solution of the linear relaxation of MILP.

$$\sum_{i=1}^{m}\sum_{j=1}^{2} u'_{(i,j)} = n - m,$$

$$\alpha u'_{(i,1)} = \beta u'_{(i,2)} \quad \forall i = 1,\dots,m,$$

$$\lambda_i u'_{(i,j)} = \lambda_k u'_{(k,j)} \quad \forall j = 1,2, \forall i = 1,\dots,m, \forall k = i,\dots,m.$$

Thus, the different values of $u'_{(i,j)}, i = \overline{1,m}, j = \overline{1,2}$ can be deduced directly according to the following formula:

- $u'_{(1,1)} = \frac{\alpha+\beta}{\alpha}(n-m)\sum_{i=1}^{m}\frac{\lambda_1}{\lambda_i}$
- $u'_{(1,2)} = \frac{\alpha u'_{(1,1)}}{\beta}$
- $u'_{(k,j)} = \frac{\lambda_1 u'_{(1,j)}}{\lambda_k}\ \forall j = 1,2; \forall k = 2,\dots,m$

It is clear that this step is done in $O(m)$ times.

6.4.1.4 Rounding Values

According to the solution of PLS obtained through the previous step, rounded integer values $u''_{(i,j)}$ can be done by: $u''_{(i,j)} = \left\lfloor u'_{(i,j)} \right\rfloor$, $i = 1,\dots,m; j = 1,2$. Thus $u''_{(i,j)}$ check all the constraints of MILP, excepting constraints (6.1). In fact, we can easily deduce that $\forall i = 1,\dots,m; k = i+1,\dots,m$:

- $n - 2m \le \sum_{i=1}^{m}\sum_{j=1}^{2} u''_{(i,j)} \le n - m$
- $\alpha u''_{(i,1)} \le \alpha u'_{(i,1)} = \beta u''_{(i,2)} < \beta(u''_{(i,2)} + 1)$
- $\beta u''_{(i,2)} \le \beta u'_{(i,2)} = \alpha u''_{(i,1)} < \alpha(u''_{(i,1)} + 1)$
- $\lambda_i u''_{(i,1)} \le \lambda_i u'_{(i,1)} = \lambda_k u'_{(k,1)} < \lambda_k(u''_{(k,1)} + 1)$
- $\lambda_k u''_{(k,1)} \le \lambda_k u'_{(k,1)} = \lambda_i u'_{(i,1)} < \lambda_i(u''_{(i,1)} + 1)$
- $\lambda_i u''_{(i,2)} \le \lambda_i u'_{(i,2)} = \lambda_k u'_{(k,2)} < \lambda_k(u''_{(k,2)} + 1)$
- $\lambda_k u''_{(k,2)} \le \lambda_k u'_{(k,2)} = \lambda_i u'_{(i,2)} < \lambda_i(u''_{(i,2)} + 1)$

With the values $u''_{(i,j),i=\overline{1,m},j=\overline{1,2}}$, we can determine an optimal schedule of $Qm|p_j = p, d_j = d$, *non-restrictive* $|\sum(\alpha E_j + \beta T_j)$ for $m + \sum_{i=1}^{m}\sum_{j=1}^{2} u''_{(i,j)}$ jobs in which m jobs are "JIT".

However, we need to determine the scheduling of $n-(m+\sum_{i=1}^{m}\sum_{j=1}^{2}u''_{(i,j)})$ remaining jobs. According to inequality (1”), this number is bounded by m. As we have: $n-(m+\sum_{i=1}^{m}\sum_{j=1}^{2}u''_{(i,j)}) \leq m$, the scheduling of these remaining jobs is carried out in $O(m^2)$ by applying algorithm 6.14.

Theorem 6.4. *: An optimal solution of $Qm|p_j=p,d_j=d$, non-restrictive $|\sum(\alpha E_j+\beta T_j)$ can be performed in constant execution time bounded by $O(m^2)$.*

According to Theorem 6.4, we can deduce the following corollary:

Corollary 6.1. *An optimal solution of $Qm|p_j=p|\sum C_j$ can be performed in constant execution time bounded by $O(m^2)$.*

Proof. The proof is given by simply replacing the decision variables $u_{(i,j)}$ by u_i, where u_i corresponds to the number of jobs executed on machine M_i. Thus, the proof remains valid. □

6.4.2 $Qm|p_j=p,d_j=d$, non-restrictive $|\sum(\alpha_j E_j+\beta_j T_j)$

Let us study the $(m\geq 1)$ uniform parallel machines $Qm|p_j=p,d_j=d$, *non-restrictive* $|\sum(\alpha_j E_j+\beta_j T_j)$ where earliness and tardiness penalties are depended on jobs.

By following the same approach as presented in Sect. 6.3.1, for a single machine scheduling problem we can determine matrix of job assignment costs and thus show that is equivalent to the assignment problem.

Theorem 6.5. *An optimal solution for the scheduling problem $1|p_i=p,d_i=d$, non-restrictive $|\sum(\alpha_i E_i+\beta_i T_i)$ with non-restrictive common due date can be found in $O(n^3)$ time.*

Proof. In this case, in matrix of job assignment costs Q 6.17, we neglect the rows corresponding to kth early position with $k>n$. □

So, we can deduce the following theorems.

Theorem 6.6. *An optimal solution for the single machine scheduling problem with l non-restrictive due dates, denoted by $1|p_j=p,d_j\in D,|D|=l,l$ fixed, non-restrictive $|\sum(\alpha_j E_j+\beta_j T_j)$ can be found in $O(l^{l+1}n^{l+2})$ time.*

Theorem 6.7. *An optimal solution for the uniform parallel machines scheduling problem with l non-restrictive due dates, denoted by $Qm|p_j=p,d_j\in D,|D|=l,l$ fixed$|\sum(\alpha_j E_j+\beta_j T_j)$ can be found in $O(m2^m l^{ml+1}n^{m(l-1)+3})$ time.*

6.5 Unknown Common Due Date and Just-in-Time Scheduling

The concepts of JIT approach in inventory management according to which jobs are to be completed as close to their due dates as possible, has stimulated research on the scheduling problems with due dates assignment and non-regular objective functions dependent on earliness and tardiness costs. In relevant models, the decision maker not only takes sequencing and scheduling decisions, but also determines the optimal values of due dates which is specified as controllable parameters. The optimal choice of the controllable values reflects a possible cooperation between various participants of a production process or supply chain, see some motivation surveys for more details [14, 15, 26, 28].

The classical due dates assignment and JIT scheduling problem aims at determining the optimal job schedule and due dates that minimize a linear combination of three costs: the cost for completing a jobs early, the cost for completing jobs tardy and the cost related to the setting of due dates. As a pioneering result, Panwalkar et al. [37] introduced an efficient $O(n\log n)$-time algorithm for the single machine problem which, using the classical three-field notation of [16], is referred to as $1|d_j = d|\sum(\alpha E_j + \beta T_j + \gamma d_j)$ where $\alpha_j = \alpha$ and $\beta_j = \beta$. In the case of identical parallel machines, Cheng and Chen [7] proved that the problem is NP-hard even for two machines. Additionally, they showed that the special case $Pm|p_j = p, d_j = d|\sum(\alpha E_j + \beta T_j + \gamma d_j)$, for which the number of machines is constant, is polynomial. Fifteen years later, Mosheiov and Sarig [36] studied the two uniform machines and show that the JIT scheduling and due-date assignment problem can be solved in constant time. The extension to m machines can be solved in polynomial time (in the number of jobs), however the computation time increases exponentially with the number of machines. Then, the practical solution procedure can solve to optimality only instances with a relatively small m value. In the case of unequal job penalties, Mosheiov and Yovel [35] showed that an optimal solution for the single machine and identical parallel machine scheduling problems with unitary-length jobs can be calculated in $O(n^4)$ time. In their study, the due date is considered as a real value, so that the approach is still valid for the case of equal-length jobs.

A challenging question refers to possible extensions of the above special case to multi-machine settings with different performances (uniform machine). Hence, in following, we propose to study the special scheduling problem with l distinct due dates (these due dates are decision variables) and equal-length jobs on uniform parallel machines. In the case of given due dates, the problem is shown to be polynomially solvable by studying all the interesting cases [23]. Each one is modeled by an assignment matrix and solved by polynomial algorithm. By following the same approach, we show that the due dates assignment and JIT scheduling problem on uniform parallel machines with equal-length jobs is polynomially solvable. The main contribution of this section is the following. First, we show that the special case of scheduling equal-length jobs on a single machine with an unknown common due date can be solved in $O(n^3)$ time, which improves upon the complexity of the algorithm proposed by Mosheiov and Yovel [35]. Next, we show how to generalize our method in the case of single machine and uniform parallel machines with l assignable due dates, l is constant.

6.5.1 Job Assignment Cost Matrix for a Given Due Date

Consider a single machine scheduling problem with a given common due date (assume that $d_j = d$ and d is constant). Let J_s be a splitting job, that is, job J_s is such that $S_s \leq d < C_s$. As it is unnecessary to minimize the total cost by inserting idle times between two adjacent jobs sharing the same due date, there are only two situations of an optimal sequence [14]: the first job starts at time zero or there is an on-time job if $S_s = d$. Let k be the maximum number of early jobs on the machine. If common due date d is a multiple of processing time p then $k = d/p - 1$; otherwise $k = \lfloor d/p \rfloor$ (Fig. 6.16). It is clear that the job scheduled at the $(k+1)$-th position noted $J_{[k+1]}$ is the splitting job J_s. Let $\delta_1 = d - kp$, $\delta_2 = p - \delta_1$ be the distances between S_s, C_s and d.

The job positions are defined as follows (see Fig. 6.16):

- Concerning the early jobs set: the last job processed just before J_s is at the first early position and is denoted by J_{e_1}. This job may be on-time. The job processed just before the previous one is at the second early position and is denoted by J_{e_2}, and so on.

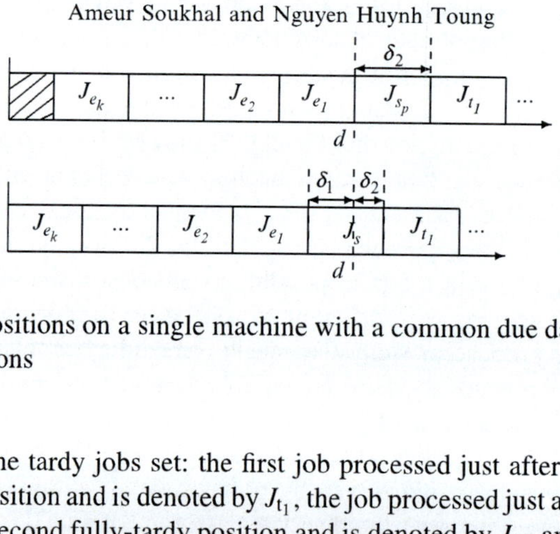

Fig. 6.16: Job positions on a single machine with a common due date: the two possible configurations

- Concerning the tardy jobs set: the first job processed just after J_s is at the first fully-tardy position and is denoted by J_{t_1}, the job processed just after the previous one is at the second fully-tardy position and is denoted by J_{t_2}, and so on.

The cost of assigning a job to the ith early position is defined by $\alpha_{[k-i+1]}((i-1)p + \delta_1)$ which corresponds to the contribution of this job to the total earliness cost. Similarly, the cost of assigning a job to the ith fully-tardy position is defined by $\beta_{[k+1+i]}(ip + \delta_2)$ and the cost of assigning a splitting job J_s is defined by $\delta_2 \beta_s$.

Consequently, we can define an assignment costs matrix Q in which each column represents all possible position costs of a job and each row represents an assignment position among early positions, splitting position and fully-tardy positions. So, for each job represented by the corresponding column, the item in a row of the matrix shows the cost incurred for the job if it is scheduled in a particular position. More precisely, the rows and columns of matrix Q are denoted as follows (details of Q are presented in Fig. 6.17).

- The j-th column corresponds to job J_j.
- The first row corresponds to the straddling job J_{s}.
- The $(2 \times i)$-th rows correspond to the costs of early jobs J_{e_i}, $1 \leq i \leq k$.
- The $(2 \times i + 1)$-th rows correspond to the costs of fully-tardy jobs J_{t_i}, $1 \leq i \leq k$.
- The last $(k+1+i)$-th rows correspond to the costs of the fully-tardy job J_{t_i}, $k+1 \leq i \leq n$.

Since there are at most k early jobs, a splitting job and n fully-tardy jobs, the number of rows is at most $k+n+1$, and then bounded by $2n$. So, this matrix is referred to as $Q_{((k+n)\times n)}$, in which each element is denoted by $(q_{i,j})_{1 \leq i \leq k+n+1; 1 \leq j \leq n}$. A set $S = (q_{k_1,1}; q_{k_2,2} \ldots ; q_{k_n,n})$ of n elements from Q is called a *feasible solution* with $Z = f(S) = \sum_{r=1}^{n} q_{k_r,r}$ if and only if no two elements are from the same row (by definition, no two elements of S are in the same column). Therefore, to determine an optimal solution S^* we must find a feasible set S from Q that minimizes Z. Hence, this problem is equivalent to the well-known assignment problem [34] when the due date is fixed. As we know, the assignment problem resolution has received particular attention from several researchers, e.g., we cite here some well-known results – Kuhn in 1955 [34], Edmonds and Karp in 1972 [10], Goldberg and Tarjan in 1988 [13], Ahuja and Orlin in 1989 [1], etc. Furthermore, the research is ongoing. Thus, in the following, for a simplification, we consider that assignment problem with matrix $(x \times y)$ can be solved in $O(xy\min(x,y))$.

Job positions	Jobs			
	J_1	J_2	$\cdots$	J_n
J_s	$\delta_2\beta_1$	$\delta_2\beta_2$	$\cdots$	$\delta_2\beta_n$
J_{e_1}	$\delta_1\alpha_1$	$\delta_1\alpha_2$	$\cdots$	$\delta_1\alpha_n$
J_{t_1}	$(\delta_2+p)\beta_1$	$(\delta_2+p)\beta_2$	$\cdots$	$(\delta_2+p)\beta_n$
J_{e_2}	$(\delta_1+p)\alpha_1$	$(\delta_1+p)\alpha_2$	$\cdots$	$(\delta_1+p)\alpha_n$
J_{t_2}	$(\delta_2+2p)\beta_1$	$(\delta_2+2p)\beta_2$	$\cdots$	$(\delta_2+2p)\beta_n$
$\cdots$	$\cdots$	$\cdots$	$\cdots$	$\cdots$
$\cdots$	$\cdots$	$\cdots$	$\cdots$	$\cdots$
J_{e_k}	$(\delta_1+(k-1)p)\alpha_1$	$(\delta_1+(k-1)p)\alpha_2$	$\cdots$	$(\delta_1+(k-1)p)\alpha_n$
J_{t_k}	$(\delta_2+kp)\beta_1$	$(\delta_2+kp)\beta_2$	$\cdots$	$(\delta_2+kp)\beta_n$
$J_{t_{k+1}}$	$(\delta_2+(k+1)p)\beta_1$	$(\delta_2+(k+1)p)\beta_2$	$\cdots$	$(\delta_2+(k+1)p)\beta_n$
$\cdots$	$\cdots$	$\cdots$	$\cdots$	$\cdots$
$J_{t_{n-1}}$	$(\delta_2+(n-1)p)\beta_1$	$(\delta_2+(n-1)p)\beta_2$	$\cdots$	$(\delta_2+(n-1)p)\beta_n$

Fig. 6.17: The matrix of job assignment costs

6.5.2 $1|p_j = p, d_j = d$ *unknown*$|\sum(\alpha_j E_j + \beta_j T_j + \gamma d)$

This section deals with the single machine problem with a common unknown due date (due date is a decision variable). Several propositions of an optimal schedule can be found in [14].

Proposition 6.2. *There exists an optimal schedule in which*

1. *There is no idle time between jobs, that is, the first job starts at time zero and all jobs are processed contiguously [4].*
2. *There exists an optimal V-shaped schedule, that is, early jobs are sequenced in a non-increasing order of p_j/α_j, and fully-tardy jobs are sequenced in a non-decreasing order of p_j/β_j [38] (notice that in the particular case where $p_j = p$, $\forall j$, a V-shaped schedule can be obtained in sequencing the early jobs in the non-decreasing α_j order and the fully-tardy jobs in the non-increasing β_j order).*
3. *If $n\gamma \geq \sum_{j=1}^{n} \beta_j$, then $d = 0$. Otherwise, the completion time of one job coincides with the common due date d [37].*

This problem can be solved by enumerating all $(n+1)$ values of the common due date d in the set $\{0, p, 2p, \ldots, np\}$. For each value of d, an assignment matrix can be built and the corresponding assignment problem solved in $O(n^3)$ time. As $(n+1)$ solutions are obtained, the optimal solution of the considering problem corresponds to the lowest cost solution. The overall time complexity is in $O(n^4)$ [35].

However, in this particular case, the necessary computation time can be improved according to the following theorem.

Theorem 6.8 ([22]). *An optimal solution of the $1|p_j = p, d_j = d$ unknown$|\sum(\alpha_j E_j + \beta_j T_j + \gamma_j d)$ scheduling problem can be obtained in $O(n^3)$.*

Proof. Instead of the due date enumeration, it is more efficient to add the due date cost as a part of the early contribution cost. Due to $d = \sum_{j;C_j \leq d} p$ and the fact that due date cost per unit time is $n\gamma$, we can consider that the due date contribution on each early job with processing time p is $n\gamma p$. In other words, the objective function can be written as follows:

$$\begin{aligned}
\sum_{j=1}^{n}(\alpha_j E_j + \beta_j T_j + \gamma d) &= \sum_{j=1}^{n}(\alpha_j E_j + \beta_j T_j) + n\gamma d \\
&= \sum_{j=1, C_j \leq d}^{n} \alpha_j E_j + \sum_{j=1, C_j > d}^{n} \beta_j T_j + n\gamma d \\
&= \sum_{j=1, C_j \leq d}^{n} \alpha_j E_j + \sum_{j=1, C_j > d}^{n} \beta_j T_j + n\gamma \sum_{j=1, C_j \leq d}^{n} p \\
&= \sum_{j=1, C_j \leq d}^{n} (\alpha_j E_j + np\gamma) + \sum_{j=1, C_j > d}^{n} (\beta_j T_j).
\end{aligned}$$

Consequently, the problem consists of determining a set of early jobs and tardy jobs where

- The cost of an early job J_j ($C_j \leq d$) is defined by $\alpha_j E_j + np\gamma$.
- The cost of a tardy job J_j ($C_j > d$) is defined by $\beta_j T_j$.

Thus, to find an optimal solution, we can build an assignment matrix as defined in Fig. 6.18: n rows and $2n$ columns. The optimal due date is a multiple of p. Thus, the value of an item (i.e., element of matrix Q) corresponding to a job J_j and a position is:

- $p\beta_j$, if it corresponds to a splitting position.
- $kp\alpha_j + np\gamma$, if it corresponds to a k-th early position.
- $(k-1)p\beta_j$, if it corresponds to a k-th fully-tardy position.

Finding the minimum feasible solution to this modified assignment matrix determines simultaneously optimal position for each job and optimal due date. Of course, the optimal due date is equal to the number of optimal early jobs k^* times p, that is, $d^* = k^* \times p$.

Job positions	Jobs: J_1	J_2	...	J_n
J_s	$p\beta_1$	$p\beta_2$	...	$p\beta_n$
J_{e_1}	$np\gamma$	$np\gamma$	...	$np\gamma$
J_{t_1}	$2p\beta_1$	$2p\beta_2$	...	$2p\beta_n$
J_{e_2}	$p\alpha_1 + np\gamma$	$p\alpha_2 + np\gamma$	...	$p\alpha_n + np\gamma$
J_{t_2}	$3p\beta_1$	$3p\beta_2$	...	$3p\beta_n$
J_{e_3}	$2p\alpha_1 + np\gamma$	$2p\alpha_2 + np\gamma$	...	$2p\alpha_n + np\gamma$
J_{t_3}	$4p\beta_1$	$4p\beta_2$	...	$4p\beta_n$
...	...	...	...	...
...	...	...	...	...
$J_{e_{n-1}}$	$(n-2)p\alpha_1 + np\gamma$	$(n-2)p\alpha_2 + np\gamma$	...	$(n-2)p\alpha_n + np\gamma$
$J_{t_{n-1}}$	$np\beta_1$	$np\beta_2$	...	$np\beta_n$
J_{en}	$(n-1)p\alpha_1 + np\gamma$	$(n-1)p\alpha_2 + np\gamma$	...	$(n-1)p\alpha_n + np\gamma$

Fig. 6.18: The modified matrix of job assignment costs

Consequently, for determining an optimal solution, we consider only $\delta_1 = 0$ and $\delta_2 = p$ with the modified matrix Q' of Fig. 6.18. Notice that this resolution method is solely valid on a single machine and a common due date case. □

6.5.3 $1|p_j = p, d_j \in D, |D| \leq l, D\ unknown, l\ fixed| \sum(\alpha_j E_j + \beta_j T_j + \gamma_j d_j)$

We now turn to the single machine problem with a set of unknown due dates $D = \{D_1, D_2, \ldots, D_l\}$, with constant l. Let n_k be the number of jobs that should be scheduled as close as possible to D_k.

Proposition 6.3. *There exists an optimal schedule in which*

1. *There is no idle time, and then the first job starts at time zero and the other jobs are processed contiguously.*
2. *Two adjacent jobs J_i and J_j are scheduled according to*
   - *The non-increasing order of p_j/α_j, if they are early jobs and $\max(C_i,C_j) \leq \min(d_i,d_j)$.*
   - *The non-decreasing order of p_j/β_j if they are fully-tardy jobs and $\min(C_i,C_j) \geq \max(d_i,d_j)$.*
3. *For each job set $\mathcal{N}_k$, if $n_k\gamma \geq \sum_{J_j \in \mathcal{N}_k} \beta_j$, then $D_k = 0$. Otherwise, the completion time of one job from the set $\mathcal{N}_k$ coincides with D_k.*

Proof. Proposition 6.3.1 is trivial. Proposition 6.3.2 can be proved by pair-wise interchange arguments. For instance, consider an optimal schedule having two adjacent early jobs J_i and J_j as illustrated in Fig. 6.19, and suppose that $d_i \leq d_j$. Let x be the distance between the maximum completion time of these two jobs and d_i $(x = \min(d_i,d_j) - \max(C_i,C_j) = d_i - max(C_i,C_j))$. And let δ be the distance between d_i and d_j.

If $C_i < C_j$, the contribution cost of these two jobs to the objective function is obtained by $\alpha_i(p_j+x)+\alpha_j(x+\delta)$; otherwise, the contribution cost should be $\alpha_j(p_i+x+\delta)+\alpha_i x$. Thus, J_i is scheduled before J_j if and only if:

$$\alpha_i(p_j+x)+\alpha_j(x+\delta) \leq \alpha_j(p_i+x+\delta)+\alpha_i x$$
$$\Leftrightarrow \alpha_i p_j \leq \alpha_j p_i$$
$$\Leftrightarrow p_j/\alpha_j \leq p_i/\alpha_i$$

Consequently, the adjacent early jobs should be scheduled according to the non-increasing order of p_j/α_j (non-decreasing order of α_j in the case of identical processing time). By a similar proof, adjacent fully-tardy jobs should be scheduled according to the non-decreasing order of p_j/β_j (non-increasing order of β_j in the case of identical processing time).

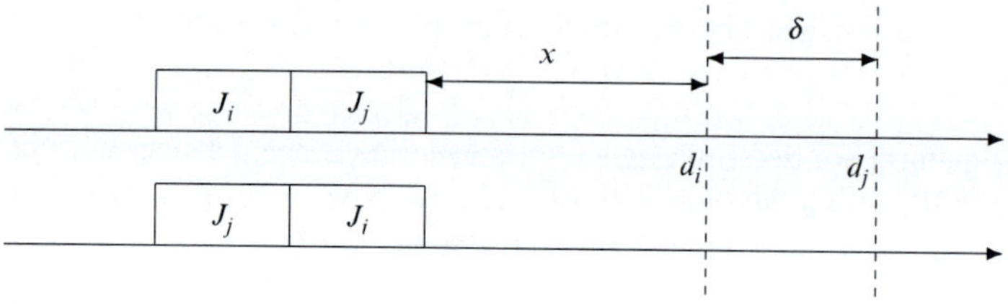

Fig. 6.19: Optimal position rule for two adjacent early jobs

For Proposition 6.3.3, the proof that D_k should be zero if $n_k\gamma \geq \sum_{J_j \in \mathcal{N}_k} \beta_j$' is similar to the common due date case presented in [37]. More precisely, let us assume that there exists a $D_k \neq 0$ and $n_k\gamma \geq \sum_{J_j \in \mathcal{N}_k} \beta_j$. Once D_k is decreased one unit time without changing the given sequence, the deviation from the new overall cost is determined by $\sum_{J_j \in \mathcal{N}_k, C_j \geq D_k} \beta_j - n_k\gamma - \sum_{J_j \in \mathcal{N}_k, C_j < D_k} \alpha_j$, which is a non-positive value (since $\sum_{J_j\in\mathcal{N}_k,C_j\geq D_k}\beta_j - n_k\gamma \leq \sum_{J_j\in\mathcal{N}_k}\beta_j - n_k\gamma \leq 0$ and $-\sum_{J_j\in\mathcal{N}_k,C_j<D_k}\alpha_j \leq 0$). Thus, D_k should be decreased and extending the same reasoning, D_k should be zero.

Now, consider a solution S with some D_k such that there is no job $J_j \in \mathcal{N}_k$ with $C_j = D_k$.

(a) If D_k is left time-shifted by one unit, the deviation cost is $\Delta = n_k\gamma + \sum_{J_j\in\mathcal{N}_k;C_j>D_k}\beta_j - \sum_{J_j\in\mathcal{N}_k;C_j\leq D_k}\alpha_j$. When $\Delta \geq 0$, solution S can be improved by left time-shifting D_k by one unit. The procedure can continue repeatedly until $D_k = 0$ or $D_k = C_j$ with any $J_j \in \mathcal{N}_k$.
(b) If D_k is right time-shifted by one unit, the deviation cost is $-\Delta$. This means that S can be improved by right time-shifting D_k until $D_k = C_j$ with any $J_j \in \mathcal{N}_k$.

Consequently, there is an optimal solution with $D_k = 0$ or $D_k = C_j$ with any $J_j \in \mathcal{N}_k$.
□

From proposition 6.3, we can derive a polynomial time algorithm for which the main steps are the following:

- For each combination of $\{D_1, D_2, \ldots, D_l\}$ where $D_k, k = 1, 2, \ldots, l$ is chosen from $\{0, p, 2p, \ldots, np\}$ Do:
  - Determine the assignment matrix Q as introduced in Sect. 6.5.1.
  - Solve in $O(n^3)$ time the corresponding assignment problem.
- The optimal solution corresponds to the best obtained solution.

Consequently, we have the following theorem.

Theorem 6.9 ([22]). *An optimal solution of the* $1|p_j = p, d_j \in D, |D| \leq l, D$ *unknown,* l *fixed*$|\sum(\alpha_j E_j + \beta_j T_j + \gamma_j d_j)$ *scheduling problem can be done in* $O(n^{l+3})$.

6.5.4 $P|p_j = p, d_j \in D, |D| \leq l, D$ *unknown*, l *fixed*$|$ $\sum(\alpha_j E_j + \beta_j T_j + \gamma_j d_j)$

In this section, we consider the case of m identical parallel machines. We first introduce some useful propositions.

Proposition 6.4. *There exists an optimal schedule in which:*

1. *Two adjacent jobs* J_i *and* J_j *are scheduled according to:*
   - *The non-increasing order of* p_j/α_j*, if they are early jobs and* $\max(C_i, C_j) \leq \min(d_i, d_j)$.
   - *The non-decreasing order of* p_j/β_j *if they are fully-tardy jobs and* $\min(C_i, C_j) \geq \max(d_i, d_j)$.
2. *For each job set* $\mathcal{N}_k$*, if* $n_k\gamma \geq \sum_{J_j\in\mathcal{N}_k}\beta_j$*, then* $D_k = 0$. *Otherwise, the completion time of one job in set* $\mathcal{N}_k$ *coincides with* D_k.

Proof. The proof is the same as in the single machine case (see Sect. 6.5.3). □

Let S be an optimal schedule and let X_i be the size of starting idle time on machine M_i, that is, the idle time between time zero and the starting time of the first job scheduled on machine M_i (X_i can be 0).

Proposition 6.5. *There exists an optimal schedule in which* $X_i \leq p$, $\forall i = 1,\ldots,m$.

Proof. Suppose that there is an optimal schedule S in which $\exists i \in \{1,\ldots,m\} : X_i > p$. Consider three types of machines (see example in Fig. 6.20):

1. M_{i_1}: machines with $X_{i_1} \neq 0$.
2. M_{i_2}: machines with $X_{i_2} = 0$ and there is no idle time after the first scheduled job.
3. M_{i_3}: machines with $X_{i_3} = 0$ and there is an idle time after the first scheduled job. Clearly, the due date of this first scheduled job should be between 0 and p.

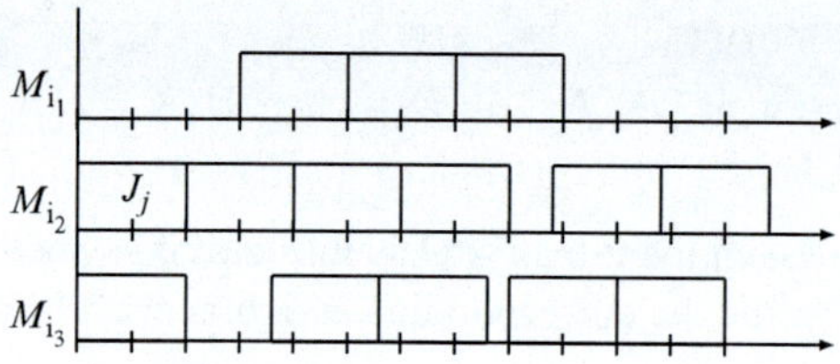

Fig. 6.20: Three types of machine according to the starting idle time

Hence, machine M_i is of type i_1. If there is no machine of type i_2, we can reduce the total cost by

- Left-shifting by one time unit the jobs scheduled on the machines of type i_1.
- Left-shifting by one time unit the jobs scheduled on the machines of type i_3 except the job in the first position.
- Decreasing one time unit all the due dates greater than p.

Which contradicts that S is an optimal solution. Consider now that there is at least one machine of type i_2 where J_j is in the first position.

- If $d_j > p$, J_j can be moved and scheduled at the first position on M_i. In order to reduce the overall costs (which contradicts that S is an optimal solution) J_j is executed at the last possible position in time without delaying the subsequent jobs. So, J_j is started later which reduces its earliness costs.
- Otherwise ($d_j \leq p$), we define S' obtained from S by moving J_j to schedule at time zero M_i. Clearly, the total objective costs of S and S' are equal but S' has no starting idle time with size greater than p.

Henceforth, there is an optimal schedule in which X_i is less than or equal to p, $\forall i = 1,\ldots,m$. □

	J_1	J_2	J_3	J_4	J_5	J_6	J_7	J_8	J_9	J_{10}
α_i	100	100	100	100	100	100	100	100	100	100
β_i	100	100	100	100	100	100	100	1	1	1
γ_i	30	30	10	10	1	1	1	1	1	1

Fig. 6.21: Penalty costs

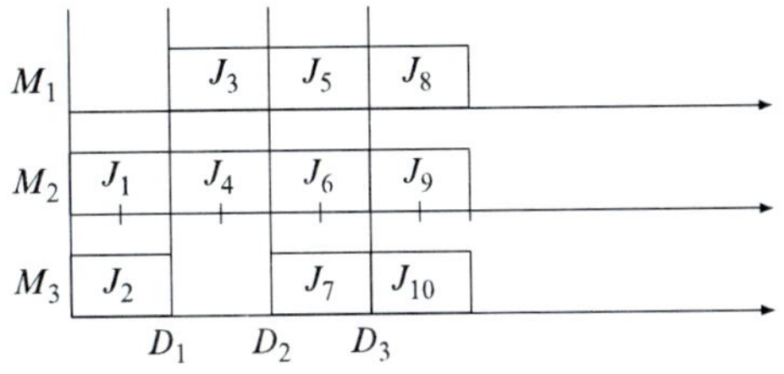

Fig. 6.22: Optimal solution

Example 6.3. Let's consider a numerical example with ten jobs to be scheduled on three machines. Jobs J_1 and J_2 should be scheduled as close as possible to D_1; J_3 and J_4 as close as possible to D_2; the remaining jobs as close as possible to D_3. Their processing times are p and their penalties are given in Fig. 6.21. Figure 6.22 illustrates an optimal solution where we have the three types of machines.

Proposition 6.6. *There exists an optimal schedule in which* $X_i \in \{0, p\}$, $\forall i = 1, \ldots, m$.

Proof. According to proposition 6.5, we have $X_i \leq p$, $\forall i = 1, \ldots, m$. Suppose that $\exists i \in \{1, \ldots, m\} : 0 < X_i < p$. From an optimal solution S, let us consider the sub-sequence σ of jobs scheduled between the first two idle times on machine M_i. Let A (resp. B) be the total earliness (resp. tardiness) penalties per time unit for the early jobs (resp. tardy jobs) formed σ.

When the sequence σ is *right time-shifted* by one time unit, the state of all jobs concerning early/tardy does not change. Perhaps, some early jobs become on-time. The deviation cost is $\Delta = A - B$.

When the sequence σ is *left time-shifted* by one time unit, the state of all jobs concerning early/tardy does not change. Perhaps, some tardy jobs become on-time. The deviation cost is $-\Delta$.

As S is optimal, Δ should be 0. So, we can build another optimal solution by left time-shifting the sequence σ and then X_i is reduced by one time unit. By repeating this procedure, we obtain an optimal solution with $X_i = 0$. □

In the following, we study the number of idle times on each machine between two adjacent due dates.

Proposition 6.7. *There exists an optimal schedule in which on each machine there is at most one idle time between two adjacent due dates.*

Proof. Assume that from an optimal solution S there are two idle times on machine M_i between two adjacent due dates. The sequence between these two idle times is denoted by σ. It is easy to show by a pair-wise interchange argument that σ can be partitioned into two sub-sequences ($\sigma = \sigma_1 // \sigma_2$):

- Sub-sequence σ_1 has only fully-tardy jobs.
- Sub-sequence σ_2 has only early jobs.

The objective cost of S can be improved by both left time-shifting σ_1 and right time-shifting σ_2 (in contradiction with S an optimal schedule). Thus, there are at most one idle time between two adjacent due dates on machine M_i. □

Proposition 6.8. *There exists an optimal schedule in which all completion times and due dates are multiples of p.*

Proof. According to optimal solution S, let $\mathscr{J}_{\text{mult}}$ ($\mathscr{J}_{\text{nomult}}$ resp.) be the set of jobs which their completion times are multiple of p (resp. are not multiple of p); Let $\mathscr{D}_{\text{mult}}$ ($\mathscr{D}_{\text{nomult}}$ resp.) be the set of due dates whose values are multiples of p (are not multiples of p resp.).

Let A_1 (B_1 resp.) be the total earliness (tardiness resp.) penalties per time unit of jobs in $\mathscr{J}_{\text{mult}}$ and having due dates belonging to $\mathscr{D}_{\text{nomult}}$.

Let A_2 (B_2 resp.) be the total earliness (tardiness resp.) penalties per time unit of jobs in $\mathscr{J}_{\text{nomult}}$ and having due dates belonging to $\mathscr{D}_{\text{mult}}$. We define $G = \gamma \times \sum_{D_k \in \mathscr{D}_{\text{nomult}}} n_k$.

It is clear that all jobs of $\mathscr{J}_{\text{nomult}}$ can be right-shifted or left-shifted by one time unit without changing the starting time of all jobs of $\mathscr{J}_{\text{mult}}$.

When all jobs of $\mathscr{J}_{\text{nomult}}$ and all due dates of $\mathscr{D}_{\text{nomult}}$ are left-shifted by one time unit, the early/tardy state of most of the jobs does not change. At most, some tardy jobs in the set $\mathscr{J}_{\text{nomult}}$ and having due dates belonging to $\mathscr{D}_{\text{mult}}$ (some early jobs in the set $\mathscr{J}_{\text{mult}}$ and having due dates belonging to $\mathscr{D}_{\text{nomult}}$ resp.) may become on-time. The deviation cost is defined by $\Delta = -A_1 + B_1 + A_2 - B_2 - G$. As S is optimal, Δ should be ≤ 0.

Similarly, by right-shifting by one time unit all jobs of $\mathscr{J}_{\text{nomult}}$ and all due dates of $\mathscr{D}_{\text{nomult}}$, the deviation cost is $(-\Delta)$. As S is optimal, $(-\Delta)$ should be ≤ 0.

Consequently, Δ should be 0. So, we can build another optimal solution by left time-shifting all jobs of $\mathscr{J}_{\text{nomult}}$ and all due dates of $\mathscr{D}_{\text{nomult}}$. This procedure can be repeated until we obtain an optimal solution with all job completion times and all due dates being multiples of p. □

Thus, we deduce that there is an optimal schedule in which the size of each idle time is a multiple of p. Furthermore, we can show that the size of each idle time is exactly p as follows.

Proposition 6.9. *There exists an optimal schedule in which the size of each idle time is equal to p.*

Proof. We will show that the size of each idle time is not greater than $2p$ nor equal to $2p$. The idle time duration is therefore equal to p.

- To show that the size of idle time is not greater than $2p$ between two adjacent due dates $D_{[k]}$ and $D_{[k+1]}$, let us consider an idle time on machine M_i with duration $3p$. If there is a job J_j scheduled in the middle of the considered idle time on machine $M_{i'}$, $i' \neq i$ (see Fig. 6.23 for an example), we can always reduce the total cost by moving J_j to be processed on M_i in the considered idle time interval: if J_j is early, it is moved at the end of the interval and if J_j is tardy, it is moved at the beginning of the interval.

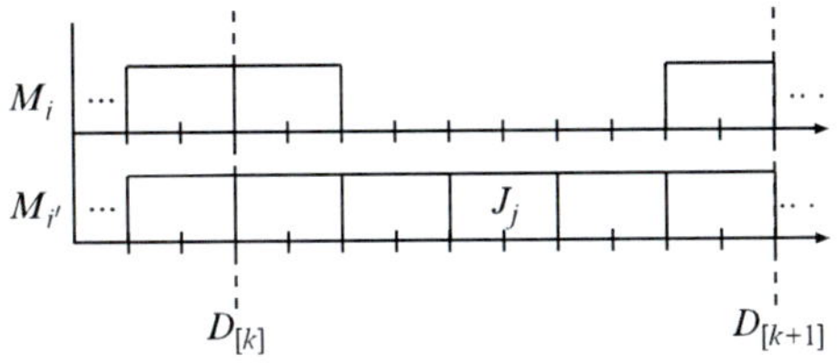

Fig. 6.23: Example with idle time $X = 3p$

If there is no job J_j scheduled in the considered idle time on another machine $M_{i'}$ $\forall i', i' = 1, \ldots, m \wedge i' \neq i$, then the objective cost can be reduced by left time-shifting all jobs executed after this idle time on all machines and left time-shifting all due dates that occurred after this idle time.
Thus, there is no optimal solution that has an idle time with duration equal to $3p$. The case where the duration of idle time is greater than $3p$ can be proved similarly.

- Consider now the case where the duration of idle time is $2p$.
If there is a machine $M_{i'}$, $i' \neq i$ which has two jobs J_{j_1} and J_{j_2} processed in the considered idle time interval (see Fig. 6.24), then J_{j_2} can be moved to be processed by machine M_i. The objective function cost is unchanged. We then have a new optimal solution with $X = p$, as pictured in Fig. 6.25.

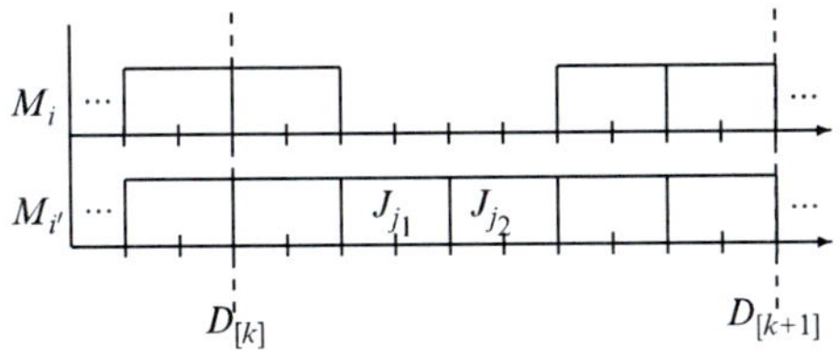

Fig. 6.24: Example with idle time $X = 2p$

Otherwise, in the case where there is no machine $M_{i'}$, $i' \neq i$ with two jobs J_{j_1} and J_{j_2} processed in the considered idle time interval, then the solution here is not optimal since the objective function cost can be reduced by left time-shifting all jobs executed after this idle time on all machines and left time-shifting all due dates that appear after the considered idle time (for instance in Fig. 6.26, left-shifting by one time unit the jobs scheduled after J_{j_3},J_{j_4} and J_{j_5} and left-shifting

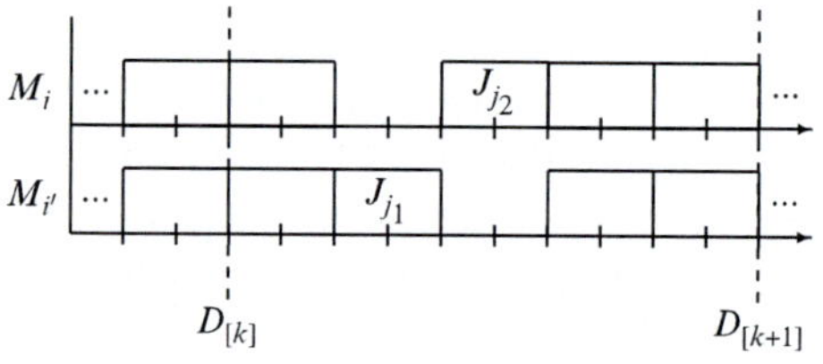

Fig. 6.25: Transformation to new solution with idle time $X = p$

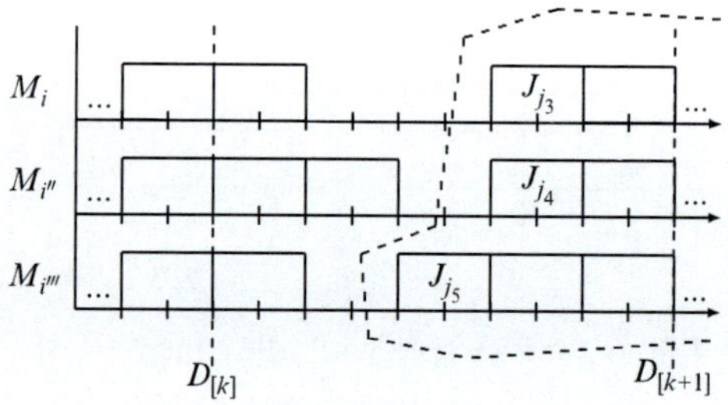

Fig. 6.26: Example for transformation to new solution with idle time $X = p$

by one time unit due dates greater than $D_{[k]}$ will reduce total due date penalties). Consequently, there are at most one idle time between two adjacent due dates on machine M_i and its size is p. $\square$

According to the above propositions, we can deduce that there is an optimal solution in which the job completion times and optimal due dates are multiples of processing time p, there is at most one idle time between two adjacent due dates on each machine and the size of each idle time is p. The maximum job completion time and maximum optimal due date are then bounded by $(n+l)p$. Moreover, these maximum values can be improved by considering $n+ml$ jobs (ml fictive jobs represented l idle times on each machine) to schedule consecutively as soon as possible on m machines. As there is no job scheduled after qp with $q = \lceil (n+ml)/m \rceil$, the job completion times and optimal due dates are then less than or equal to qp. Thus, we derive the following two-step solution for $Pm|p_j = p, d_j \in D, |D| \leq l, D\ unknown,$ $l\ constant|\sum(\alpha_j E_j + \beta_j T_j + \gamma_j d_j)$ scheduling problem.

The main steps of the algorithm are the following:

- For each combination of $\{D_1, D_2, \ldots, D_l\}$ where $D_k, k = 1, 2, \ldots, l$ is chosen from $\{0, p, 2p, \ldots, qp\}$ (i.e., for each due date we have exactly $n+1$ values and bear in mind that a job J_j can be scheduled on one of m machines). Do:
  - Determine the assignment matrix Q as introduced in Sect. 6.5.1.
  - Solve in $O(n^3)$ time the corresponding assignment problem.
- The optimal solution corresponds to the best obtained solution.

The dimension of Q is then $mq \times n$. Consequently, the complexity to determine a solution to the assignment problem is $O(mn^3)$ (because $q \leq \lceil (n+mn)/m \rceil < 2n$). The overall computation time is then in $O(mn^{l+3})$. Consequently, we have the following theorem.

Theorem 6.10 ([22]). *An optimal solution to the $Pm|p_j = p, d_j \in D, |D| \leq l$, D unknown, l constant$|\sum(\alpha_j E_j + \beta_j T_j + \gamma_j d_j)$ scheduling problem can be obtained in $O(mn^{l+3})$.*

As the complexity is polynomial as far as l is constant, we deduce the following for the case where the number of machines is unknown as follows.

Theorem 6.11. *The $P|p_j = p, d_j \in D, |D| \leq l, D$ unknown, l fixed$|\sum(\alpha_j E_j + \beta_j T_j + \gamma_j d_j)$ scheduling problem is polynomial.*

6.5.5 Additional Constraint and Not Depending Weights on Jobs

In this section, additional constraint is considered where an optimal solution does not contain voluntary idle time. In particular, in this section, the problem consists on determining an optimal solution where starting-idle-time on machines is forbidden. According to the classical notation of scheduling problem, the term *nmit* is added. Hence, the studied scheduling problem is denoted by $Qm|p_j = p, d_j = d, nmit|\sum(\alpha E_j + \beta T_j + \gamma d)$ for which an optimal solution does not contain starting-idle-time $\forall j = 1, \ldots, m$. We denote by d^{opt} the optimal value of the common due date.

The single machine scheduling problem $1|p_j = p, d_j = d$ *unknown*, $nmit|\sum(\alpha E_j + \beta T_j + \gamma d)$ with an unknown common due date which is also to be minimized is studied in [37]. The authors show that an optimal solution for this problem exists which does not contain voluntary idle time. They propose a polynomial time algorithm to determine an optimal solution in $O(nlgn)$ times. Remark that if $\gamma \geq \beta$ the problem is simple to solve since the reduction of one time unit of d enables us to reduce the global cost by $(n\gamma - n\beta)$. Hence, an optimal solution is obtained by setting $d^{\text{opt}} = 0$ where jobs are scheduled according to the SPT rule; else, if $\gamma < \beta$, an optimal solution exists in which d^{opt} is given by the completion time of the job scheduled at the position $\lceil (n(\beta - \gamma))/(\alpha + \beta) \rceil$.

In the case of uniform parallel machines to minimize $\sum(\alpha E_j + \beta T_j + \gamma d)$, we have the following theorem.

Theorem 6.12. *For the problem $Qm|p_j = p, d_j = d, nmit|\sum(\alpha E_j + \beta T_j + \gamma d)$, an optimal solution of the set of feasible solution where all machines have no starting-idle-time can be computed in $O(m^2)$.*

Proof. It is easy to show that an optimal solution of $Qm|p_j = p, d_j = d, nmit|$ $\sum(\alpha E_j + \beta T_j + \gamma d)$ is also an optimal solution of problem $Qm|p_j = p|\sum C_j$ which can computed in $O(m^2)$ (see Theorem 6.4.1.4). Hence, the optimal value of the due date is calculated with a greedy algorithm in two steps:

1. If the due date is not null then there is a JiT job. In this case we have at most $(mn+1)$ possible values of d^{opt}.
2. For a given value of due date, we determine the number of early jobs and tardy jobs behind each machine. So the cost will be calculated in $O(m)$ (cf. remark 6.1).

Now we should determine the value of d^{opt}. However, we will show that the number of all possibility values of d^{opt} is bounded by $O(m)$. Thus, the calculation of d^{opt} can be done in $O(m^2)$ as follows.

Let n_1, n_2 and n_3 the number of early jobs, JIT jobs and tardy jobs according to a given d. Therefore, $n = n_1 + n_2 + n_3$. For a given value of due date d, the cost will be increased if due date is moved either on the left or on the right a unit of time. So, we have

$$\begin{cases} -(n_1+n_2)\alpha + n_3\beta - \gamma \leq 0 \\ n_1\alpha - (n_2+n_3)\beta + \gamma \quad \leq 0 \end{cases} \Rightarrow \begin{cases} n_3 \leq \frac{n\alpha+\gamma}{\alpha+\beta} \\ n_1 \leq \frac{n\beta-\gamma}{\alpha+\beta} \end{cases}.$$

Consequently, $n_2 \leq m \Rightarrow \frac{n\beta-\gamma}{\alpha+\beta} \leq n_1 + n_2 \leq \frac{n\beta-\gamma}{\alpha+\beta} + m$. Hence, the sum n_1+n_2 is at most $m+1$ values. We verify then only at most m corresponding values of d^{opt}. The procedure will be then as follows:

- Using the same approach for proving Proposition 1, the scheduling problem with a given $d = 0$ can be solved in $O(m^2)$.
- For the least value of (n_1+n_2), on each machine, the number of jobs including early jobs and JIT jobs would be determined in $O(m^2)$. The objective function cost is then calculated in $O(m)$.
- For the other values of (n_1+n_2) and on each machine, the number of jobs including early jobs and JIT jobs would be determined in $O(m)$. The objective function cost is calculated in $O(m)$.

The execution time is then bounded by $O(m^2)$. The optimal value d^{opt} in the case is noted by d_0. □

6.6 Conclusion

This chapter deals with single machine and parallel machines scheduling problems with a limited set of due dates and equal-size jobs. The first part of this chapter is dedicated to the case where a due date is given where the objective is to minimize the total cost defined by the weighted sum of the earliness cost and the tardiness cost. Both cases have been studied: restrictive due date and non-restrictive due date. The main results are summarized in Table 6.6.

Table 6.6: JIT with given common due date and equal-size jobs

Problem	Method and Complexity status	Reference
$1\|p_j = p\|\sum(E_j + T_j)$	Polynomial $O(n\log n)$	[12]
$1\|p_j = 1, d_j = d\|\sum w_j(E_j + T_j)$	Polynomial $O(n\log n)$	[18]
$1\|p_j = 1, d_j\|\sum(\alpha_j E_j + \beta_j T_j)$	Polynomial $O(n^4)$	[31]
$1\|p_j = p, d_j = d\|\sum(\alpha_j E_j + \beta_j T_j)$	Polynomial $O(n^3)$	Sect. 6.3.1
$1\|p_j = p, d_j \in D, \|D\| = l, l\ fixed\|\sum(\alpha_j E_j + \beta_j T_j)$	Polynomial $O(l^{l+1}n^{l+2})$	[24,25] Sect. 6.3.2
$Qm\|p_j = p, d_j \in D, \|D\| \leq l, l\ fixed\|\sum(\alpha_j E_j + \beta_j T_j)$	$O(m2^m l^{ml+1} n^{m(l-1)+3})$	Sect. 6.3.3
$1\|p_j = p, d_j = d$, *non-restrictive* $\|\sum(\alpha_j E_j + \beta_j T_j)$	Polynomial $O(n^3)$	[24,25,35]
$Qm\|p_j = p, d_j = d$, *non-restrictive* $\|\sum(\alpha E_j + \beta T_j)$	Polynomial $O(m^2)$	Sect. 6.4.1
$Qm\|p_j = p, d_j = d$, *non-restrictive* $\|\sum(\alpha_j E_j + \beta_j T_j)$	Polynomial $O(n^3)$	Sect. 6.4.2

Table 6.7: JIT with assignment common due date and equal-size jobs

Problem	Method and Complexity status	Reference
$P\|p_i = p, d_i = d\|\sum(\alpha E_i + \beta T_i + \gamma d)$	Polynomial	[7]
$Qm\|p_i = p, d_i = d\|\sum(\alpha E_i + \beta T_i + \gamma d)$	Polynomial $O(3^m)$	[7]
$Qm\|p_j = p, d_j = d, nmit\|\sum(\alpha E_j + \beta T_j + \gamma d)$	Polynomial $O(m^2)$	Sect. 6.5.5
$1\|p_j = p, d_j = d\|\sum(\alpha_j E_j + \beta_j T_j + \gamma d)$	Polynomial $O(n^3)$	Sect. 6.5.2
$1\|p_j = p, d_j \in D, \|D\| \leq l, l\ fixed\|\sum(\alpha_j E_j + \beta_j T_j + \gamma d)$	Polynomial $O(n^{l+3})$	Sect. 6.5.3
$P\|p_j = p, d_j \in D, \|D\| \leq l, l\ fixed\|\sum(\alpha_j E_j + \beta_j T_j + \gamma d)$	Polynomial $O(mn^{l+3})$	Sect. 6.5.4

The second part of this chapter is dedicated to the case where a due date corresponds to a decision variable where the objective is to minimize the total cost defined by the weighted sum of the earliness cost, the tardiness cost, and a cost related to due date assignment. The main results are summarized in Table 6.7.

References

1. Ahuja, R.K., Orlin, J.B.: A fast and simple algorithm for the maximum flow problem. Operations Research **37**, 748–759 (1989)
2. Alidaee, B., Panwalkar, S.S.: Single stage minimum absolute lateness problem with a common due date on non-identical machines. Journal of the Operational Research Society **44**(1), 29–36 (1993)
3. Bagchi, U., Sullivan, R.S., Chang, Y.L.: Minimizing absolute and squared deviations of completion times with different earliness and tardiness penalties about a common due-date. Naval Research Logistics **34**(5), 739–751 (1987)
4. Baker, K.R., Scudder, G.D.: Sequencing with earliness and tardiness penalties: a review. Operations Research **38**, 22–36 (1990)
5. Baptiste, P., Brucker, P.: Scheduling equal processing time jobs. In JY Leung (Ed.): *Handbook of scheduling: algorithms, models and performance analysis*, CRC Press, Boca Raton, FL, USA (2004)
6. Bruno, J., Coffman, Jr.E.G., Sethi, R.: Scheduling independent tasks to reduce mean finishing time. Communicaiton of the ACM **17**, 382–387 (1974)

7. Cheng, T.C.E., Chen, Z.L.: Parallel-machine scheduling problems with earliness and tardiness penalties. Journal of the Operational Research Society **45**, 685–695 (1994)
8. Cole, R., Ost, K., Schirra, S.: Edge-coloring bipartite multigraphs in $O(E \log D)$ time. Combinatorica **21**, 5–12 (2001)
9. De, P., Ghosh, J.B., Wells, C.E.: Due-date assignment and early/tardy scheduling on identical parallel machines. Naval Research Logistics **41**(1), 17–32 (1994)
10. Edmonds, J., Karp, R.M.: Theoretical improvements in algorithmic efficiency for network flow problems. Journal of the Association for Computing Machinery **19**, 248–264 (1972)
11. Emmons, H: Scheduling to a common due-date on parallel uniform processors. Naval Research Logistics Quarterly **34**, 803–810 (1987)
12. Garey, M.R., Tarjan, R.E., Wilfong, G.T.: One-processor scheduling with assymmetric earliness and tardiness penalties. Mathematics of Operations Research **13**, 330–348 (1988)
13. Goldberg, A.V., Tarjan, R.E.: A new approach to the maximum-flow problem. Journal of the Association for Computing Machinery **35**, 921–940 (1988)
14. Gordon, V., Proth, J.M., Chu, C.: A survey of the state-of-art of common due date assignment and scheduling research. European Journal of Operational Research **139**, 1–25 (2002)
15. Gordon, V. S., Proth, J.-M., Strusevich, V. A.: Scheduling with due date assignment. In Handbook of scheduling, J.Y.T Leung eds., CRC Press, Boca Raton, FL, USA (2004)
16. Graham, R.L., Lawler, E.L., Lenstra, J.K., Rinnooy Kan, A.H.G.: Optimization and approximation in deterministic sequencing and scheduling: a survey. Annals of Discrete Mathematics **5**, 287–326 (1979)
17. Hall, N.G., Kubiak, W., Sethi, S.P.: Earliness-tardiness scheduling problem, II: Deviation of completion times about a restrictive common due date. Operations Research **39**, 847–856 (1991)
18. Hall, N.G., Posner, M.E.: Earliness-tardiness scheduling problem, I: Weighted deviation of completion times about a common due date. Operations Research **39**,836–846 (1991)
19. Hoogeveen, J.A., Van de Velde, S.L.: Scheduling around a small common due date. European Journal of Operational Research **55**(2), 237–242 (1991)
20. Hoogeveen, J.A., Oosterhout, H., Van de Velde, S.L.: New lower and upper bounds for scheduling around a small common due date. Operations research **42**(1), 102–110 (1994)
21. Hopcroft, J.E., Karp, R.M.: An $n^{5/2}$ algorithm for maximum matchings in bipartite graphs. SIAM Journal on Computing **2**, 225–231 (1973)
22. Huynh Tuong, N., Soukhal, A.: Due dates assignment and JIT scheduling with equal-size jobs. European Journal of Operational Research **205**(2), 280-289 (2010)
23. Huynh Tuong, N., Soukhal, A.: Some new polynomial cases in just-in-time scheduling problems with multiple due dates. In Proceeding of the 6th International Conference on Research, Innovation & Vision for the Future in Computing & Communications Technologies (RIVF'08), Ho Chi Minh (Vietnam), 36–41 (2008)
24. Huynh Tuong, N., Soukhal, A.: Ordonnancement juste-à-temps sur une seule machine avec date de fin souhaitée commune. 9ème Congrés de la Société Française de Recherche Opérationnelle et d'Aide à la Décision (ROADEF'08), Febuary 25–27, Clermont-Ferrand (France), 146–148 (2008)
25. Huynh Tuong, N., Soukhal, A.: Polynomial cases and PTAS for just-in-time scheduling on parallel machines around a common due date. 11th International Workshop on Project Management and Scheduling (PMS'08), April 28-30, Istanbul (Turkey), 152–155 (2008)
26. Józefowska, J.: Just-in-time Scheduling : Models and Algorithms for Computer and Manufacturing Systems. Springer-Verlag New York Inc. (2007)
27. Jurisch, B., Kubiak, W., Józefowska, J.: Algorithms for minclique scheduling problems. Discrete Applied Mathematics **72**, 115–139 (1997)
28. Kaminsky, P., Hochbaum, D.: Due date quotation models and algorithms. In Handbook of scheduling, J.Y.T Leung eds., CRC Press, Boca Raton, FL, USA (2004)
29. Kanet, J.J.: Minimizing the average deviation of job completion times about a common due date. Naval Research Logistics Quarterly **28**, 643–651 (1981)
30. Kao, M.-Y., Lam, T.-W., Sung, W.-K., Ting, H.-F.: A decomposition theorem for maximum weight bipartite matchings. SIAM Journal on Computing **31**, 18–26 (2002)

31. Kovalyov, M.Y.: Soft and negotiable release and due dates in supply chain scheduling. Rapport de projet ORDO-COO-OC de GDR RO (2006)
32. Kovalyov, M.Y., Kubiak, W.: A fully polynomial approximation scheme for the weighted earliness-tardiness problem. Operations Research **47**, 757–761 (1999)
33. Kubiak, W., Lou, S., Sethi, S.: Equivalence of mean flow time problems and mean absolute deviation problems. Operations Research Letters **9**, 371–374 (1990)
34. Kuhn, W.H.: The Hungarian Method for the assignment problem. Naval Research Logistics Quarterly **2**, 83–97 (1955)
35. Mosheiov, G., Yovel, U.: Minimizing weighted earliness–tardiness and due date cost with unit processing–time jobs. European Journal of Operations Research **172**, 528–544 (2006)
36. Mosheiov, G., Sarig, A.: Due-date assignment on uniform machines. European Journal of Operational Research **193**(1), 49–58 (2009)
37. Panwalkar, S.S., Smith, M.L., Seidmann, A.: Common due-date assignment to minimize total penalty for the one machine scheduling problem. Operations Research **30**, 391–399 (1982)
38. Smith, W.E.: Various optimizers for single-stage production. Naval Research Logistics Quarterly **3**, 59–66 (1956)
39. Sourd, F.: New exact algorithms for one-machine earliness-tardiness scheduling. INFORMS Journal of Computing **21**(1), 167–175 (2009)
40. Sun, H., Wang, G.: Parallel machine earliness and tardiness scheduling with proportional weights. Computers & Operations Research **30**(5), 801–808 (2003)
41. T'kindt, V., Billaut, J.-C.: Multicriteria Scheduling: Theory, Models and Algorithms. Springer, Heidelberg (2006)
42. Webster, S.: The complexity of scheduling job families about a common due date. Operations Research Letters **20**(2), 65–74 (1997)
43. Xiao, W.-Q., Li, C.-L.: Approximation algorithms for common due date assignment and job scheduling on parallel machines. IIE Transactions **34**(5), 467–477 (2002)
44. Yuan, J.: The NP-hardness of the single machine common due date weighted tardiness problem. Systems Science and Mathematical Sciences **5**, 328–333 (1992)

Chapter 7
No-Wait Scheduling Problems with Batching Machines

A. Oulamara

Abstract Scheduling problems with batching operation are largely considered in the literature of the last decade. Batching means that sets of jobs which are processed on the same machine are grouped together into batches. One distinguish in the literature of scheduling two types of batching, namely, parallel batching and serial batching. In parallel batching, jobs are processed simultaneously so that the processing time of batch is given by the largest processing times of jobs of this batch. In serial batching, jobs are processed sequentially so that the processing of batch is given by the total processing times of jobs of this batch. This chapter focuses on several problems of no-wait scheduling problem including batching machines. Theoretical and experimental results are presented and some future researches are discussed.

7.1 Introduction

Scheduling problems with batching operation are largely considered in the literature of the last decade. Batching means that sets of jobs which are processed on the same machine are grouped together into batches. The motivation for batching jobs is in gain in efficiency to process jobs together in a batch rather than to process them individually. Several application involve the batching operation, for instance, the "burn in" operations in the manufacture of circuit boards, which are performed in ovens that can process several jobs [22], production in metal sheet industry in witch metal sheets pass through a multi-head hole-punching machine [13], chemical industry in which processing products consisting of photolytical, galvanic or any other chemical baths [18], tire manufacturing in which the vulcanization operations are made on press-machines [17].

A. Oulamara
LORIA laboratory, Nancy University, Ecole des Mines de Nancy, Parc de Saurupt, 54042 Nancy, France, e-mail: Ammar.Oulamara@loria.fr

R.Z. Ríos-Mercado and Y.A. Ríos-Solís (eds.), *Just-in-Time Systems*, Springer Optimization and Its Applications 60, DOI 10.1007/978-1-4614-1123-9_7,

In the literature of scheduling, batching occurs in two versions: parallel batching and serial batching. In parallel batching, several jobs are processed together in batch on specific machine. This machine is called parallel batching machine abbreviated as *p-batching machine*. On this machine jobs are processed simultaneously so that the processing time of batch is given by the largest processing times of jobs of this batch, and jobs of the same batch start and finish their processing at the same time. An upper bound on the batch size may be imposed, that is, the capacity of batch is bounded by an integer value, which reflects the maximum number of jobs that can be gathered in batch. Sometimes, this limit reflect the total weight or total volume of jobs. In serial batching, jobs may be batched if a setup time occurs on a machine when it process several jobs and the setup time is required between the processing of jobs with different characteristics. This machine is called serial batching machine, abbreviated as *s-batching machine*. The jobs in the same batch are processed sequentially, so that the processing time of a batch is equal to the sum of the processing times of jobs in this batch. Furthermore the batch availability notion is used to describe the availability of jobs after their processing on s-batching machine. The Batch availability in the case of a single s-batching machine or in parallel s-batching machines means that all jobs of the same batch are available together when the last job in this batch has completed its processing. Similarly, for shop problems, batch availability means that all jobs of a batch become available for processing on the next machine, or are completed, at the time that the last jobs in that batch completes its processing.

Scheduling problems with batching machines are extensively considered in the literature. Potts and Van Wassenhove [21], Albers and Bruckers [1], Webster and Backer [23], and Potts and Kovalyov [19] provide surveys and analysis of both types of scheduling with batching operation. For the single p-batchine machine problem, Brucker et al. [5] consider cases of unbounded and bounded batching machine and several regular criteria (makespan, total completion tame, maximum lateness, total tardiness, number of late jobs). They provides a classification of problems as polynomially or pseudopolynomially solvable, binary or unary NP-hard, or open. When jobs have release dates, except the makespan criterion which is polynomial [5] all other regular criteria are NP-hard or open problems. For instance, Chen et al. [7] prove that the maximum lateness is NP-hard in the strong sense, whereas the weighted total completion time is NP-hard in ordinary sense [9]. Sometimes jobs cannot be batched together, due to technological constraints or to physical and chemical properties of jobs that form a batch; for instance, in [2] authors describe an application in tire manufacturing in which the tire curing operation are made by p-batching machine, and jobs are processed together in batch only if they share a same processing time. In that case, we call this constraint as compatibility constraint, that is, only compatible jobs are batched together in same batch. Thus, a compatibility graph is defined for that problem in which vertices represent jobs and two vertices are connected by an edge if they are compatible. Several compatibility graphs are considered in the literature of batch scheduling problems, for instance, Boudhar and Finke [4] consider the makespan minimization in scheduling with p-batching machine and general graph compatibility between jobs, they show that the problem is NP-hard in the strong sense if the capacity of batching machine $b \geq 3$ and it is polynomial solved in $O(n^3)$ when $b = 2$. Special cases of graphs are considered in the literature, for instance, interval graph [17], bipartite graph [4, 10], split graphs [3, 10].

Concerning shop scheduling, Potts et al. [20] study the complexity of minimizing the makespan in openshops, jobshops and fowshops with two p-batching machines. For the fowshop and job shop, they show that problems minimizing the makepsan are solvable in $O(nlogn)$ time when the size of batch is unbounded, whereas when the size of batches is restricted, authors showed that minimizing the makespan is binary NP-hard for both problems. For the openshop, Potts et al. [20] show that the problem the minimizing the makespan is solvable in $O(n)$ when the size of batches is unbounded and the problem is binary NP-hard when the size is bounded.

Scheduling problems with serial batching machines are considered in the literature for the case of single machine problems and shop scheduling problems. For single machine problem, Cofmann et al. [8] propose a dynamic programming algorithm that solve the total completion time criterion in $O(nlogn)$ time, Albers and Brucker [1] show that the weighted completion time is NP-hard in the strong sense. Webster and Baker [23] consider the maximum lateness criterion they propose a dynamic programming algorithm that minimize the maximum lateness in $O(n^2)$. For the criterion of the number of late jobs, Brucker and Kovalyov [6] propose a dynamic programming algorithm that gives an optimal solution in $O(n^3)$. Shop scheduling problems are considered in Glass et al. [15], authors provide a complexity results and approximation algorithms. Specifically, they show that minimizing the makespan in two s-batching machines under batch availability is NP-hard in the strong sense and they show that the batch are consistent in the optimal solution. Concerning the openshop scheduling problem, authors in [15] show that minimizing the makespan is NP-hard in the ordinary sense, and the optimal solution contains one, two or three consistent batches on each machine. Although several studies are focused on shop scheduling with batching machines in the literature, there are still interesting problems that are not considered, such as flowshop scheduling with mixed batching machine.

In this chapter, we consider two interesting problems of scheduling in flowshop scheduling with two batching machines integrating the no-wait constraints. In the first problem, we consider the makespan minimization in two flowshop scheduling problem, where the first machine is p-batch machine and the second machine is p-batch machine. In the second problem, we consider the makespan minimization in flowshop with two s-batching machines and no-wait constraint between the processing of batches on machines.

The remainder of this chapter is organized as follows. Section 7.2 considers the problem of flowshop scheduling with two mixed batching machines, a complexity of makespan minimization is presented and several special cases are studied. Section 7.3 presents the no-wait flowshop with two s-batching, the complexity of the makespan minimization is presented together with a Lagrangian relaxation based heuristic. Finally, some concluding remarks are contained in Sect. 7.4.

7.2 No-Wait Flowshop with Mixed Batching Machines

In this section, we consider the no-wait flowshop scheduling problem. There are n jobs to be batched and scheduled for processing on two batching machines. Each job must be processed on batching machines M_1 and M_2 and visit them in this order. The processing time for job j on machines M_1 and M_2 is at least p_j and q_j time units respectively. Machine M_1 is *p-batching machine* with capacity b ($b < n$), that is, can handle up to b jobs simultaneously. The batch processing time on this machine is equal to the maximum processing time of the jobs in this batch. Machine M_2 is *s-batching machine*, and handles jobs sequentially. The batch processing time on this machine is equal to the sum of processing times of the jobs in this batch. A setup time s is required on this machine prior to batch execution. The batches are processed in a no-wait fashion, so that each batch's completion time on machine M_1 will coincide with its starting time on machine M_2.

A schedule is characterized by partitioning the jobs into batches and then sequencing the batches. Due to the no-wait constraints, batch partition and batch sequence are the same for both machines, and the objective is to find a schedule that will minimize the batch completion time. By using the general notation for scheduling problems introduced by Graham et al. [16] we write $F2 \mid p-batch(1)$, $s-batch(2), b, no-wait \mid C_{\max}$ to refer to this problem.

7.2.1 NP-Hardness

In this section, we show that the problem $F2|p-batch(1), s-batch(2), b, no-wait|C_{\max}$, abbreviated in the following as problem $P(b)$, is strongly NP-hard. The following decision problem, which is known as NP-complete in the strong sense (see Garey and Johnson [12]) is used for the reduction.

Numerical Matching with Target Sums (NMTS). Given two sets of integers U and V, where $U = \{u_1, \ldots, u_n\}$ and $V = \{v_1, \ldots, v_n\}$, and a target vector $W = \langle W_1, \ldots, W_n \rangle$ with positive integers. Can $U \cup V$ be partitioned into n disjoint sets $N_1, \ldots, N_n$, with each N_i ($i = 1, \ldots, n$) containing exactly one element from each of U and V, such that $u_{N_i} + v_{N_i} = W_i$?

Theorem 7.1. *Problem $P(b = 2)$ is strongly NP-hard.*

Proof. Given an arbitrary instance I of the NMTS problem, first we propose a modification of I as follows:
Let $w = \max\{W_1, \ldots, W_n\}$. We build a new instance I' of NMTS such that $U' = \{u'_1, \ldots, u'_n\}$, $V' = \{v'_1, \ldots, v'_n\}$ and the target vector $W' = \langle W'_1, \ldots, W'_n \rangle$ where, $u'_i = u_i$, $u'_i = u_i + 3w$, $W'_i = W_i + 3w$, $i = 1, \ldots, n$. Clearly, the instance I' has a solution, if and only if the instance I has a solution.

From an arbitrarily modified instance I of the NMTS problem, we construct the following instance of problem $P(b = 2)$ with $2n^2 + 2(n+1)$ jobs. The jobs are then partitioned into three groups:

- X-jobs: denoted $X_{i,j}$,$(i,j=1,\ldots,n)$
- Y-jobs: denoted $Y_{i,j}$, $(i,j=1,\ldots,n)$
- Z-jobs: denoted $Z_{i,j}$, $(i=1,\ldots,n+1,\ j=1,2)$

The job processing times on both machines are given in Table 7.1, with a setup time of $s=0$ on the second machine.

Table 7.1: Processing times of jobs

	X - Jobs	Y - Jobs	Z - Jobs					
	$X_{i,j}$ $(i,j=1,\ldots,n)$	$Y_{i,j}$ $(i,j=1,\ldots,n)$	$Z_{1,1}$	$Z_{1,2}$	$Z_{i,1}$ $(i=2,\ldots,n)$	$Z_{i,2}$ $(i=2,\ldots,n)$	$Z_{n+1,1}$	$Z_{n+1,2}$
p_j	W_i	W_i	0	0	W_{i-1}	W_{i-1}	W_n	W_n
q_j	u_i	v_i	$2w$	W_1-2w	$2w$	W_i-2w	0	0

Thus, the NMTS problem has a solution, if and only if $P(b=2)$ has a feasible sequence S, such that the completion time for S is $C_{\max}(S)\leq(n+1)\sum_{i=1}^{n}W_i$.

First, suppose that NMTS has a solution. Let $N_1,\ldots,N_n$ be n disjoint sets, such that each set N_i contains exactly one element of U and one element of V, and $u_{N_i}+v_{N_i}=W_i$, $i=1,\ldots,n$. Thus, a schedule S exists for the problem $P(b=2)$, such that the completion time $C_{\max}(S)$ is equal to $(n+1)\sum_{i=1}^{n}W_i$. In this schedule, the jobs are grouped into two groups of batches, each one contains exactly two jobs. The groups are constructed as follows:

- $G1$ contains batches $D_i=\{Z_{i,1};Z_{i,2}\}$, $i=1,\ldots,n$, with a processing time of W_{i-1} on the first machine, except for batch $\{Z_{1,1};Z_{1,2}\}$ which has a zero processing time on the first machine.
- $G2$ contains batches $D_{l,i}=\{X_{l,j};Y_{l,k}\}$, $l,i=1,\ldots,n$, $j\in N_i$ and $k\in N_i$, with a processing time of W_l on the first machine and $u_j+v_k=W_i$ on the second machine. This class contains n^2 batches.

The first batch $D_1=\{Z_{1,1};Z_{1,2}\}$, from $G1$, is scheduled in the first position in S. Let r_i be the position of the batch $D_i=\{Z_{i,1};Z_{i,2}\}$ in S, where $\pi_1=1$, and $\pi_i=\pi_{i-1}+2(n-i+2)$, $i=2,\ldots,n+1$. Let $\pi_{l,i}$ be the position in S of the batch $D_{l,i}=\{X_{i,j};Y_{i,k}\}$, $l,i=1,\ldots,n$, from group $G2$, where $j\in N_i$ and $k\in N_i$. Then $\pi_{l,i}=\pi_i+2(l-i)+1$ if $i\leq l$ and $\pi_{l,i}=\pi_l+2(i-l)$ if $i>l$. In the sequence S, all batches are sequenced in no-wait fashion and there is no idle-time on either machine. Therefore,

$$C_{\max}(S)=\sum_{i=1}^{n}q(D_i)+\sum_{l-1}^{n}\sum_{i=1}^{n}q(D_{l,i})=\sum_{i-1}^{n}W_i+n\sum_{i=1}^{n}W_i=(n+1)\sum_{i=1}^{n}W_i.$$

Conversely, suppose that there is a schedule S^* such that $C_{\max}(S^*)\leq(\text{n+1})\sum_{i=1}^{n}W_i$. Let d be the number of batches in S^*, where $d\geq n^2+n+1$ (batch size on the first machine is bounded by 2). Denote by $p(B_i)$ the processing time of batch B_i, where $i = 1,\dots,d$, on the first machine. For each B_i, $p'(B_i)$ is defined as follows:

$$p'(B_i) = \begin{cases} 0 & \text{if } B_i \text{ contains exactly one job} \\ \min\{p_j \,/\, j \in B_i\} & \text{otherwise} \end{cases}.$$

The batch size on the first machine is bounded by 2, so

$$\sum_{i=1}^{d} p(B_i) + \sum_{i=1}^{d} p'(B_i) = \sum_{i=1}^{2n^2+2(n+1)} p_j = 2(n+1)\sum_{i=1}^{n} W_i,$$

and $C_{cmax}(S^*) \leq (n+1)\sum_{i=1}^{n} W_i$, yielding,

$$\sum_{i=1}^{d} p(B_i) = \sum_{i=1}^{d} p'(B_i) = (n+1)\sum_{i=1}^{n} W_i.$$

Thus, (1) each batch B_i, where $i = 1,\dots,d$, contains exactly two jobs, (i.e. $d = n^2+n+1$), (2) the jobs which have the same indices are grouped in the same batch (i.e. a batch containing the job $X_{i,j}$ contains also a job $X_{i,k}$ or $Y_{i,l}$ or $Z_{i+1,t}$), and (3) There is no idle time on the first machine.

On the second machine, the sum of the job processing times is given by

$$n\sum_{i=1}^{n} u_i + n\sum_{i=1}^{n} v_i + \sum_{i=1}^{n} W_i = (n+1)\sum_{i=1}^{n} W_i = C_{\max}(S).$$

Because there is no idle time on the second machine, only the batch $\{Z_{1,1};Z_{1,2}\}$ can be scheduled in the first position, with a zero processing time on the first machine and an W_1 processing time on the second.

As mentioned above, the jobs which have the same indices are grouped in the same batch, (i.e. a job $X_{i,j}$ of the set $X_i = \{X_{i,1},\dots,X_{i,n}\}$ can be grouped either with a job $X_{i,k}$, where $j \neq k$, of set X_i or with a job $Y_{i,k}$ of set $Y_i = \{Y_{i,1},\dots,Y_{i,n}\}$ or with a job $Z_{i+1,k}$, where $k = 1,2$). This leads to the following cases:

(a) If $Z_{i+1,1}$ is grouped with a given job $X_{i,j}$ in batch D_i, then the processing time for D_i on the second machine will be $q(D_i) = u_i + 2w \leq 3w < W_k$, where $k = 1,\dots n$. This creates idle time on the second machine between the batch D_i and the following batch in the sequence S^*.

(b) If $Z_{i+1,1}$ is grouped with a given job $Y_{i,j}$ in batch D_i, then the processing time for D_i on the second machine will be $q(D_i) = v_i + 2w > W_k$, where $k = 1,\dots n$. This creates idle time on the first machine between the batch D_i and the following batch in the sequence S^*.

Thus, $Z_{i+1,1}$ can only be grouped with $Z_{i+1,2}$. In addition, when choosing the NMTS instance, the jobs in set X_i can be batched only with the jobs in set Y_i. Otherwise, idle time is caused on either the first or the second machine. Thus, all the batches in the sequence S^* have the form $\{X_{i,j};Y_{i,k}\}$ and $\{Z_{i,1};Z_{i,2}\}$.

Without loss of generality, it can be assumed that the batch $D = \{X_{i,j};Y_{i,k}\}$ is sequenced at position l. Let D' be the batch sequenced at position $l+1$. The processing time for D' on the first machine is equal to some W_l, where $W_l \in W$. But given that D' is sequenced in a no-wait fashion, and given that there is no idle time on either machine, $q(D) = q(X_{i,j}) + q(Y_{i,k}) = W_l$ for each $W_l \in W$. Thus, the set $N_l = \{u_j, v_k \ / \ u_i + v_j = W_l\}$ contains exactly one element of U and one element of V. Repeating this last operation for all values of l ($l = 1,\ldots,n$), produces a solution for the instance I of the NMTS problem. □

7.2.2 *Equal Processing Times on* M_1

Let consider the special case where all jobs have the same processing time on the first machine, $p_j = p$, $j = 1,\ldots,n$. We distinguish two cases: (1) the case where the capacity of p-batching machine is equal to two, denoted by $P(p,b=2)$, and (2) the case where this capacity is greater than or equal to three, denoted by $P(p,b \geq 3)$.

7.2.2.1 A Capacity $b = 2$

Lemma 7.1. *There exists an optimal schedule for the problem $P(p,b=2)$ in which all batches, excepting the last, are full if the number of jobs is not a multiple of two.*

Proof. Let S be an optimal schedule for the problem $P(p,b=2)$, containing the batches $(B_1,\ldots,B_u)$. Let B_j be the first non full batch of S (B_j contains only one job). Let B_k, the non full batch of S, follow B_j in the schedule. A new schedule S' is built from S by completing B_j with a job from B_k and shifting all the batches that follow B_k in schedule S to the left. Clearly, $C_{\max}(S') \leq C_{\max}(S)$. Therefore, S' is an optimal schedule. Repeating such modifications yields a schedule whose batches (excepting the last) are full, if the number of batches is not a multiple of two. □

Without loss of generality, it can be assumed that the number of jobs is a multiple of two; otherwise a dummy job can be added, with a processing time p on M_1 and a zero processing time on M_2. Thus, for a given instance of the problem $P(p,b=2)$ with n jobs, the optimal sequence contains $n/2$ batches.

Theorem 7.2. *The problem $P(p,b=2)$ reduces to a maximum weight matching problem.*

Proof. Consider a complete graph $G = (U,E)$ where U is the set of vertices $U = \{u_1,\ldots,u_n\}$ corresponding to the jobs, and E is the set of edges $E = \{e_1,\ldots,e_m\}$ corresponding to the possible batching of jobs in a given batch. Let w_j be the weight of edge e_j, such that $w_j = \max\{s + q_{u_j} + q_{v_j} - p, 0\}$ where $e_j = (u_j, v_j)$, $j = 1,\ldots,m$. Let A be the vertex-edge incidence matrix of the complete graph $G = (U,E)$ where $a_{i,j} = 1$ if the edge j is incident to the vertex i and 0 otherwise. Let x_j be a decision variable such that $x_j = 1$ if the jobs connected by the edge j are in the same batch and 0 otherwise.

The integer program corresponding to minimize the makespan is

$$\begin{cases} \min C_{\max} = s + \frac{n}{2}p + \sum_{j=1}^{m} w_j x_j + \sum_{i=1}^{n}\left(1 - \sum_{j=1}^{m} a_{i,j}x_j\right) q_i \\ s.t.\ \sum_{j=1}^{m} a_{i,j}x_j \le 1 \quad i = 1,\dots,n \\ \sum_{j=1}^{m} x_j = \frac{n}{2} - 1 \end{cases}$$

We have,

$$\begin{aligned} C_{\max} &= s + \frac{n}{2}p + \sum_{j=1}^{m} w_j x_j + \sum_{i=1}^{n}\left(1 - \sum_{j=1}^{m} a_{i,j}x_j\right) q_i \\ &= s + \frac{n}{2}p + \sum_{j=1}^{m} w_j x_j + \sum_{i=1}^{n} q_i - \sum_{i=1}^{n}\sum_{j=1}^{m} a_{i,j} q_i x_j \end{aligned}$$

$$\begin{aligned} &= s + \frac{n}{2}p + \sum_{i=1}^{n} q_i - \sum_{j=1}^{m}\left(\sum_{i=1}^{n} a_{i,j} q_i - w_j\right) x_j \\ &= s + \frac{n}{2}p + \sum_{i=1}^{n} q_i - \sum_{j=1}^{m} (q_{u_j} + q_{v_j} - w_j) x_j \\ &= s + \frac{n}{2}p + \sum_{i=1}^{n} q_i - \sum_{j=1}^{m} (\min\{p - s, q_{u_j} + q_{v_j}\}) x_j. \end{aligned}$$

Hence $s + (n/2)p + \sum_{i=1}^{n} q_i$ is constant and greater than $\sum_{j=1}^{m}(\min\{p - s, q_{u_j} + q_{v_j}\})x_j$. Then minimizing $C_{\max}$ is equivalent to maximizing $\sum_{j=1}^{m} d_j x_j$ where $d_j = \min\{p - s, q_{u_j} + q_{v_j}\}$ if the edge e_j is incident to the vertices u_j and v_j. Consequently, the integer model can be reduced to a weight matching problem in complete a graph $G = (V,E)$ with the weight d_j on the edge e_j, $j = 1,\dots,m$. This problem can be solved optimally using Edmonds's algorithm [11] in $O(n^3)$. □

The following algorithm gives the optimal schedule for the problem $P(p, b = 2)$.

Algorithm Best-Batch

1. Construct a valued complete graph $G = (V,E)$ where each edge in E is valued by $\min\{p - s, q_i + q_j\}$ if the edge is incident to the vertices i and j.
2. Find a maximum weight matching in a complete graph G.
3. For each edge e_i of the matching, process the corresponding two jobs incident to e_i in the same batch, and execute the batches in an arbitrary order, except for the two jobs that are not part of the matching, process this two jobs together in the same batch at the end of the schedule.

7.2.2.2 A Capacity $b \geq 3$

We show that minimizing the makespan is NP-hard in the strong sense when the capacity of the p-batching machine is greater than or equal to three. We use the reduction of the 3-PARTITION problem, which is known to be strongly NP-hard [12].

3-PARTITION. Given $3n$ positive integers $a_1, \ldots, a_{3n}$, where $\sum_{i=1}^{3n} a_i = nW$ for some W and $W/4 < a_i < W/2$ for $i = 1, \ldots, 3n$, can the index set $N = \{1, \ldots, 3n\}$ be partitioned into n disjoint subsets $N_1, \ldots, N_n$ such that $\sum_{i \in N_j} a_i = W$ and $|N_j| = 3$ for $j = 1, \ldots, n$?

Theorem 7.3. *The problem $P(p, b \geq 3)$ is strongly NP-hard.*

Proof. First we consider the case in which the capacity of first machine is equal to three, then we extend the proof to the case in which the capacity is greater than three. Given an arbitrary instance of 3-PARTITION, we construct the following instance of problem $P(p, b = 3)$, with the set $N = \{1, \ldots, 3n+3\}$ of jobs. The job processing times on the first machine are equal to W and are equal $q_j = a_j$ for $j = 1, \ldots, 3n$ and $q_{3n+1} = q_{3n+2} = q_{3n+3} = 0$ on the second machine. The capacity of the first machine is equal to three and the setup time on the second machine is equal to zero.

To prove the theorem, we show that for the constructed instance of the problem $P(p, b = 3)$, there is a schedule S, such that the makespan is $C_{\max}(S) \leq (n+1)W$, if and only if the 3-PARTITION instance has a solution.

First, suppose that 3-PARTITION has a solution, and $N_1, \ldots, N_n$ are the required subsets of N. The schedule S contains following $n+1$ batches, which are defined by $N_1, \ldots, N_n$ and N_{n+1} where N_{n+1} contains the jobs $3n+1$, $3n+2$ and $3n+3$. The jobs in batch N_1 are processed on machine one in the interval $[0, W]$ and are processed on machine two in the interval $[W, 2W]$. In the interval $[jW, (j+1)W]$ for $j = 1, \ldots, n$, the jobs in batch N_{j+1} are processed on machine one, and the jobs in batch N_j are processed on machine two. At time $(n+1)W$, machine one has completed all of its processing. Since the jobs in N_{n+1} have zero processing times on machine two, they are processed on machine two in the interval $[(n+1)W, (n+1)W]$. Thus, $C_{\max}(S) = (n+1)W$.

Suppose now that there is a schedule S, such that $C_{\max}(S) \leq (n+1)W$. Let v be the number of batches in S. Since all jobs have the same processing times W on the machine one, it is easy to show that, (1) each batch B_i, $i = 1, \ldots, v$, contains exactly three jobs, (*i.e.* $v = n+1$) and (2) there is no idle time on the first machine.

On the second machine, the sum of the processing times are given by $\sum_{i=1}^{3n+3} a_i = nW$. Therefore, there is no idle time on machine two, except at the beginning. This idle time W corresponds to the time machine two spends waiting to process the first batch.

Without loss of generality, it can be assumed that the batch B_l, composed of jobs i, j and k, is sequenced in position l, and that the batch B_{l+1} is sequenced in the position $l+1$. The processing time for B_{l+1} on machine one is equal to W. But since B_{l+1} is sequenced in a no-wait fashion, and since there is no idle time on either machine prior to batch B_{l+1}, then $a_i + a_j + a_k = W$. Thus, the set $N_l =$

$\{i, j, k\}$ contains exactly three jobs. Repeating this last operation for all values of l ($l = 1, \ldots, n$) results in a solution for the 3-PARTITION instance.

To complete the proof, the scheduling problem instance can be extended to the problem $P(p,b)$, where $3 < b < n$. In order to make it easier to reduce the 3-PARTITION problem to $P(p,b)$, the scheduling instance, with $(b-3)(n+1)$ jobs and W processing times on machine one and zero processing times on machine two, has been added. Using the same principle, it can be shown that the problem $P(p,b)$ allows a schedule S with a makespan less than or equal to $(n+1)W$, if and only if the 3-PARTITION problem has a solution. □

7.2.3 Equal Processing Times on M_2

Let consider the special case in which all jobs have the same processing time on the second machine, $q_j = q$, $j = 1, \ldots, n$. In the following we denote this problem by $P(q,b)$.

Lemma 7.2. *There exists an optimal solution for problem $P(q,b)$ where the jobs are assigned to the batches on the first machine according to the non-decreasing order of their processing times p_j, $j = 1, \ldots, n$.*

Proof. Let S^* be an optimal solution for the problem $P(q,b)$ consisting of the batch $\langle B_1, \ldots, B_u \rangle$. Suppose that the jobs are indexed in the non-decreasing order of their processing times on the first machine. Let J_i and J_{i-k} ($k > 1$) be two jobs of the same batch B_s and J_{i-1}, a job of another batch B_r where $r > s$. Consider a new schedule S' in which the jobs J_{i-k} and J_{i-1} are exchanged. Since all the jobs have the same processing time on the second machine and the batches are scheduled in a no-wait fashion, S' is feasible. Moreover, the new completion time for the batch B_r in S' is not greater than that of B_r in S^*, because $p_{i-k} \leq p_{i-1}$. Thus, the new schedule is also optimal. This transformation yields a schedule with the desired properties. □

According to Lemma 7.2, jobs are assigned to batches in the non-decreasing order of their processing times on the first machine. The number of batches in the optimal solution and the first jobs assigned to each batch remain to be determined.

A polynomial algorithm for the problem $P(q,b)$ is detailed below. Jobs are reindexed in the non-decreasing order of their processing times on the first machine. Consider the optimal schedule S with k batches, each containing i_j jobs, where $1 \leq i_j \leq b, i = 1, \ldots, k$.

As shown in Fig. 7.1, the makespan can be written as follows:

$$C_{\max}(S) = k \times s + n \times q + \sum_{i=1}^{k} \Delta_i.$$

The makespan is given as a function of the idle time for machine two. Thus, minimizing the makespan is equivalent to minimizing the total idle time, $\sum_{i=1}^{k} \Delta_i$ for

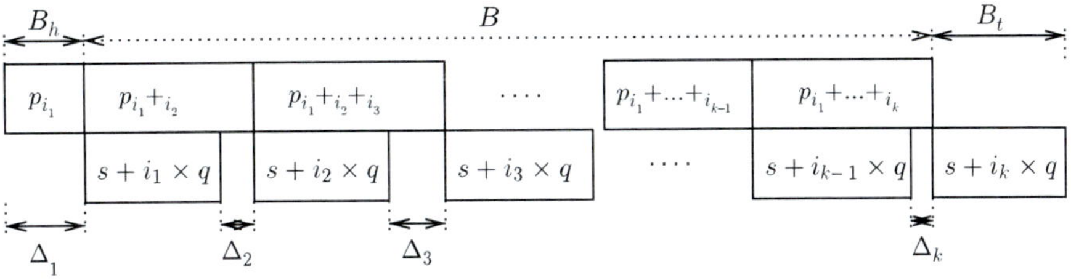

Fig. 7.1: Structure of the optimal schedule

the second machine. Note that when index k and vector $I = (i_1, i_2, \ldots, i_k)$ change, the profile of schedule S, defined in Fig. 7.1 by B_h, B and B_t, also changes, B_h and B_t are called *heads* and *tails*, respectively. Our algorithm is based on the idea that for each profile of the schedule S defined by B_h, there is an optimal schedule that minimizes the total idle time on the second machine. From all the possible profiles B_h, the one that best minimizes the makespan is chosen.

Let h $(1 \leq h \leq b)$ be the number of jobs in the first batch in schedule S. In order to find an optimal schedule S_h for the profile defined by h, a dynamic programming was used to build an optimal schedule with a forward scheme in which the jobs are added to the end of the current (partial) schedule.

For each value of h denote by K_h the maximum number of batches (not counting the first) that can be contained in the schedule S_h is $K_h = n - h$. Let $\Delta_k(i_k, j, l)$ denote the total minimum idle time added to the total setup time, on the second machine for the partial schedule containing the batches $i_{k+1}, \ldots, i_{K_h}$, given that the kth batch contains i_k jobs and that there are j jobs and l batches left to allocate. C_h denotes the value of the makespan corresponding to the optimal schedule S_h with the profile B_h.

Algorithm Dyn-Batch

1. Initialization: $\Delta_k(i, 0, l) = 0; \quad 1 \leq k \leq n, \quad 1 \leq i \leq n, \quad 0 \leq l \leq n$
2. Calculate C_h, $1 \leq h \leq b$, such that $C_h = p_h + s + n \times q + \Delta_0(h, n-h, K_h)$ where $K_h = n - h$ and

$$\Delta_k(i_k, j, l) = s + \min_{1 \leq i_{k+1} \leq \min\{b, j\}} \{\max\{0, p_{n-j+i_{k+1}} - s - i_k \times q\} + \Delta_{k+1}(i_{k+1}, j - i_{k+1}, l - 1)\},$$

3. $C_{\max} = \min\{C_h \mid h = 1, \ldots, b\}$

Theorem 7.4. *Algorithm Dyn-Batch gives the optimal solution for the problem $P(q)$ in $O(n^3)$.*

Proof. Lemma 7.2 states that the jobs are assigned to the batches according to the non-decreasing order of their processing times on the first machine. The optimal partition of the jobs remains to be determined. For a given value of h $(1 \leq h \leq b)$, algorithm Dyn-Batch chooses the best sequence with the minimum value for the total idle time plus setup time for the set of jobs $N_h = \{h+1, \ldots, n\}$, divided into all possible values l of the batches with $l \leq \bar{n}_h$. Since algorithm Dyn-Batch considers all values of h, an optimal schedule is produced in step 3 (above). Algorithm Dyn-Batch calculates C_h for each value h, for a total of b values. For a given h, the total number of possible values for j and l is $O(n^2)$. The number of possible values for k and i_k is $O(n)$ and $O(b)$ respectively. Since the recurrence relation is computed in $O(n^3)$, the algorithm's computation time is $O(n^3)$. □

7.2.4 Approximation

In this section, we propose an approximation algorithm that creates a schedule with a makespan that is, at most, twice the optimum values and run in $O(n^3)$ for the problem $P(b)$ described in Sect. 7.2. This approximation algorithm is based on a generalization of Algorithm Dyn-Batch (presented in Sect. 7.2.3). The jobs are reindexed in the non-decreasing order of their processing times on the first machine. Using the same idea of *schedule profile* that was developed in Sect. 7.2.3, the Algorithm Dyn-Batch is generalized as follows:

Heuristic H

1. Initialization: $\Delta_k(i,0,l) = 0; \quad 1 \leq k \leq n, \quad 1 \leq i \leq n, \quad 0 \leq l \leq n$
2. Calculate C_h, $1 \leq h \leq b$, such that $C_h = p_h + s + \sum_{i=1}^{n} q_i + \Delta_0(h, n-h, K_h)$ where
   $K_h = n - h$ and
   $$\Delta_k(i_k,j,l) = s + \min_{1 \leq i_{k+1} \leq \min\{b,j\}} \left\{ \max\left\{0, p_{n-j+i_{k+1}} - s - \sum_{r=n-j-i_k+1}^{n-j} q_r \right\} + \Delta_{k+1}(i_{k+1}, j-i_{k+1}, l-1) \right\},$$
3. $C_{\max} = \min\{C(h,t) \mid t,h = 1,\ldots,b\}$

Theorem 7.5. *Heuristic H produces a schedule S_{h} with a guaranteed performance ratio of two and this bound is tight.*

Proof. Let S_{h} be the schedule obtained by the heuristic H containing batches $(B_1,\ldots,B_u)$. Consider only the one-machine batching problem on first machine. According to Brucker et al. [5], the optimal schedule is obtained by saturated batches $(B'_1,\ldots,B'_v)$ in SPT-order with processing times of $P(B'_j)$ $j = 1,\ldots,v$ on the first machine (jobs of zero duration can be added if necessary, in order to obtain a total number of jobs that is a multiple of b). The processing is completed on the second machine, using consistent batches$(B'_1,\ldots,B'_v)$ in a no-wait fashion. The resulting schedule is denoted S_{spt} and its corresponding makespan is $C_{\max}(S_{\mathrm{spt}})$. Since heuristic H also enumerated the partition $(B'_1,\ldots,B'_v)$, one has

$$C_{\max}(S_{\mathrm{H}}) \leq C_{\max}(S_{\mathrm{spt}}) \leq \sum_{i=1}^{v} P(B'_j) + v.s + \sum_{i=1}^{n} q_i.$$

Let S^* be the optimal schedule of the problem $P(b)$, and $C_{\max}(S^*)$ its completion time, yielding,

$$C_{\max}(S^*) \geq \sum_{i=1}^{v} P(B'_j) \quad \text{and} \quad C_{\max}(S^*) \geq v.s + \sum_{i=1}^{n} q_i$$

Consequently,

$$C_{\max}(S_{\mathrm{H}}) \leq \sum_{i=1}^{v} P(B'_j) + v.s + \sum_{i=1}^{n} q_i \leq 2C_{\max}(S^*)$$

and

$$\frac{C_{\max}(S_{\mathrm{H}})}{C_{\max}(S^*)} \leq 2$$

Thus, heuristic H provides a solution with a performance ratio equal to two.

An instance for which this ratio was reached is presented now. Let I be an instance for the problem $P(b)$, composed of four job classes U, V, W and X, each containing b jobs. The setup time for the second machine is equal to zero. The job processing times are given in the following table (Table 7.2):

Table 7.2: Processing times of jobs

Classes	Jobs	p_j	q_j
U	$U_j,\ j=1,\ldots,b$	ε	ε
V	V_1,	$(b-1)\times\varepsilon$	$L+\varepsilon$
	$B_j\ j=2,\ldots,b$	$b\times\varepsilon$	ε
W	$W_j,\ j=1,\ldots,b$	$b\times\varepsilon$	ε
X	$X_j,\ j=1,\ldots,b$	$L+b\times\varepsilon$	ε

The makespan $C_{\max}(S_{\mathrm{H}})$ for the schedule S_{H} produced by the heuristic H is equal to $C_{\max}(S_{\mathrm{H}}) = 2L + (4b+1)\varepsilon$. The optimal value of the makespan for the instance I is $C_{\max}(S^*) = L + (4b+1)\varepsilon$ because the lower bound $LB = \min_j p_j + \sum_{j=1}^{4b} q_j$ for $C_{\max}$ is reached by $C_{\max}(S^*)$.

Thus,

$$\frac{C_{\max}(S_{\mathrm{H}})}{C_{\max}(S^*)} = \frac{2L + (4b+1)\varepsilon}{L + (4b+1)\varepsilon}$$

and

$$\lim_{L\to+\infty} \frac{2L + (4b+1)\varepsilon}{L + (4b+1)\varepsilon} = 2.$$

This completes the proof. □

7.3 No-Wait Flowshop with Two s-Batching Machines

In this section, we consider the second problem of no-wait flowshop scheduling problem with the makespan criterion to be minimized. There are n jobs to be processed on two s-batching machines M_1 and M_2. Each job must be processed on machine M_1 and M_2 and visiting them in this order. The processing time of jobs j on machines M_1 and M_2 is p_j and q_j time units respectively. Machines M_1 and M_2 are s-batching machines, that can handle jobs sequentially. The batch processing time on these machines is equal to the sum of processing times of jobs in this batch. Before a batch is processed, a setup time s_1 and s_2 are required on machine M_1 and M_2 respectively. The setup times can be either anticipatory, denoted *as* or non-anticipatory, denoted *ns*. Furthermore, when a batch is completed on the first machine is immediately transferred and processed on the second machine, that is, the no-wait constraints are effective for the batches processing. Due to these no-wait constraints batches are identical on both machines. According to the classification scheme introduced by Graham et al. [16], we denote the no-wait flowshop problem with sequential batching machines by F2 $|s-batch(1,2),\, no-wait|\, C_{\max}$.

7.3.1 NP-Hardness

In this section, we show that problem $F2|s-batch(1,2), no-wait,\ as or ns|C_{\max}$ is NP-hard in the strong sense. The 3-PARTITION problem, which is known to be NP-complete in the strong sense (see Garey and Johnson [12]), is used for the reduction.

Theorem 7.6. *Problem* $F2|s-batch(1,2),\ no-wait, as|C_{\max}$ *is NP-hard in the strong sense.*

Proof. Given an arbitrary instance of 3-PARTITION, we construct the following instance of problem $F2|sum-batch(1,2),\ no-wait, as|C_{\max}$ with $3n$ jobs. The processing times of jobs are $p_j = 3a_j$ and $q_j = a_j$, $j = 1,\ldots,3n$, and the setup times are $s_1 = 0$ and $s_2 = 2W$.

We show that for the constructed instance of the problem $F2|sum-batch(1,2)$, $no-wait, as|C_{\max}$, there exists a schedule S such that $C_{\max} \leq (3n+1)W$, if and only if 3-PARTITION has a solution.

Assume that 3-PARTITION has a solution, and $N_1,\ldots,N_n$ are the required subsets of N. Then the problem $F2|sum-batch(1,2), no-wait, as|C_{\max}$ has a schedule S with makespan $C_{\max}(S) = (3n+1)W$. The feasible schedule S is constructed by forming n batches $B_1,\ldots,B_n$ where $B_i = \{a_j | j \in N_i\}$ for $i = 1,\ldots,n$. The jobs of B_1 are processed on machine M_1 in the interval $[0,\ 3W]$, and the first setup time on machine M_2 is performed in the interval $[W,\ 3W]$. In the interval $[3jW,\ 3(j+1)W]$, for $j = 1,\ldots,n-1$, the jobs of batch B_{j+1} are processed on machine M_1, and machine M_2 processes jobs of batch B_j and the setup of batch B_{j+1}. At time $3nW$, machine M_1 has completed the processing of all batches, and the second machine processes jobs of batch B_n in the interval $[3nW,\ 3(n+1)W]$. Thus the makespan $C_{\max}(S) = 3(n+1)W$.

Conversely, suppose that the desired schedule S exists with makespan $C_{\max} \leq (3n+1)W$. Due to the no-wait constraints, schedule S has consistent batches. Furthermore, there are no more than n batches, otherwise, the makespan of schedule S is greater than the total setup time plus the processing on machine M_2, that is, $C_{\max}(S) \geq 2(n+1)W + nW > (3n+1)W$.

Let $B_1, B_2, \ldots, B_v$ be the batches of schedule S where $v \leq n$. Let $B_{(j)}$ be the batch sequenced at position j of schedule S, then the makespan of S is greater than the sum of the total setup time and processing time on machine M_1 of batches that are processed before batch $B_{(j)}$, plus the setup time on machine M_1 plus the processing time on machine M_1 and M_2 of batch $B_{(j)}$ plus the total setup time plus processing times on machine M_2 of batches processed after $B_{(j)}$. Thus, we have

$$\sum_{i=1}^{j-1}\left(s_1 + \sum_{k\in B_{(i)}} p_k\right) + s_1 + \sum_{k\in B_{(j)}} (p_k + q_k) + \sum_{i=j+1}^{v}\left(s_2 + \sum_{k\in B_{(i)}} q_k\right) \leq (3n+1)W.$$

Substituting the values of setup times and replacing $\sum_{k\in B_{(i)}} p_k$ by $3\sum_{k\in B_{(i)}} q_k$ since $\sum_{k\in B_{(i)}} p_k = 3\sum_{k\in B_{(i)}} q_k$ for $i = 1,\ldots,v$, we obtain

$$3\sum_{i=1}^{j}\sum_{k\in B_{(i)}} q_k + 2(v-j)W + \sum_{i=j}^{v}\sum_{k\in B_{(i)}} q_k \leq (3n+1)W,$$

$$\Rightarrow \quad 3\sum_{i=1}^{v}\sum_{k\in B_{(i)}} q_k - 3\sum_{i=j+1}^{v}\sum_{k\in B_{(i)}} q_k + \sum_{i=j}^{v}\sum_{k\in B_{(i)}} q_k \leq (3n+1-2v+2j)W,$$

since $\sum_{i=1}^{v}\sum_{k\in B_{(i)}} q_k = nW$, we deduce

$$\sum_{k\in B_{(j)}} q_k \leq (1-2v+2j)W + 2\sum_{i=j+1}^{v}\sum_{k\in B_{(i)}} q_k.$$

For $j = v$, the latter inequality reduces to $\sum_{k\in B_{(v)}} q_k \leq W$. For $j = v-1$, we have $\sum_{k\in B_{(v-1)}} q_k \leq -W + 2\sum_{k\in B_{(v)}} q_k \leq W$, thus $\sum_{k\in B_{(v-1)}} q_k \leq W$. Similarly, we obtain $\sum_{k\in B_{(j)}} q_k \leq W$ for $j = v-2, v-3, \ldots, 1$. Furthermore, we have $\sum_{i=1}^{v}\sum_{k\in B_{(i)}} q_k = nW$, thus we obtain $v = n$ and $\sum_{k\in B_{(j)}} q_k = W$, for $j = 1,\ldots,n$. Thus batches $B_{(j)}$, $j = 1,\ldots,n$, form solution to the 3-PARTITION problem. □

Corollary 7.1. *The problem $F2|sum-batch(1,2), no-wait, ns|C_{\max}$ is NP-hard in the strong sense.*

$$\begin{cases} \min C_{(n)} & & \\ \text{s.t.} \;\; C_{(i)} - C_{(i-1)} \geq \sum_{j=1}^{n} q_j x_{i,j} & (i = 1,\ldots,n), & (7.1) \\ \quad\;\; C_{(i)} - C_{(i-1)} \geq \sum_{j=1}^{n} (p_j + q_j) x_{i,j} - \sum_{j=1}^{n} q_j x_{i-1,j} & (i = 1,\ldots,n), & (7.2) \\ \quad\;\; \sum_{i=1}^{n} x_{i,j} \quad = 1 & (j = 1,\ldots,n), & (7.3) \\ \qquad\qquad C_{\pi(i)} \geq 0 \,;\, x_{i,j} \in \{0,1\}, & (i,j = 1,\ldots,n), & \\ \qquad\qquad C_{\pi(0)} = 0 \,;\, x_{0,j} = 0, & (j = 1,\ldots,n). & \end{cases}$$

7.3.2 Integer Programming Formulation

In this section, we propose an integer linear programming formulation for the problem $F2|sum - batch(1,2), no - wait, as|C_{\max}$, then from this formulation we deduce a second compact integer linear formulation that will be used to solve the problem by Lagrangian relaxation. In the rest of this section denote by $P(n)$ the problem $F2|sum - batch(1,2), no - wait, as|C_{\max}$.

Let I be an instance of the problem $P(n)$, each feasible schedule S of instance I contains at most n batches. The problem consists of partitioning the set of jobs into at most n batches that minimize the makespan. Let $B_{(i)}$ the batch sequenced at position i of schedule S, and let $C_{(i)}$ be the completion time of batch $B_{(i)}$ on the second machine.

Let $x_{i,j}$ be a decision variable that equal to 1 if job j is assigned to batch $B_{(i)}$ and 0 otherwise, $i, j = 1,\ldots,n$. W.l.o.g. we assume that the setup times are equal to zero on both machines. In the following, an integer linear programming formulation of the problems $P(n)$ is given

Constraints (7.1) and (7.2) ensure the no-wait processing of batches on machines M_1 and M_2. Constraint (7.3) ensure the assignment of each job to one batch.

Given the formulation of problem $P(n)$, the total processing time of schedule S is given by

$$\begin{aligned} C_{\pi(n)} &= C_{\pi(0)} + \sum_{i=1}^{n} \max\left\{ \sum_{j=1}^{n} q_j x_{i,j} \,;\, \sum_{j=1}^{n} (p_j + q_j) x_{i,j} - \sum_{j=1}^{n} q_j x_{i-1,j} \right\} \\ &= \sum_{i=1}^{n} \sum_{j=1}^{n} q_j x_{i,j} + \sum_{i=1}^{n} \max\left\{ 0 \,;\, \sum_{j=1}^{n} p_j x_{i,j} - \sum_{j=1}^{n} q_j x_{i-1,j} \right\} \\ &= \sum_{j=1}^{n} q_j + \sum_{i=1}^{n} \max\left\{ 0 \,;\, \sum_{j=1}^{n} p_j x_{i,j} - \sum_{j=1}^{n} q_{2,j} x_{i-1,j} \right\}. \end{aligned}$$

Let $w_i = \max\left\{ 0 \,;\, \sum_{j=1}^{n} p_j x_{i,j} - \sum_{j=1}^{n} q_j x_{i-1,j} \right\}$. Then the new integer linear program is given by (IP) program

$$(IP)\begin{cases} \min \sum_{i=1}^{n} w_i \\ s.t.\ \ w_i \geq \sum_{j=1}^{n} p_j x_{i,j} - \sum_{j=1}^{n} q_j x_{i-1,j} & (i=1,\ldots,n), \\ \sum_{i=1}^{n} x_{i,j} = 1 & (j=1,\ldots,n), \\ \quad w_i \geq 0,\ \ x_{0,j}=0,\ \ x_{i,j}\in\{0,1\},\ (i,j=1,\ldots,n). \end{cases}$$

7.3.3 Lagrangian Relaxation Heuristic

In this section, we derive a Lagrangian relaxation of IP formulation and we use the subgradient heuristic to solve the relaxed problem. Consider the following linear formulation obtained by Lagrangian relaxation of constraints $w_i \geq \sum_{j=1}^{n} p_j x_{i,j} - \sum_{j=1}^{n} q_j x_{i-1,j}$, $i=1,\ldots,n$, with Lagrange multipliers λ_i, $i=1,\ldots,n$.

$$\begin{cases} \text{Min}\ \ L_\lambda(x,w) = \sum_{i=1}^{n}(1-\lambda_i)w_i + \sum_{i=1}^{n}\sum_{j=1}^{n} \lambda_i p_j x_{i,j} - \sum_{i=1}^{n}\sum_{j=1}^{n} \lambda_i q_j x_{i-1,j} \\ s.c.\ \ \sum_{i=1}^{n} x_{i,j} = 1,\ \ (j=1,\ldots,n) \\ \quad x_{0,j}=0,\ 0\leq x_{i,j}\leq 1,\ \ (i,j=1,\ldots,n) \end{cases}$$

For given vector of multipliers $\lambda \in \mathbb{R}^n_+$, denote $L(\lambda)$ the Lagrangian function,

$$L(\lambda) = \min\left\{ \sum_{i=1}^{n}(1-\lambda_i)w_i + \sum_{i=1}^{n}\sum_{j=1}^{n} \lambda_i p_j x_{i,j} - \sum_{i=1}^{n}\sum_{j=1}^{n} \lambda_i q_j x_{i-1,j}\ : \right.$$

$$\left. \sum_{i=1}^{n} x_{i,j} = 1,\ \ (j=1,\ldots,n), x_{0,j}=0,\ 0\leq x_{i,j}\leq 1\ \ i,j=1,\ldots,n \right\}.$$

Since $L(\lambda)$ is lower bound of the optimal solution of the problem (IP), then the best value of vector λ is obtained by solving the so-called Lagrangian dual problem L^* defined by:

$$L^* = \max_{\lambda\geq 0} L(\lambda).$$

Remark that for given vector $\lambda \in \mathbb{R}^n_+$, the Lagrangian function $L(\lambda)$ is decomposable function of $x_{i,j}$ and w_i, $(i,j=1,\ldots,n)$, that is, $L(\lambda) = L_x(\lambda) + L_w(\lambda)$ where,

$$L_w(\lambda) = \min_{\lambda\geq 0}\left\{ \sum_{i=1}^{n}(1-\lambda_i)w_i\ :\ \ w_i\geq 0,\ \ i=1,\ldots,n \right\}$$

and

$$L_x(\lambda) = \min_{\lambda \geq 0} \left\{ \sum_{j=1}^{n} \left(\lambda_n p_j x_{n,j} + \sum_{i=1}^{n-1} (\lambda_i p_j - \lambda_{i+1} q_j) x_{i,j} \right) : \right.$$

$$\left. \sum_{i=1}^{n} x_{i,j} = 1, \ \ j = 1,\ldots,n; \ \ x_{0,j} = 0, \ 0 \leq x_{i,j} \leq 1, \ i,j = 1,\ldots,n \right\}.$$

Given this decomposition, we have the following result.

Lemma 7.3. *The optimal value of the Lagrangian dual problem L^* is given by the maximum value of $L_x(\lambda)$, where $0 \leq \lambda_i \leq 1$ ($i = 1,\ldots,n$).*

Proof. If there exist some λ_i, $i = 1,\ldots,n$, where $\lambda_i > 1$, then the optimization problem $\min\{\sum_{i=1}^{n}(1-\lambda_i)w_i, \ w_i \geq 0, \ i = 1,\ldots,n\}$ is unbounded and $L_w(\lambda) = -\infty$. Furthermore, for $\lambda_i \leq 1$, $i = 1,\ldots,n$ the optimal value of w_i is equal to zero and $L^*_w(\lambda) = 0$. Thus, the optimal value of Lagrangian dual is given by the optimal value of $L_x(\lambda)$, that is, $L^* = \max_{0 \leq \lambda \leq 1} L_x(\lambda)$. □

Denote $i^*(j) = \arg\min\{\min_i\{\lambda_i p_j - \lambda_{i+1} q_j\} \ ; \ \lambda_n p_j\}$, it is easy to verify that $L_x(\lambda)$ have the optimal solution the set $X = \{x_{i,j} \,|\, x_{i,j} = 1$ if $i = i^*(j)$ and 0 otherwise, $i,j = 1,\ldots,n\}$, and this solution is given in $O(n)$.

7.3.3.1 Solving the Dual Lagrangian Problem

In order to compute the optimal value of dual Lagrangian problem, we use the subgradient optimization algorithm to determine the vector of Lagrangian multipliers, and thus the best solution for the problem IP. Let *LB* be the best lower bound of the dual Lagrangian problem and let *UB* be the best feasible solution to problem *IP*. Initially, *LB* and *UB* are set to $-\infty$ and $+\infty$, respectively.

Given a vector of multipliers λ^k of the kth iteration, let x^k_λ be the solution of Lagrangian function $L_x(\lambda)$ at the kth iteration. A feasible solution of problem *IP* is obtained by *Gilmore-Gomory algorithm* [14] applied on processing times of batches constructed by the feasible assignment solution x^k_λ of jobs to batches. Let $GG(x^k_\lambda)$ be the makespan of the obtained schedule.

At the kth iteration, if x^k_λ satisfy all relaxed constraints, that is,

$$\forall i, \quad \sum_{j=1}^{n} p_j x^k_{i,j} - \sum_{j=1}^{n} q_j x^k_{i-1,j} \leq 0$$

then x^k_λ is optimal solution of the problem *IP*, otherwise, the vector of multipliers of the $(k+1)$th iteration is calculated by a subgradient R^k_i defined as $R^k_i = \sum_{j=1}^{n} p_j x^k_{i,j} - \sum_{j=1}^{n} q_j x^k_{i-1,j}$, then we calculate the value of Lagrangian function $L_x(\lambda)$ with a new the vector multipliers. Two criteria are used to stop the subgradient optimization method, namely, the total number of iterations without improving of the best solution, and when the variation of vector of multipliers does not improve the solution already found.

7.3.3.2 Lower Bound

Since the lower bound for the problem IP, obtained by solving the Lagrangian function $L_x(\lambda)$, may take negative value, we propose a new lower bound described as the largest value between: the shortest processing time of jobs on the first machine plus the total processing times of all jobs on the second machine, and the total processing time of all jobs on the first machine plus the shortest processing time of jobs on the second machines, that is,

$$LB = \max\left\{\min_j\{p_{1,j} \,:\, j \in N\} + \sum_{k=1}^{n} p_{2,k}\,;\, \min_j\{p_{2,j} \,:\, j \in N\} + \sum_{k=1}^{n} p_{1,k}\right\}.$$

Finally the subgradient optimization algorithm is described in Algorithm Subgradient optimization.

7.3.4 Computational Experiments

We performed a series of computational experiments to evaluate the efficiency of the subgradient optimization algorithm. We generated the number of jobs n in the set $\{10,20,50,200,500\}$. For each value of n, we generated 20 instances. The processing times are random integers uniformly distributed in $[10,100]$.

We compared the subgradient algorithm with the lower bound LB as described above. We calculated the average distance between the solution obtained by subgradient algorithm, denoted $C_{\max}(SG)$, and the lower bound LB. This distance is given by:

$$D_{\text{avg}} = \frac{C_{\max}(SG) - LB}{LB}.$$

We also calculated the average CPU time T_{avg} for which the solution of subgradient algorithm is obtained. Table 7.3 reports the result of computational experiments.

From the Table 7.3, we can see that the subgradient algorithm gives good results. For all tested instances, we note that the relative distance between solutions provided by subgradient algorithm and lower bound are between 7% and 10%. We also note that the subgradient algorithm is very fast. Indeed, for instances of 500 jobs, the CPU time does not exceed 17 s. We can say in conclusion that this subgradient algorithm gives efficient results.

Algorithm Subgradient optimization

1. **Initialization.** $\lambda = \lambda^0$, $LB = -\infty$, $UB = +\infty$, $\rho^0 = 0.01$, $Nb\_iter = 50$, $iter = 1$, $Total\_iter = 0$, $Total = 200$, $\rho = \rho^0$, $k = 1$.
2. **Iteration** k.

   i. Calculate the solution of Lagrangian function $L_x(\lambda^k)$.
   $x^k_{i,j} = 1$ if $i = \arg\min\left\{\min_i\{\lambda^k_i p_j - \lambda^k_{i+1} q_j\}\ ;\ \lambda^k_n p_j\right\}$ and 0 otherwise.

   ii. **If** x^k is feasible solution for problem (IP), that is, $\forall i$,

$$\sum_{j=1}^{n} p_j x^k_{i,j} - \sum_{j=1}^{n} q_j x^k_{i-1,j} \leq 0$$

   then x^k is an optimal solution of the problem (IP), go to (4).
   Otherwise,

   i. Update lower bound: if $LB < L^*_{x^k}(\lambda^k)$ then, $LB = L^*_{x^k}(\lambda^k)$, $iter = 0$
   ii. Construct a feasible solution to problem (IP): calculate $GG(x^k)$. If $UB > GG(x^k)$ then $UB = GG(x^k)$ and $iter = 0$,
   iii. Update the vector of multipliers:

$$R^k_i = \sum_{j=1}^{n} p_{1,j} x^k_{i,j} - \sum_{j=1}^{n} p_{2,j} x^k_{i-1,j} \quad i = 1,\ldots,n,$$

$$\theta = \rho \frac{UB - L^*_x(\lambda^k, x^k)}{\sum_{i=1}^{n}(R^k_i)^2},$$

$$\lambda^{k+1} = \min\left\{1\ ;\ \max\left\{0\ ;\ \lambda^k + \theta \sum_{i=1}^{n} R^k_i\right\}\right\},$$

   $Total\_iter = Total\_iter + 1$, if $iter \geq Nb\_iter$ then $\rho = \frac{\rho}{2}$ and $iter = 0$, otherwise $iter = iter + 1$. Go to (3).

3. **Stopping criteria**
   If $\rho \leq \rho^0$ or $Total\_iter \geq Total$ go to (4), otherwise go to (2).
4. **Termination.**
   x^k is the best solution of the problem (IP) and $GG(x^k)$ is the value of its makespan.

Table 7.3: Computational results

	D_{avg}	T_{avg} in seconds
$n = 10$	0.0773	2.11
$n = 20$	0.0794	4.68
$n = 50$	0.0915	5.78
$n = 200$	0.0957	12.34
$n = 500$	0.1091	16.16

7.4 Conclusion

Scheduling problems with batching operation becomes more considered problems in the literature due to several applications. In this chapter, the scheduling problems with batching operation are presented for both types of batching models (serial batching and parallel batching) and the literature is reviewed, in which the most problems are NP-hard. In this chapter, we concentrate on the extension of flowshop scheduling problems with batching machines. Two problems are considered. In the first problem, we considered the flowshop scheduling consisting of two batching machines. The first machine is p-batching machine whereas the second one is s-batching machine. In the second problem, we considered the flowshop scheduling consisting of two s-batching machines. In both problems, the no-wait constraint is imposed on the processing of batches between machine under the makespan performance measure. We showed that both problems are NP-hard, and we focused on development of polynomial algorithms for spacial cases, and heuristics for general problems. There are various interesting topics related to shop scheduling with batching, for future research. For instance, developing approximation algorithms for problems with more than two machines and integrating no-wait constraints. Another interesting research topic is to connect no-wait batch scheduling with travel salesman problem. In the classical no-wait flowshop the problem is modeled as TSP problem. It will be challenging to model no-wait batch scheduling as special case of TSP problem.

References

1. Albers, S., Brucker, P.: The complexity of one-machine batching problems. Discrete Applied Mathematics, **47**, 87–107 (1993)
2. Bellanger, A., Oulamara, A.: Scheduling hybrid flowshop with parallel batching machines and compatibilities. Computers and Operations Research **36**(6), 1982–1992 (2009)
3. Boudhar, M.: Dynamic scheduling on a single batch processing machine with split compatibility graphs. Journal of Mathematical Modelling and Algorithms **2**, 17–35 (2003)
4. Boudhar, M., Finke, G.: Scheduling on a batch machine with job compatibilities. Belgian journal of operations research, statistics and computer science **40**, 69–80 (2000)
5. Brucker, P., Gladky, A., Hoogeveen, J.A., Kovalyov, M.Y., Potts, C.N., Tautenhahn, T., Van de Velde, and S.L.: Scheduling a batch processing machine. Journal of Scheduling **1**, 31–54 (1998)
6. Brucker, P., Kovalyov, M.Y.: Single machine batch scheduling to minimize the weighted number of late jobs. Mathematical Methods of Operations Research **43**, 1–8 (1996)
7. Cheng, T.C.E., Liu, Z., Yu, W.: Scheduling jobs with release dates and deadlines on a batch processing machine. IIE Transactions **33**(8), 685–690 (2001)
8. Coffman Jr., E.G., Yannakakis, M., Magazine, M.J., Santos, C.A.: Batch sizing and job sequencing on a single machine. Annals of Operations Research **26**, 135–147 (1990)
9. Deng, X., Feng, H., Zhang, P., Zhang, Y., Zhu, H.: Minimizing mean completion time in a batch processing system. Algorithmica **38**(4), 513–528 (2004)
10. De Werra, D., Demange, M., Escoffier, B., Monnot, J., Paschos, V. Th.: Weighted coloring on planar, bipartite and split graphs : Complexity and approximation. Discrete Applied Mathematics **157**(4), 819–832 (2009)

11. Edmonds, J.: Paths, trees, and flowers. Canadian Journal of Mathematics **17**, 449–467 (1965)
12. Garey, M.R., Johnson, D.S.: Computers and Intractability: A Guide to NP-completeness. W.H. Freeman, San Francisco (1979)
13. Gavranovic, H., Finke, G.: Graph partitioning and set covering for optimal design of production system in the metal industry. In Proceedings of the second Conference on Management and Control of Production and Logistics - MCPL'00, Grenoble (2000)
14. Gilmore, P.C., Gomory, R. E.: Sequencing a one state-variable machine: A solvable case of the traveling salesman problem. Operations Research **12**, 655–679 (1964)
15. Glass, C.A., Potts, C.N., Strusevich, V.A.: Scheduling Batches with Sequential Job Processing for Two-Machine Flow and Open Shops. Informs Journal on computing **13**(2), 120–137 (2001)
16. Graham, R.L., Lawler, E.L., Lenstra, J.K., Rimooy Kan, A.H.G.: Optimization and approximation in deterministic sequencing and scheduling : a survey. Annals of Discrete Mathematics **5**, 287–326 (1979)
17. Oulamara, A., Finke, G., Kamgaing Kuiten, A.: Flowshop scheduling problem with batching machine and task compatibilities. Computers & Operations Research **36**(2), 391–401 (2009)
18. Oulamara, A., Kovalyov, M.Y., Finke, G.: Scheduling a no-wait flowshop with unbounded batching machines. IIE Transactions on Scheduling and logistics **37**(8), 685–696 (2005)
19. Potts, C.N., Kovalyov, M.Y.: Scheduling with batching : A review. Journal of Operational Research **120**(2), 228–249 (2000)
20. Potts, C.N., Strusevich, V.A., Tautenhahn, T.: Scheduling batches with simultaneous job processing for two-machine shop problems. Journal of Scheduling **4**(1), 25–51 (2001)
21. Potts, C.N., Van Wassenhove, L.N.: Integrating scheduling with batching and lotsizing : A review of algorithms and complexity. Operations Research Society **42**, 395–406 (1992)
22. Uzsoy, R., Yang, Y.: Minimizing total weighted completion time on a single batch processing machine. Production and Operations Management **6**(1), 57–73 (1997)
23. Webster, S., Baker, K.R.: Scheduling groups of jobs on a single machine. Operations Research **43**, 692–704 (1995)

Part II
Just-in-Time Philosophy: Interaction with Other Areas

Chapter 8
Just-in-Time Scheduling in Modern Mass Production Environment

Joanna Józefowska

Abstract The main goal of just-in-time production planing is the reduction of the in-process inventory level. This goal may be achieved by completing the items as close to their further processing (or shipment) dates as possible. In the mass production environment it is too costly to define and control due dates for individual items. Instead, the model proposed in Toyota is applied that assumes monitoring the actual product rate of particular products. The objective is to construct schedules with minimum deviation from an ideal product rate. In the approach aimed at minimization of the Product Rate Variation (PRV), the control process concentrates on product types, not individual items. In this chapter we discuss the PRV model and scheduling algorithms developed to solve this problem with two types of objectives: to minimize the total or maximum deviation from the ideal product rate. We present algorithms proposed in the context of the just-in-time production scheduling as well as in other areas, adopted later to solve the PRV problem. One of the most interesting problems discussed in this context is the apportionment problem. Originally, the PRV problem was defined as a single machine scheduling problem. We show that some algorithms can be generalized to solve the parallel-machine scheduling problem as well.

8.1 Introduction

The problem of just-in-time scheduling has been considered since the new production planning concept called just-in-time was introduced in Toyota [17]. The main goal of this approach was to improve the business return of investment by reducing the in-process inventory cost. Just-in-time system is often characterized as a system where only the necessary products are produced in the necessary quantities at the necessary time.

Joanna Józefowska
Poznań University of Technology, Poznań, Poland,
e-mail: joanna.jozefowska@cs.put.poznan.pl

R.Z. Ríos-Mercado and Y.A. Ríos-Solís (eds.), *Just-in-Time Systems*, Springer Optimization and Its Applications 60, DOI 10.1007/978-1-4614-1123-9_8,

As a result of introducing this approach new optimization problems occurred and, consequently, new models were introduced to the theory of scheduling. In general, two different approaches to this problem may be observed in the literature. In the first approach due dates are defined explicitly for each task and the objective is to minimize the total cost resulting from scheduling tasks before or after their due dates. The second approach aims at constructing a schedule which at any time maintains a certain given proportion between the number of items of all products completed up to that time. A survey of models and algorithms in this field is presented in [9].

Mass production is defined as the production of large amounts of standardized products especially on the production or assembly lines. The concept of mass production is applied to various kinds of products, in particular, to discrete solid parts and to assemblies of such parts such as household appliances and automobiles. Mass production is often characterized as extremely inflexible, while modern markets demand diverse products available in relatively small amounts. In consequence, the production and assembly lines had to become more flexible. There are two main ways to achieve this flexibility while maintaining some benefits of mass production. The first way is the design of products based on a modular construction. A good example is the automotive industry where many diverse models of a car may be obtained by simply exchanging the engine (Diesel or gasoline engine), gearbox (manual or automatic), and various kinds of additional equipment. Since the assembly of modules is very similar all different types may be assembled on the same line. Another way that allows more flexibility in production and assembly lines is the flexibility of machines. The machine flexibility is a result of technological development and reflects the system's ability to be changed (or set-up) to produce new product types, and ability to change the order of operations executed on a part. As a result the production and assembly lines are capable to produce relatively small amounts of diverse products maintaining the efficiency of mass production. They are called flexible production lines or mixed-model assembly lines.

The just-in-time concept may also be addressed in the environment of mixed-model assembly lines, as for example in [16]. One of the optimization problems to be solved in order to effectively utilize this type of assembly lines is determination of the sequence for producing different products on the line [18]. Let us consider the Bill of Materials (BOM) of a product completed on the line. It defines the number of items of each part type required to complete the final product. The number of items may be quite big in case of complex products. In a just-in-time system, we would like to have low inventories of particular part types and keep the line operating without any idle time caused by the shortage of the components. One could build a schedule of products and then, following the BOM assign due dates to all items of various part types. However, assigning each item a separate due date following from the schedule of products on the line may be too complex from the organizational point of view. Therefore, the objective considered in case of flexible assembly lines, is to keep a constant rate of usage of every part used by the line. The problem is to find a sequence of products completed on the line such that the product rate (i.e. the average number of products of a given type completed in a time unit) remains constant over time. Observe that this objective reflects the general just-in-time goal, since if parts are delivered at a constant rate and product rate variation is minimized then the in-process inventory stays below a given threshold. The scheduling problem resulting from this approach is called the Product Rate Variation (PRV) problem.

The PRV problem occurs also in supply chain management as well as in computer systems like real-time systems [11] or packet scheduling [12] and may be also considered as a *fair division problem*. The fair division problem is defined in the context of discrete resources which should be allocated among users proportionally to the weights of particular users. One of the most common problems examined in this area is the apportionment problem. The oldest apportionment problems relate to representative democracy, and, in particular, deal with deciding on the number of representatives a state is entitled to in a legislative body. Balinski and Young [3] developed a systematic analysis of this problem and proved some important properties of the solution methods. Their *theory of apportionment* provides tools for analysis of the fair division algorithms, in particular the PRV solution methods. An extensive survey of results and applications of the fair division problem may be found in [13].

In this chapter, we discuss the Product Rate Variation problem and we propose an extension of the original single-machine model to the case of multiple identical parallel machines. We show that this generalization opens a new interesting research area in the field of just-in-time scheduling.

In Sect. 8.2, we introduce the basic mathematical model of the single machine case and a transformation of the just-in-time scheduling problem to the apportionment problem. Section 8.3 contains a short review of algorithms existing for the single machine case. Finally, in Sect. 8.4, we formulate the multi-machine problem and propose solution algorithms based on two known approaches to the minimization of total and maximum deviation. A summary of results and directions for further research are provided in Sect. 8.5.

8.2 Mathematical Model

We consider a scheduling problem that occurs in an assembly line where n different products are manufactured. The number of items of each product to be completed in the considered period is known and it is assumed that processing time of each item takes one unit of time. The goal is to find such a schedule that at each time unit the proportion of items of particular products completed by that time unit remains constant. The problem was formulated as a mixed integer optimization problem by Miltenburg [16]. It was named the Product Rate Variation Problem by Kubiak and Sethi [14]. Below we present the basic formulation with the most often considered objective functions. This problem may be transformed to the problem of apportionment examined by Balinski and Young [4]. This transformation leads to interesting observations and provides a systematic characteristics of solution approaches to the considered problem. This similarity was observed by Balinski and Shahidi [2] and Balinski and Ramirez [1] and further developed by Jzefowska, Jzefowski, and Kubiak in [10]. We present these results in Sect. 8.2.2.

8.2.1 Basic Model

Let us assume that n products are planned for production in the considered time period. For each product the number of items to be produced d_i is given by the Master Production Schedule. We assume that processing time of a single item of any product takes one time unit. Thus, the time period required to complete all items equals $D = \sum_{i=1}^{n} d_i$, and so $t \in \{1,\ldots,D\}$. The proportion of each product, called *product rate*, equals $r_i = \frac{d_i}{D}$. The objective is to find a schedule on a single machine such that for each product i at each time unit t the ratio of the number of items of product i completed by time unit t to the total number of items produced up to time t is as close as possible to r_i. Under these assumptions, an ideal production volume of product i completed up to time t is calculated as tr_i. Let us denote by x_{ik} the actual number of items of product i completed by time unit t, $i = 1,\ldots,n$, $t = 1,\ldots,D$. The objective function may be formulated in several ways. The most popular ones are listed below:

- Total squared deviation, introduced by Miltenburg [16],

$$\text{minimize} \quad \sum_{i=1}^{n}\sum_{t=1}^{D}(x_{it} - tr_i)^2. \tag{8.1}$$

- Total deviation, also introduced by Miltenburg [16],

$$\text{minimize} \quad \sum_{i=1}^{n}\sum_{t=1}^{D}|x_{it} - tr_i|. \tag{8.2}$$

- Maximum deviation, proposed by Steiner and Yeomans [19],

$$\text{minimize} \quad \max_i \max_t |x_{it} - tr_i|. \tag{8.3}$$

A feasible solution is obtained if the following constraints (8.4)–(8.7) are satisfied:

$$\sum_{i=1}^{n} x_{it} = t, \quad t = 1,\ldots,D, \tag{8.4}$$

$$0 \le x_{it+1} - x_{it} \le 1, \quad i = 1,\ldots,n, \; t = 1,\ldots,D-1, \tag{8.5}$$

$$x_{iD} = d_i, \quad i = 1,\ldots,n, \tag{8.6}$$

$$x_{it} \in \mathbf{Z}^+ \cup \{0\}, \quad i = 1,\ldots,n, \; t = 1,\ldots,D. \tag{8.7}$$

Constraints (8.4) guarantee that exactly t items are completed in the time interval from 1 through t. Constraints (8.5) express the cumulative character of variables (x_{it} may not decrease while t increases) and prevent from assigning more than one item in a single time unit (x_{it} may increase at most by 1 in a single time unit). Finally, constraints (8.6) ensure that the demand of all products following from the production plan will be fulfilled. Finally, constraints (8.7) restrict the set of decision variables to nonnegative integers $x_{it}, i = 1,\ldots,n, t = 1,\ldots,D$.

In case of the objective functions (8.1)–(8.3) an optimal solution may be found in $O(nD)$ time. We present the relevant algorithms in Sect. 8.3.

8.2.2 Just-in-Time Scheduling and the Apportionment Problem

The most common apportionment problems relate to representative democracy, and, in particular, deal with deciding on the number of representatives a state is entitled to in a legislative body. The number of seats a state receives in the body should be proportional to the population of the state. This problem was first encountered in the context of deciding the representations of states in the US parliament. The problem is formulated as follows. Consider n states and a vector of nonnegative integers (populations) p_i, $i = 1,\ldots,n$, and a house of size $h \geq 0$. The problem is to find an allocation for h, i.e. a vector of nonnegative integers a_i, $i = 1,\ldots,n$, such that $\sum_{i=1}^{n} a_i = h$. The first method approved by the US House of Representatives was the Hamilton method. Its author, Alexander Hamilton, was the first US Secretary of Treasury. The Hamilton method is defined as follows.

Hamilton method of apportionment

> **begin**
> $SD := \frac{\sum_{i=1}^{n} p_i}{h}$; % Calculate the Standard Divisor
> **for** i:=1 **to** n **do**
>   **begin**
> $SQ[i] := \frac{p_i}{SD}$; % Calculate the Standard Quotas
> $a[i] := \lfloor SQ[i] \rfloor$;
>   **end**
> renumber the states so that:
> $SQ[1] - a[1] \geq SQ[2] - a[2] \geq \ldots \geq SQ[n] - a[n] \geq 0$;
> **while** $\sum_{i=1}^{n} a[i] < h$ **do**
>   **begin**
> $a[i] := a[i] + 1$;
> $i := i + 1$;
>   **end**
> **end**

Although the Hamilton method looks reasonably fair it has some disadvantage that was discovered in 1880 and is known as the Alabama Paradox. Namely, the chief clerk of the U.S. Census Office, C. W. Seaton computed apportionments for all house sizes between 275 and 350 members. He then pointed out that according to the Hamilton method in a house of 299 members Alabama has 8 seats, while in a house of 300 members Alabama receives only 7 seats. Such method is clearly unfair. Even before this evidence was published, Webster method was adopted in 1842. Initially the Webster method was defined as follows.

Webster method

begin
$SD := \frac{\sum_{i=1}^{n} p_i}{h}$; % Calculate the Standard Divisor
for $i := 1$ **to** n **do** $a[i] := round\left(\frac{p_i}{SD}\right)$;
while $\sum_{i=1}^{n} a[i] \neq h$ **do**
  begin
  $SD := MD$;
  for $i := 1$ **to** n **do** $a[i] := round\left(\frac{p_i}{SD}\right)$;
  end;
end

The value of the so called *modified divisor MD* was found in a trial and error procedure.

Only in the second half of the twentieth century Balinski and Young [3] provided formal analysis of the apportionment problem and extensive characteristics of the apportionment methods. A method of apportionment $M(p,h)$ is a rule that for any vector p and house size h returns a vector of apportionments a. We write in this case $a \in M(p,h)$. Balinski and Young define several properties of apportionment methods. One of them is called *house monotonicity*. Any apportionment method such that $a \in M(p,h)$ and $a' \in M(p,h+1)$ with $a' \geq a$ is house monotone. Notice, that the Alabama paradox does not occur if a house monotone method is applied. Obviously, the Hamilton method is not house monotone. Another important property is *staying within the quota*. A method of apportionment M stays within the quota if all its allocations satisfy the following inequalities:

$$\left\lfloor \frac{p_i h}{\sum_{i=1}^{n} p_i} \right\rfloor \leq a_i \leq \left\lceil \frac{p_i h}{\sum_{i=1}^{n} p_i} \right\rceil . \tag{8.8}$$

An important class of apportionment methods applicable to solve scheduling problems are divisor methods. A divisor method is based on a so-called *divisor function* $d(a)$ that is any increasing real valued function defined for all integers $a \geq 0$ such that $a \leq d(a) \leq a+1$ and such that there exists no pair of integers $b \geq 0$ and

$c \geq 1$ with $d(b) = b+1$ and $d(c) = c$. Balinski and Young [3] proposed the following iterative procedure finding an apportionment according to a divisor method with divisor function $d(a)$.

Divisor method

> **begin**
> **for** $i := 1$ **to** n **do** $a[i] := 0$; % Initialize
> **while** $h > 0$ **do**
>   **begin**
>   find k such that $\frac{p_k}{d(a_k)} \geq \frac{p_i}{d(a_i)}$ for any i;
> $a[k] := a[k] + 1$; % assign a seat to state k
> $h := h - 1$; % reduce the number of unassigned seats
>   **end**
> **end**

Balinski and Young [3] showed that the Webster method is a divisor method of apportionment with divisor function $d(x) = x + 0.5$.

The most important result proved by Balinski and Young [3] is the so called *Impossibility Theorem* saying that there exists no method of apportionment that is population monotone and stays within the quota. Since any divisor method of apportionment is house monotone, none of them stays within the quota.

It is quite easy to notice that the problem of apportionment is very similar to the PRV problem. Namely, the products may be considered states, the time period of length t may be considered the house size and the demand vector d_i corresponds to the population vector p_i. Then an apportionment a_i corresponds to x_{it}, i.e. the number of items of product i completed up to time t. There is one important issue: a method $M(p,h)$ finds a feasible solution of the corresponding PRV problem only if it is a house monotone method.

8.3 Solution Algorithms

Solution algorithms for the problem of minimization of the Product Rate Variation were proposed by Miltenburg [16], Kubiak and Sethi [14], and Steiner and Yeomans [19]. Similar problems were considered by Tijdeman [21] and Inman and Bulifin [8] and the algorithms they proposed may be also used to solve the PRV problem. Finally, since Bautista, Companys and Corominas [5] proved that the Inman–Bulfin algorithm is equivalent to the Webster method of apportionment some research began to examine the properties of the scheduling algorithms as apportionment methods. Summary of results in this field may be found in [10].

8.3.1 Minimizing the Total Deviation

Minimization of the total squared deviation and the total deviation were the first objectives considered in the context of the just-in-time scheduling by Monden [17]. However, the solution approach proposed in [17] was very time consuming. A more efficient optimization algorithm for the problem (8.2), (8.4)–(8.7) was developed by Kubiak and Sethi [14]. This algorithm is polynomial in D. Moreover, the Webster method of apportionment also minimizes the total and the total squared deviation. It is worth mentioning that the Hamilton method minimizes the total deviation as well; however, it is not a house monotone method, so it cannot be applied to solve a scheduling problem.

8.3.1.1 Miltenburg Algorithms

Miltenburg [16] proposed an enumerative algorithm for the problems of minimizing the total and total squared deviation. This algorithm has exponential worst case complexity. Thus, he also proposed two heuristics with complexity $O(nD)$ and $O(n^2D)$.

8.3.1.2 Webster Method of Apportionment

The Webster method of apportionment finds optimal solution for any objective function being an l_p norm of $|x_{it} - tr_i|$, thus also function (8.1). Since no divisor method of apportionment stays within the quota the Webster method does not always find an optimal solution for the objective function (8.3).

8.3.1.3 Kubiak–Sethi Algorithm

The algorithm proposed by Kubiak and Sethi in [14] and [15] finds an optimal solution of the problem with objective function (8.1) and constraints (8.4)–(8.7). The main idea is to reduce the problem to a corresponding assignment problem, which is polynomially solvable. Let us now present the reduction. For each item j of model i an ideal position t in the schedule is calculated as follows from Fig. 8.1.

Let us consider the curves corresponding to the deviation $|k - tr_i|$ of item $k = j - 1$ and $k = j$ of the same product i. In the interval $[0, Z^i_j]$, a smaller deviation is obtained for $j - 1$ items scheduled, while in the interval $[Z^i_j, D]$ it is desirable to schedule j items of the product what is shown in a bold line. The ideal moment to schedule the j-th copy is calculated as the intersection point of curves $tr_i - j$

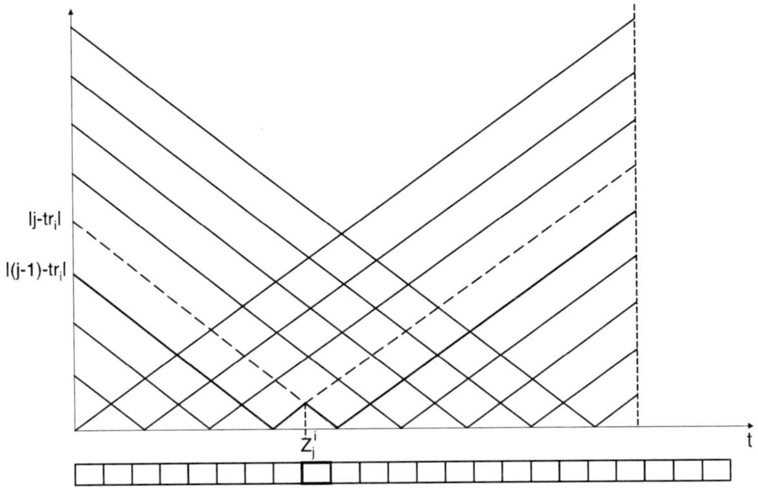

Fig. 8.1: The idea of calculating the ideal completion time of item j

and $j-1-tr_i$. Obviously, it may not be an integer. Thus, the best feasible completion time of item j is calculated as:

$$Z_j^{i*} = \left\lceil \frac{2j-1}{2r_i} \right\rceil. \tag{8.9}$$

Now, let us denote by C_{jt}^i the cost of completing the j-th item of product i in time unit t, $i = 1,\dots,n$, $j = 1,\dots,d_i$, $t = 1,\dots,D$. If $t < Z_j^{i*}$ then item j is completed too early and at each time unit l such that $t \le l < Z_1^{i*}$ we add to the objective function a deviation exceeding the optimal one by $(j-lr_i)^2 - (j-1-lr_i)^2$. Symmetrically, if $t > Z_j^{i*}$ then item j is completed too late and at each time unit l such that $Z_1^{i*} \le l < t$ we add to the objective function a deviation exceeding the optimal one by $(j-1-lr_i)^2 - (j-lr_i)^2$. This additional deviation is denoted by ψ_{jl}^i and calculated as follows:

$$\psi_{jl}^i = |(j-1-lr_i)^2 - (j-lr_i)^2| \tag{8.10}$$

for $i = 1,\dots,n$, $j = 1,\dots,d_i$, $t = 1,\dots,D$. Finally, the cost C_{jt}^i of scheduling the j-th copy of product i in time unit t is calculated by the following formula:

$$C_{jt}^i = \begin{cases} \sum_{l=t}^{Z_j^{i*}-1} \psi_{jl}^i & \text{if } t < Z_j^{i*} \\ 0 & \text{if } t = Z_j^{i*} \\ \sum_{l=Z_j^{i*}}^{t-1} \psi_{jl}^i & \text{if } t > Z_j^{i*} \end{cases}. \tag{8.11}$$

Since only each copy may be scheduled only in one time unit and each time unit may be assigned to only one copy, an optimal schedule is found by solving the following assignment problem.

$$\text{minimize} \quad \sum_{t=1}^{D} \sum_{(i,j)\in J} C^i_{jt} x^i_{jt} \tag{8.12}$$

subject to

$$\sum_{i=1}^{n} \sum_{j=1}^{d_i} x^i_{jt} = 1, \quad t = 1,\ldots,D, \tag{8.13}$$

$$\sum_{t=1}^{D} x^i_{jt} = 1, \quad (i,j) \in J, \tag{8.14}$$

where $J = \{(1,1),\ldots,(1,d_1),\ldots,(n,1),\ldots,(n,d_n)\}$ and (i,j) denotes the j-th item of product i,

$$x^i_{jt} = \begin{cases} 1 \text{ if } (i,j) \text{ is assigned to period } t \\ 0 \qquad\qquad \text{otherwise} \end{cases}. \tag{8.15}$$

The Kubiak–Sethi algorithm does not stay within the quota; however, it is house-monotone. The coefficients ψ^i_{jt} can be modified so that the solution of the assignment problem gives an optimal schedule for the problem with objective function (8.2). Kubiak and Sethi [14] show that in fact, the method works for any objective function of the form

$$\text{minimize} \quad \sum_{i=1}^{n} \sum_{t=1}^{D} F_i(x_{it} - tr_i), \tag{8.16}$$

where F_i is a nonnegative convex function satisfying $F_i(0) = 0$, $F_i(y) > 0$ for $y \neq 0$, $i = 1,\ldots,n$.

8.3.1.4 Inman–Bulfin Algorithm

Inman and Bulfin [8] consider the problem with objective functions (8.1) and (8.2) under constraints (8.4)–(8.7). Their approach consists in calculating ideal completion times for particular items of each product $Z^i_j = \frac{2j-1}{2r_i}$ and minimizing the following objective functions, respectively:

$$\text{minimize} \quad \sum_{i=1}^{n} \sum_{j=1}^{d_i} (y_{ij} - Z^i_j)^2, \tag{8.17}$$

$$\text{minimize} \quad \sum_{i=1}^{n} \sum_{j=1}^{d_i} |y_{ij} - Z^i_j|, \tag{8.18}$$

where y_{ij} is the completion time of item j of product i. Notice, that Z_j^i may be considered a due date of item j of product i. The problem with objective function (8.18) is thus equivalent to the problem of scheduling n unit time jobs on a single machine so that the total absolute deviation between respective completion times and due dates is minimized. Garey et al. [7] proved that this problem is solved optimally by the *Earliest Due Date Rule*. A simple interchange argument may be used to prove the same result for the objective function (8.17). The complexity of the Inman–Bulfin algorithm is $O(nD)$.

8.3.2 Minimizing the Maximum Deviation

The problem of minimization of the maximum deviation is defined by objective function (8.3) and constraints (8.4)–(8.7). This problem was in fact considered by Tijdeman [21], however, in the context of the so called *chairmen assignment problem*. Tijdeman proposed an algorithm that finds a solution with $|x_{it} - tr_i| < 1$, $i = 1,\dots,n$, $t = 1,\dots,D$, but this solution does not necessarily minimize (8.3). It is easy to observe that house monotone methods of apportionment that stay within the quota may be also applied to find solutions with $|x_{it} - tr_i| < 1, i = 1,\dots,n$, $t = 1,\dots,D$. Later, Steiner, and Yeomans [19] proposed an algorithm that is polynomial in D and minimizes (8.3).

8.3.2.1 Apportionment Methods Staying Within the Quota

The first population monotone apportionment method staying within the quota was developed by Balinski and Young [3]. Few years later, Still [20] proposed a more general algorithm. We present the Still approach as a scheduling algorithm that finds solutions with $\max_{it} |x_{it} - tr_i| < 1$. In order to define the set of items eligible to be scheduled in time unit t two tests are performed. The first one, the *upper quota test* consists in checking if $|x_{it-1} + 1 - tr_i| < 1$. The second test, the *lower quota test* is more sophisticated. Let t_i denote the first time unit when the next item of product i may be scheduled without violating the lower quota. The value of t_i can be calculated as follows:

$$t_i = \left\lceil \frac{D(x_{it-1} + 1)}{d_i} \right\rceil . \tag{8.19}$$

For each time unit g such that $t \le g \le t_i$ we define

$$s_k(g,i) = \begin{cases} x_{it-1} + 1 & \text{for } k = i \\ \max\left\{x_{kt-1}, \frac{d_k}{D} g\right\} & \text{for } k \neq i \end{cases} . \tag{8.20}$$

If $\sum_{k=1}^{n} s_k(g,i) \le g$ for each $t \le g \le t_i$, then product i satisfies the lower quota test.

Still algorithm

begin
$t := 1$;
while $t \leq D$ **do**
 begin
 $J_t := \{i :$ satisfies the lower and upper quota tests$\}$;
 find $k \in J_t$ such that $\frac{d_k}{d(x_{kt})} \geq \frac{d_i}{d(x_{it})}$ for any $i \in J_t$;
 schedule the next item of product k in time unit t;
 $t := t + 1$;
 end
end

The divisor function $d(x)$ may be selected arbitrarily, so the Still method defines in fact a class of algorithms. Still proved that J_t is never empty and that each house monotone method that stays within the quota belongs to this class.

8.3.2.2 Tijdeman Algorithm

Tijdeman [21] formulated the following problem called the *chairmen assignment problem*. Consider n states that form a union and every year a chairmen of the union is to be appointed in such a way that the cumulative number of chairmen from state i is proportional to its weight r_i, $i = 1, \ldots, n$, such that $\sum_{i=1}^{n} r_i = 1$. Let us denote by x_{it} the total number of chairmen from state i in the periods $1, \ldots, t$. The objective is to minimize $\sup_{i,t} |x_{it} - tr_i|$.

It is clear that this problem is equivalent to the scheduling problem (8.3)–(8.7). Consecutive years correspond to time units in the schedule and an assignment of a chairmen from state i corresponds to assignment of an item of product i. The objective function was originally defined as a supremum of the deviation, but it is easy to see that in the PRV problem this supremum belongs to the set of feasible solutions so we can write maximum in this place.

Tijdeman developed an algorithm that always finds a feasible sequence with

$$\max_{it} |x_{it} - tr_i| \leq 1 - \frac{1}{2n-2}. \tag{8.21}$$

Another property of the Tijdeman algorithm was proved in [10]. Namely, the Tijdeman algorithm is a quota divisor method of apportionment with divisor function defined as $d(x) = x + (1 - 1/(2n-2))$. Notice, that although the bound (8.21) is quite attractive, the algorithm does not guarantee finding an optimal solution with respect to (8.3).

The algorithm proposed by Tijdeman works as follows.

Tijdeman algorithm

begin
$t := 1$;
while $t \leq D$ **do**
 begin
 $J_t := \{i : tr_i - x_{it-1} \geq \frac{1}{2n-2}\}$;
 find $k \in J_t$ such that for any $i \in J_t$

$$\frac{1 - \frac{1}{2n-2} - tr_k + x_{kt-1}}{r_k} \leq \frac{1 - \frac{1}{2n-2} - tr_i + x_{it-1}}{r_i};$$

 schedule the next item $(x_{kt-1} + 1)$ of product k in time unit t;
 $t := t + 1$;
 end
end

8.3.2.3 Steiner–Yeomans Algorithm

The Steiner–Yeomans approach uses the fact that for a given maximum deviation δ the problem may be reduced to a problem of scheduling jobs with ready times and due dates. The transformation is based on the following observation illustrated in Fig. 8.2.

Assume that we look for a schedule with

$$\max_{it} |x_{it} - tr_i| \leq \delta. \tag{8.22}$$

Let us consider the j-th item of product i, $i = 1,\ldots,n$, $1 \leq j \leq d_i$. The deviation $|j - tr_i| \leq \delta$ not earlier than at point A. Thus, the earliest processing time of item (i, j) can be calculated as follows

$$E(i, j) = \left\lceil \frac{j - \delta}{r_i} \right\rceil. \tag{8.23}$$

On the other hand after point B we have $|(j-1) - tr_i| > \delta$; so, item j cannot be scheduled later than at

$$L(i, j) = \left\lceil \frac{j - 1 + \delta}{r_i} \right\rceil. \tag{8.24}$$

After calculating $E(i, j)$ and $L(i, j)$ for each $i = 1,\ldots,n$, and $j = 1,\ldots,d_i$, the following procedure is performed in order to check if a schedule exists such that each item (i, j) is scheduled in the interval $[E(i, j), L(i, j)]$.

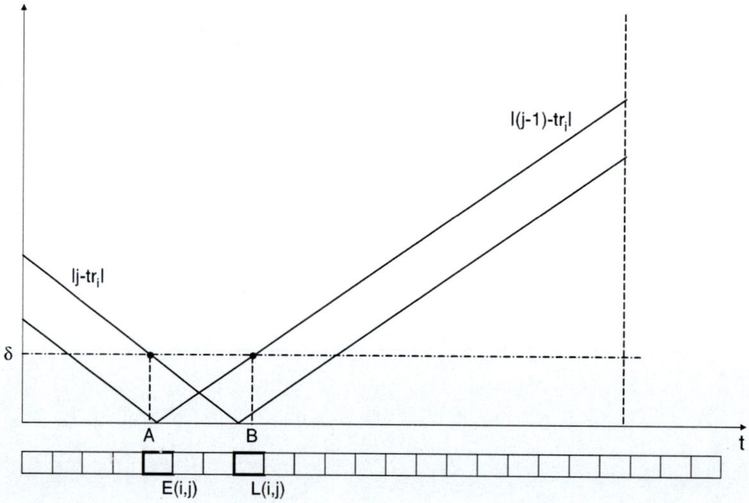

Fig. 8.2: The idea of the earliest and the latest completion times of item j with bound T

Steiner–Yeomans algorithm

begin
$t := 1$;
while $t \leq D$ **do**
  begin
  $J_t := \{(i,j) : E(i,j) \leq t \leq L(i,j)\}$;
  if J_t is empty **then** STOP: a schedule with $\max_{it} |x_{it} - tr_i| \leq \delta$ does not exist;
  find $(k,l) \in J_t$ such that $L(k,l) \leq L(i,j)$ for any $(i,j) \in J_t$;
  schedule the l-th item of product k in time unit t;
  $t := t+1$;
  end
end

In order to find a schedule with the minimum value of bound δ the algorithm is repeated for all $\delta \in \Delta$ where

$$\Delta = \left\{ \frac{D - d_{\max}}{D}, \frac{D - d_{\max} + 1}{D}, \ldots, \frac{D-1}{D} \right\}. \tag{8.25}$$

Brauner and Crama [6] show that for at least one of the values in set Δ a feasible schedule exists. Obviously, the computational effort may be reduced by applying the binary search procedure. Another property of the Steiner–Yeomans algorithm was proved in [10]. Namely, it is a quota divisor method of apportionment with divisor function defined as $d(x) = x + T$.

8.4 Multiple Parallel Machines

In the classical formulation of the PRV problem, it is assumed that tasks are scheduled on a single machine. However, in practice, it may happen that several (usually two or three) parallel identical machines (or assembly lines) are available. It is possible to extend the results obtained for a single machine to the case with multiple identical parallel machines.

We assume the same notation as in Sect. 8.2 and denote by m the number of available machines. Obviously, the minimum schedule length equals $T = \lceil D/m \rceil$. Recall that the idea following form the just-in-time approach is to keep constant usage of parts used for assembly. Thus, it is justified to relate the product rate ρ_i of product i to available time T, and calculate as follows:

$$\rho_i = \frac{d_i}{T} = \frac{d_i}{\lceil \frac{D}{m} \rceil}. \tag{8.26}$$

In other words, the goal is to distribute the production of particular items of each product evenly over the time horizon T. We will call ρ_i the *time related product rate* of product i. Notice that $0 < \rho_i \leq m$ and for $m = 1$ we have $r_i = \rho_i$, $i = 1, \ldots, n$.

The solution approach is very similar to the single machine case. For each item we calculate its deviation from optimum as $|x_{it} - t\rho_i|$, where $i = 1, \ldots, n, t = 1, \ldots, T$. The main difference is that in each time unit at most m items may be scheduled. Notice further, that if D/T is not an integer then some machines remain idle. It is easy to see that the position of the idle time may influence the value of the objective function. Let us illustrate this conjecture with the following example.

Example 8.1. Consider three products with the vector of demands $[2,3,5]$ to be scheduled on three identical parallel machines. Since we have $D = 10$, we obtain $T = 4$, and in any schedule for two time units a machine remains idle. The vector of time related product rates is $[0.5, 0.75, 1.25]$. Let us compare the schedules presented as Gantt charts in Figs. 8.3 and 8.4. For each time period and machine number the index of a scheduled product is given. In the first schedule, where two machines are idle in the last time unit, the maximum deviation equals 0.75, and the total deviation equals 3.5. In the second schedule, one machine is idle in the time units 3 and 4 and we have the deviations equal 0.5 and 3.0, respectively. In conclusion, the second schedule is a better solution with respect to both criteria. This observation should not be surprising, however, because idle time occurs in optimal schedules also in other models of just-in-time scheduling, even on a single machine [9].

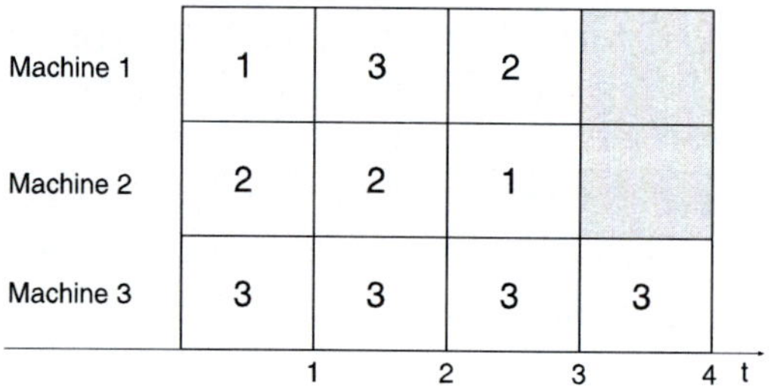

Fig. 8.3: Idle time scheduled in the last time unit

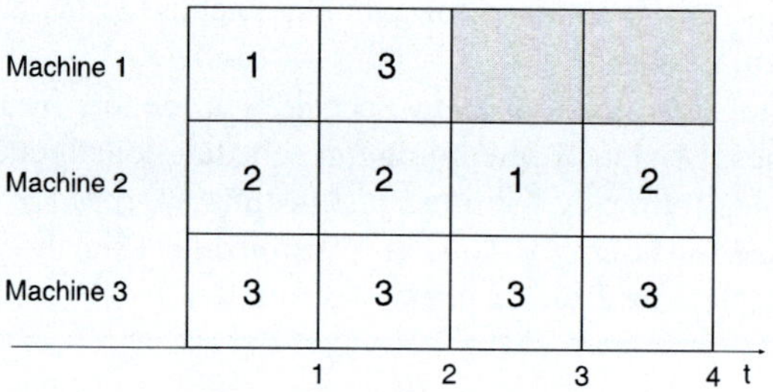

Fig. 8.4: Idle time scheduled before the last time unit

8.4.1 Total Deviation

We consider the objective functions (8.1) and (8.2) modified as follows:

$$\text{minimize} \quad \sum_{i=1}^{n}\sum_{t=1}^{T}(x_{it} - t\rho_i)^2, \tag{8.27}$$

$$\text{minimize} \quad \sum_{i=1}^{n}\sum_{t=1}^{T}|x_{it} - t\rho_i|, \tag{8.28}$$

subject to the following constraints, which correspond to constraints, (8.4)–(8.7)

$$\sum_{i=1}^{n} x_{it} \le mt, \quad t = 1,\ldots,T, \tag{8.29}$$

$$0 \le x_{it+1} - x_{it} \le m, \quad i = 1,\ldots,n, \; t = 1,\ldots,T-1, \tag{8.30}$$

$$x_{iT} = d_i, \quad i = 1,\ldots,n, \tag{8.31}$$

$$x_{it} \in \mathbf{Z}^+ \cup \{0\}, \quad i = 1,\ldots,n, \; t = 1,\ldots,T. \tag{8.32}$$

We show that the Kubiak–Sethi algorithm may be easily modified to solve the multi-machine problem. We introduce binary decision variables

$$x_{ij}^t = \begin{cases} 1 \text{ if the } j\text{–th copy of product } i \text{ is scheduled in time } t \\ 0 \text{ otherwise} \end{cases}. \tag{8.33}$$

Applying the same reasoning as in Sect. 8.3.1.3 we calculate the cost of scheduling the j-th item of product i, $i = 1, \dots, n$, $j = 1, \dots, d_i$, in time unit t as

$$C_{jt}^i = \begin{cases} \sum_{l=t}^{Z_j^{i*}-1} \psi_{jl}^i & \text{if } t < Z_j^{i*} \\ 0 & \text{if } t = Z_j^{i*} \\ \sum_{l=Z_j^{i*}}^{t-1} \psi_{jl}^i & \text{if } t > Z_j^{i*} \end{cases}, \tag{8.34}$$

where

$$\psi_{jl}^i = |(j - 1 - l\rho_i)^2 - (j - l\rho_i)^2|. \tag{8.35}$$

The objective functions (8.27) and (8.28) are minimized by solving the following transportation problem:

$$\text{minimize} \quad \sum_{(i,j)\in J} \sum_{t=1}^{T} C_{ij}^t x_{ij}^t, \tag{8.36}$$

$$\text{subject to} \quad \sum_{(i,j)\in J} x_{ij}^t \le m, \quad t = 1, \dots, T, \tag{8.37}$$

$$\sum_{t=1}^{T} x_{ij}^t = 1, \quad (i,j) \in J, \tag{8.38}$$

$$x_{ij}^t \in \{0,1\}, \quad (i,j) \in J, t = 1, \dots, T, \tag{8.39}$$

where $J = \{(1,1), \dots, (1,d_1), \dots, (n,1), \dots, (n,d_n)\}$. Obviously, the transportation problem can be solved in polynomial time.

8.4.2 Maximum Deviation

We consider the objective function (8.3) modified as follows:

$$\text{minimize} \quad \max_i \max_t |x_{it} - t\rho_i|, \tag{8.40}$$

and constraints (8.29)–(8.32).

In the case of maximum deviation, we show that the Steiner and Yeomans algorithm may be used to solve the multi-machine problem. Let us assume that we look for a schedule with maximum deviation not exceeding a given value δ.

For each copy j of product i, $i = 1,\ldots,n; j = 1,\ldots,d_i$, we may easily calculate that $|j - t\rho_i| \leq \delta$ iff:

$$E^m(i,j) = \left\lceil \frac{j-\delta}{\rho_i} \right\rceil, \tag{8.41}$$

$$L^m(i,j) = \left\lceil \frac{j-1+\delta}{\rho_i} \right\rceil. \tag{8.42}$$

Further on, we calculate the earliest and latest completion time of each item (i,j) and apply the algorithm described below.

Steiner–Yeomans algorithm for m parallel machines

begin
$t := 1$;
while $t \leq T$ **do**
if J_t is empty **then** STOP: a schedule with $\max_{it} |x_{it} - t\rho_i| \leq \delta$ does not exist
else
  begin
  $J_t := \{(i,j) : E(i,j) \leq t \leq L(i,j)\}$;
  for $h := 1$ **to** m **do**
  find $(k,l) \in J_t$ such that $L(k,l) \leq L(i,j)$ for any $(i,j) \in J_t$ and schedule the l-th item of product k in time unit t on machine h;
  $t := t+1$;
  end;
If there are still unscheduled items then a schedule with $\max_{it} |x_{it} - t\rho_i| \leq \delta$ does not exist;
end

Notice that although some J_t is not empty it may still contain less than m items. In such a case, idle time is scheduled on $m - |J_t|$ machines in time unit t. Thus, some items may remain unscheduled in time $T+1$ and in such situation a feasible schedule for δ does not exist.

Theorem 8.1. *A feasible solution exists for $\delta = 1$.*

Proof. We will prove that at each time unit $1 \leq t \leq T$ there are at least mt items with $E^m(i,j) \leq t$. Assume first that $T = \lceil D/m \rceil = D/m$. It was proved by Steiner and Yeomans [19] that a feasible solution for $\delta = 1$ exists on a single machine, so there are at least mt items with $E(i,j) \leq mt$. For each item (i,j) such that $E(i,j) \leq mt$ we have

$$E^m(i,j) = \left\lceil \frac{j-1}{\rho_i} \right\rceil = \left\lceil \frac{1}{m} \frac{j-1}{r_i} \right\rceil \leq \left\lceil \frac{1}{m} mt \right\rceil = t. \tag{8.43}$$

Thus, at each time unit there are at least m items ready to be scheduled. Similar argument shows that at each t we have at least $(T-t)m$ items with $L^m(i,j) \geq t$. For each item (i,j) such that $L(i,j) \geq mt$ we have

$$L^m(i,j) = \left\lceil \frac{j}{\rho_i} \right\rceil = \left\lceil \frac{1}{m}\frac{j}{r_i} \right\rceil \geq \frac{1}{m}\left\lceil \frac{j}{r_i} \right\rceil = \frac{1}{m}L(i,j) \geq t. \tag{8.44}$$

Now, let us assume that $T = \lceil D/m \rceil > D/m$. Obviously, in this case idle time occurs in any feasible schedule. Let us add a dummy product with $d_{n+1} = \lceil D/m \rceil - D/m$. We obtain the case considered in the first step of the proof, so we can built a feasible schedule. Notice that $T = \lceil D/m \rceil = \lceil (D + d_{n+1})/m \rceil$ and hence ρ_i for products $i = 1, \ldots, n$, does not change. Thus, in the obtained schedule we have

$$|x_{it} - t\rho_i| \leq 1, \tag{8.45}$$

for each $i = 1, \ldots n; t = 1, \ldots, T$. This ends the proof. □

The lower bound of the objective function is calculated as:

$$\delta \geq max_{i,t}\{min(\lceil t\rho_i \rceil - t\rho_i, t\rho_i - \lfloor t\rho_i \rfloor)\}. \tag{8.46}$$

This bound can be achieved as may be shown in Example 8.1 (where $T \geq 0.5$), hence the bound is tight.

8.5 Conclusions

We have discussed the problem of scheduling flexible assembly lines in the just-in-time environment. We considered the PRV problem as a model suitable to schedule systems where large amounts of similar products are manufactured with the goal to minimize the in-process inventory level. Properties of several algorithms developed for the original single machine PRV problem are discussed. Generalization of the scheduling problem to the case of multiple parallel machine system is developed. A generalization of the Kubiak–Sethi algorithm is proposed for solving the PRV problem with the objective to minimize the total deviation. The Steiner–Yeomans algorithm is generalized to solve the PRV problem with the objective to minimize the maximum deviation. In both cases, the complexity of the algorithm remains $O(nD)$.

The proposed generalization relates the product rate to the elapsed time. Another approach is possible, where the product rate is related to the production volume. Such an approach may be justified when it is important to maintain a constant proportion between production volume of various final products. This approach seems to be an interesting topic for further research in scheduling multiple machine just-in-time systems.

References

1. Balinski, M., Ramirez, V.: Parametric methods of apportionment, rounding and production. Mathematical Social Sciences **37**(2), 107–122 (1999)
2. Balinski, M., Shahidi, N.: A simple approach to the product rate variation problem via axiomatics. Operations Research Letters **22**(4-5), 129–135 (1998)
3. Balinski, M., Young, H.: The quota method of apportionment. The American Mathematical Monthly **82**(7), 701–730 (1975)
4. Balinski, M., Young, H.: Fair Representation: Meeting the Ideal of One Man, One Vote. Yale University Press (1982)
5. Bautista, J., Companys, R., Corominas, A.: A note on the relation between the product rate variation (prv) problem and the apportionment problem. Journal of the Operational Research Society **47**(11), 1410–1414 (1996)
6. Brauner, N., Crama, Y.: The maximum deviation just-in-time scheduling problem. Discrete Applied Mathematics **134**, 25–50 (2004)
7. Garey, M.R., Tarjan, R.E., Wilfong, G.T.: One-processor scheduling with symmetric earliness and tardiness penalties. Mathematics of Operations Research **13**, 330–348 (1988)
8. Inman, R.R., Bulfin, R.L.: Sequencing jit mixed-model assembly lines. Management Science **37**, 901–904 (1991)
9. Józefowska, J.: Models and Algorithms for Computer and Manufacturing Systems. Springer, New York (2007)
10. Józefowska, J., Józefowski, L., Kubiak, W.: Characterization of just in time sequencing via apportionment. In: H. Yan, G. Yin, Q. Zhang (eds.) Stochastic Processes, Optimization, and Control Theory: Applications in Financial Engineering, Queueing Networks, and Manufacturing Systems, *International Series in Operations Research & Management Science*, vol. 94, pp. 175–200. Springer US (2006)
11. Józefowska, J., Józefowski, L., Kubiak, W.: Apportionment methods and the liu-layland problem. European Journal of Operational Research **193**(3), 857–864 (2009)
12. Józefowska, J., Józefowski, L., Kubiak, W.: Dynamic divisor-based resource scheduling algorithm. In: Proceedings of the 12th International Workshop on Project Management and Scheduling. Tours, France (2010)
13. Kubiak, W.: Proportional Optimization and Fairness, *International Series in Operations Research & Management Science*, vol. 127. Springer, New York (2009)
14. Kubiak, W., Sethi, S.: A note on "level schedules for mixed-model assembly lines in just-in-time production systems". Management Science **37**(1), 121–122 (1991)
15. Kubiak, W., Sethi, S.P.: Optimal just-in-time schedules for flexible transfer lines. International Journal of Flexible Manufacturing Systems **6**, 137–154 (1994)
16. Miltenburg, J.: Level schedules for mixed-model assembly lines in just-in-time production systems. Management Science **35**(2), 192–207 (1989)
17. Monden, Y.: Toyota Production Systems. Industrial Engineering and Management Press, Norcross (1983)
18. Okamura, K., Yamashina, H.: A heuristic algorithm for the assembly line model-mix sequencing problem to minimize the risk of stopping the conveyor. Journal of Production Research **17**(3), 233–247 (1979)
19. Steiner, G., Yeomans, S.: Level schedules for mixed-model, just-in-time processes. Management Science **39**, 728–735 (1993)
20. Still, J.W.: A class of new methods for congressional apportionment. SIAM Journal on Applied Mathematics **37**(2), 401–418 (1979)
21. Tijdeman, R.: The chairman assignment problem. Discrete Mathematics **32**(3), 323–330 (1980)

Chapter 9
Just-in-Time Planning and Lot-Sizing

Nabil Absi, Stéphane Dauzère-Pérès, and Safia Kedad-Sidhoum

Abstract The goal of this chapter is to show that production planning and lot-sizing models are not contradictory with a Just-In-Time (JIT) approach. Characteristics of these models are discussed to explain how they fit with JIT concepts. We first underline the importance of reducing setup times/costs in both lot-sizing and JIT models. We show that the definition of setup time/cost reduction can be different depending on the problem. We then discuss when demands cannot be delivered at their due dates (backlogs or lost sales). The relevance of time windows in a JIT environment is presented, followed by bounded production and inventory. The importance of capacity constraints and their impact on JIT systems are also discussed. Finally, the last section of the chapter is devoted to the interest of integrating lot-sizing and scheduling in a JIT approach.

9.1 Introduction

Production planning aims at determining production quantities per period of a planning horizon to satisfy demands at minimal cost. Lot-sizing denotes production planning where setups costs or times need to be taken into account, that is when lot sizes are critical. Studying and solving lot-sizing problems seems contradictory with the Just-in-Time (JIT) philosophy whose primary objective is to produce demands only when needed to avoid unnecessary inventories. The concepts and techniques in JIT are multiple while aiming at the same goal to deliver the right product (quantity and quality) at the right time [39]. JIT production, or equivalently lean

Nabil Absi and Stéphane Dauzère-Pérès
CMP, Ecole des Mines de Saint-Étienne, 880 route de Mimet, F-13541 Gardanne, France,
e-mail: absi@emse.fr; dauzere-peres@emse.fr

Safia Kedad-Sidhoum
LIP6, Université Pierre et Marie Curie, 4 Place Jussieu, F-75252 Paris Cedex, France,
e-mail: safia.kedad-sidhoum@lip6.fr

R.Z. Ríos-Mercado and Y.A. Ríos-Solís (eds.), *Just-in-Time Systems*, Springer Optimization and Its Applications 60, DOI 10.1007/978-1-4614-1123-9_9,

production, aims at reducing seven waste types (overproduction, waiting, transportation, processing itself, stocks, motion, and defective products) in production processes. Organizations should continuously improve (*Kaizen*) their productivity, quality and flexibility, together with their suppliers and customers (JIT partnership) [53].

Inventory costs in production planning and lot-sizing models naturally tend to push demands to be produced just in time. Several reasons explain why inventories are necessary and cannot be totally removed. One of the primary reasons is the importance of setup costs and times (see Sect. 9.2). Even reduced through Single Minute Exchange of Die (SMED), one of the JIT techniques, producing or delivering a given demand just in time can be too expensive (setup costs) or time-consuming (setup times). Many examples inducing large setup costs and times can be found in production and in distribution. In process industries, major setups of several hours are often required to configure critical machines (such as setting temperature in ovens) before starting a new process type. In transportation, full truckload is often the rule to cover the fixed cost of a container. Another important reason that prevents demands to be produced just in time is capacity constraints (see Sects. 9.6 and 9.7). Products with highly seasonal demands such as toys at Christmas are typical examples of cases where capacity constraints cannot cover demands at all times. The most well-known JIT technique to better coordinate different manufacturing stages is the Kanban system. A Kanban system helps to ensure that production at one stage is only authorized if there is a demand in a downstream stage. A major constraint of such a system is that it is designed to process products at a given rate. The number of kanbans at a stage is usually calculated based on average throughput time at the stage, which is associated to its real or estimated capacity. Kanban systems are thus not very flexible and cannot handle demand variations very well. Demands that are too large cannot be met and, when demands are too small, there will be overcapacity. This is why a key tool in JIT is Production Leveling (called *Heijunka* in lean) that aims at leveling the variety and/or volume of items produced at a stage over a period of time. Production leveling means that demands may have to be advanced or postponed. Production planning models can be used to perform production leveling at minimal cost, and are thus consistent with JIT.

Production planning decisions optimize the trade-off between various conflicting economic objectives such as inventory costs and capacity related costs. In this context, lot-sizing decisions consider setup costs and times, and determine the size of the lots of different items to be produced or distributed and the time periods in which these lots have to be produced or distributed. Various constraints need to be considered depending on the problem. The basic objective function to minimize in lot-sizing is a combination of production, setup, and inventory costs. In this chapter, we will discuss the models usually considered in lot-sizing and show that many JIT aspects are embedded in these models.

This chapter does not aim at studying the relevance of JIT systems, which is widely discussed in the literature. We mention the paper of Bonney et al. [11] who examine, by means of simulation, the effect that push and pull information flows have on system performance under a variety of conditions. Christensen et al. [17]

also study the effects of distinct strategies on applied supply chain knowledge and market performance. We focus in the sequel on analyzing the relationships between JIT and lot-sizing models.

Some works should be mentioned in this scope. Jaber et al. [30] address lot-sizing with learning, forgetting and entropy costs. The analysis shows that poor control increases costs and suggests that ordering in larger lots may be an appropriate policy to counter entropy effects. Therefore, consistent with lean principles, organizations should invest in systems that reduce costs by having efficient optimization-based systems. The importance of the trade-off between setup costs and inventory costs captured in lot-sizing models are also underlined in a recent work addressing the relationship between lean production and information technology [43]. The problem of determining lot sizes of multiple items that are manufactured in a JIT environment by a single capacitated facility has been addressed in the early 1990s [37], based on available capacity and four attributes of each item: Demand rate, holding cost, setup time, and processing time. The importance of reducing product lot sizes in converting traditional job-shops systems into JIT manufacturing systems has also been addressed for closed stochastic production systems in [60] based on queuing lot size models.

A single-item lot-sizing problem can be defined as a production planning problem in which there is a time-varying demand over a planning horizon of T periods. The objective is to determine periods where production should take place to satisfy demands as well as the quantities that have to be produced in these periods subject to industrial constraints. The total production should satisfy demands while minimizing a cost function. The seminal paper on lot-sizing problems of Wagner and Whitin [58] in the late 1950s introduces one of the first lot-sizing problems addressed in the literature, that is the dynamic version of the Economic Lot-Size model (ELS). A fundamental property of the ELS model is that there exists an optimal solution that satisfies the zero-inventory property, that is a new order is placed only if the inventory drops down to zero. This implies that, in any order period, the production quantity must cover an integral number of consecutive demands. The zero-inventory property is already a JIT strategy. In fact, no over-production is done at a given period t to satisfy the demand of a given period t' if another production period exists between period t and period t'. This strategy aims at reducing inventory levels when production periods are fixed.

In the sequel of the chapter, we use the basic single-item uncapacitated lot-sizing parameters [42] defined as follows: A demand d_t, that is often a mix of forecasts and customer orders, has to be satisfied in each period t. The variable or unit production cost in period t is p_t, h_t is the unit cost for holding one unit in inventory at the end of period t and s_t is the fixed setup cost that is incurred if there is a strictly positive production in period t. In Sects. 9.2–9.6, we study some extensions of the ELS model. In Sects. 9.7 and 9.8, we study lot-sizing problems dealing with multiple items and additional constraints.

This chapter is organized as follows. In Sect. 9.2, we underline the importance of reducing setup times/costs in both lot-sizing and JIT models. We show that the definition of setup times/costs reduction can be different depending on the problem.

In Sects. 9.3 and 9.4, we discuss when demands cannot be delivered at their due dates (backlogs, overtimes and lost sales). We show the relationship between these lot-sizing concepts and some JIT principles. In Sect. 9.5, the relevance of time windows in JIT environment is presented. In Sect. 9.6, we show that lot-sizing problems with bounded inventory and production can be used to formulate different JIT concepts. Section 9.7 discusses the importance of capacity constraints and their impact on JIT systems. The last section of this chapter is devoted to the interest of integrating lot-sizing and scheduling in a JIT approach.

9.2 Setup Optimization

The importance and benefits of incorporating setup times and costs in production planning decisions is now well established, see for instance Allahverdi and Soroush [7]. As mentioned by these authors, the benefits of reducing setup times and costs include: Reduced expenses, increased production speed, increased output, reduced lead times, faster changeovers, increased competitiveness, increased profitability and satisfaction, enabling lean manufacturing, smoother flows, broader range of lot sizes, lower total cost curve, fewer stockouts, lower inventory, lower minimum order sizes, higher margins on orders above minimum, faster deliveries, and increased customer satisfaction. The reduction of setup times and costs is thus one of the most effective means to reduce inventory levels and to improve customer service which is particularly important in modern production management systems such as JIT.

There is a difference between the definitions of setup optimization in JIT and lot-sizing. Setup optimization in JIT usually means reducing structural setup times/-costs. This is done for example through SMED which is one of the JIT methods that aims at reducing wastes in manufacturing processes (e.g. changeovers). Optimizing setups in lot-sizing problems means reducing the overall setup costs/times resulting from the planning decisions. The structural setup times/costs resulting from JIT can be embedded into lot-sizing models by considering setup, start-up, and changeover times/costs. Changeover and start-up principles are described later in this section.

The overall setup optimization is also considered in JIT since we try to reduce inventory levels when smoothing production quantities over the horizon. This is done through Production Leveling which is one of the approaches used in lean manufacturing to reduce wastes. Production leveling tries to smooth workload and human resources when several products are performed by the same system in order to reduce production, setup and inventory holding costs. This approach aims at minimizing two conflicting objectives: The reduction of production costs and setup times on one hand, and inventory holding costs on the other hand. These are also the objectives of classical lot-sizing models.

The problem of cost balancing between setup and holding costs was originally formulated in the Economic Order Quantity (EOQ) model and has received significant attention since then. The original version of the EOQ model reflects the emphasis on high-volume low-cost standardized products, job specification, and hierarchical control systems prevailing in the 1970s. It can be recalled that the EOQ model assumes a constant demand rate over a continuous and infinite horizon in contrast with the ELS problem, for which the horizon is discrete and the demands and costs are time varying.

New models have been developed to deal with modern manufacturing systems such as JIT. An example is given in Vörös [56] who considers the impact of process quality in a lot-sizing model. It is stated that there is a positive correlation between process quality deterioration and lot sizes and that the holding and setup costs decrease as process quality deteriorates. A recent work from van den Heuvel and Wagelmans [49] shows that, in an optimal solution, the total holding cost in an order interval is upper bounded by a quantity proportional to the setup cost and the logarithm of the number of periods in the interval.

The ELS model can be considered as a JIT model. In fact, if setup costs are not considered and unitary production costs are not time dependent, production is performed as late as possible. When considering setup and unitary production costs, the model proposes the best trade-off between producing at the lowest cost, that is minimizing production costs, and as late as possible, that is minimizing holding costs. This trade-off is embedded in the mathematical model of the ELS below. Let X_t denote the amount produced in period t and I_t the ending inventory of period t (I_0 is the inventory level at the beginning of the first period). The binary variable Y_t is equal to 1 if $X_t > 0$ and 0 otherwise.

$$\text{Minimize} \sum_{t=1}^{T} (p_t X_t + s_t Y_t + h_t I_t), \tag{9.1}$$

Subject to:

$$I_{t-1} + X_t = d_t + I_t \qquad \forall\, t = 1, \cdots, T, \tag{9.2}$$

$$X_t \leq \left(\sum_{l=t}^{T} d_l \right) Y_t \qquad \forall\, t = 1, \cdots, T, \tag{9.3}$$

$$Y_t \in \{0, 1\} \qquad \forall\, t = 1, \cdots, T, \tag{9.4}$$

$$X_t,\ I_t \geq 0 \qquad \forall\, t = 1, \cdots, T. \tag{9.5}$$

The objective function (9.1) minimizes total production, setup and holding costs. Constraints (9.2) are the inventory balance equations. Constraints (9.3) impose that the quantity produced must not exceed the total demand requirement on the interval $[t, T]$. Constraints (9.4) and (9.5) characterize the variable domains: X_t and I_t are non-negative and Y_t is a binary variable.

The ELS problem can be represented by a minimum cost fixed charge network flow model (see Fig. 9.1) that can be formulated as follows. The set of vertices is defined by the set of periods 1 to T of the problem. Vertex 0 is a dummy vertex that represents the main source which transfers the total demand. For each node $t \in \{1, 2, \ldots, T\}$, we associate a demand d_t. Arcs $(0, t)$ are the production arcs for $t \in \{1, 2, \ldots, T\}$. The flow in arc $(0, t)$ represents the quantity X_t produced at

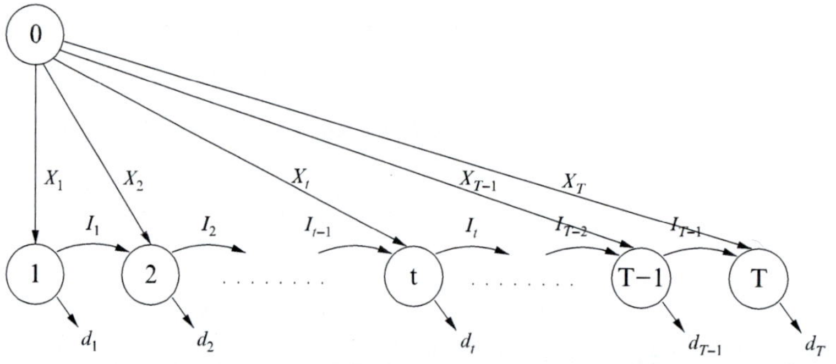

Fig. 9.1: Fixed-charge network flow model for ELS

period t. Arcs $(t, t+1)$ are the inventory arcs for $t \in \{1, 2, \ldots, T-1\}$ with a flow I_t representing the ending stock of period t and the entering stock in period $t+1$.

The seminal paper on lot-sizing problems of Wagner and Whitin [58] introduces one of the first JIT cost condition, which is often referred to as the absence of *speculative motive for early production*. The condition imposes that, if setups occur in both periods t and $t+1$, then it is more costly to produce in period t and keep the quantities in stock until period $t+1$ than to produce directly in period $t+1$. This condition illustrates the fact that it is better to produce as late as possible. Formally, we have $s_t + p_t - p_{t+1} \geq 0$.

The ELS problem is polynomial, it was originally solved in $O(T^2)$ [58]. This complexity was improved to $(O(T \log T))$ by Wagelmans et al. [57], Aggarwal and Park [4], Federgruen and Tzur [24], and Van Hoesel et al. [50]. For a comprehensive review on models and solution approaches for the single-item lot-sizing problem, the reader is referred to [13].

The classical ELS problem assumes that periods are long enough to ignore the relation between setups of consecutive periods, that is a setup is incurred at each period an item is produced. In this case, we talk about big time bucket lot-sizing problems. These problems are characterized by the fact that several items can be produced at the same period. When periods are short (generally some hours), the relation between setups of consecutive periods must be considered to model setup costs and setup times more precisely. In this case, we talk about small time buckets and the problem is called Discrete Lot-Sizing and Scheduling Problem (DLSP). The main idea behind the DLSP is that resource capacity constraints are tight and at most one item can be produced at the same period. The classical DLSP was shown to be NP-hard by Salomon et al. [44]. The general formulation considers sequence-dependent changeovers, that is the setup for processing an item at a given period depends on the product that was produced at the previous period (e.g., in some industries, the time and the cost of cleaning a machine depends on the sequence of items). Start-up costs are independent of the item sequence and are only incurred if the item was not produced at the previous period. For a recent literature review on using DLSP with sequence-dependent setups, the reader can refer to [63]. The goal of better modeling setups is to minimize the overall setup times and costs, which is one of the main objectives of JIT. Nevertheless, the DLSP is characterized by a large number of binary variables necessary to formulate changeovers which makes the problem more difficult to solve [8].

Another way to better model setups is to consider the relation between the last product processed in a given period and the first product processed in the next period, that is to allow the production run to continue over consecutive time periods without incurring multiple setups. Sox and Gao [45] study such extension called setup carry-over in big time bucket problems. Setup carry-over is also known as linked lot-sizes [46]. They can be viewed as a compromise between a fixed time bucket and a continuous time axis approach. The aim is to get the best possible flexibility of the production resources.

9.3 Demand Backlogs

In JIT systems, production is supposed to be only driven by customer demands. One of the basic requirements is to try to reach a good service level to satisfy customers. However, when capacity constraints are tight, it may not be possible to deliver all demands at their due dates. Therefore, it is interesting to evaluate the possibility of delaying production of some orders to allow a better use of the resources for high-priority orders that must meet their due dates. Backlogs on demands capture this possibility. Backlogs can also be considered in contradiction with JIT principles. They might be viewed as a degree of freedom to achieve the best possible service level. Indeed, in some situations, not considering backlogs will lead to demand losses.

It is difficult for a firm to implement a JIT system independently of its partners; Cooperation with suppliers and customers is required. This is called JIT partnership strategy. Backlogs can be allowed when a firm has a good partnership with its customers, which can be suitable to reduce inventory costs. In fact, producers may create large inventories to avoid backlogs, while small delays in the deliveries are acceptable.

Zangwill [62] proposes a lot-sizing model where demand must still be satisfied, but it is possible to satisfy a demand later than required with a penalty cost proportional to the amount backlogged and to the duration of the backlog. This model is particularly useful when capacity constraints are considered and the resource capacity is tight, and can come with a condition on the maximum number of periods until the end of which demand can be backordered. This is illustrated in Hsu and Lowe [29] who study the problem were backlogging costs are age dependent.

The ELS problem with backlogs is formulated by introducing new variables B_t that represent the quantity backlogged at the end of period t and a new parameter b_t which is a unitary penalty cost. This model can be represented by a minimum cost fixed charge network flow. Figure 9.1 can be modified by introducing new arcs for $(t, t-1)$, $t \in \{2, 3, \dots, T\}$, in the opposite direction of the inventory arcs, with a flow B_t representing the quantity backlogged at period t and used at or before period $t-1$. The mathematical model of the ELS problem with backlogs is obtained from the model (9.1)–(9.5) by introducing the new non-negative variables B_t, adding the cumulative backlogging cost $\left(\sum_{t=1}^{T} b_t B_t\right)$ to the objective function and replacing Constraints (9.2) by the following constraints:

$$I_{t-1} - B_{t-1} + X_t = d_t + I_t - B_t \qquad \forall\, t = 1, \cdots, T. \tag{9.6}$$

The ELS problem with backlogging is also polynomial. The first dynamic programming algorithm to solve it was proposed by Zangwill [62]. Wagelmans et al. [57] demonstrate that the ELS problem with backlogging has the same complexity as the classical ELS problem, that is $O(T \log T)$.

9.4 Lost Sales and Overtimes

The traditional trade-off in the classical lot-sizing problem is between the inventory holding cost and the setup cost [5]. If a holding cost incurs in a given period, a setup cost can be avoided. On the other hand, a large number of setups helps to avoid holding costs using a lot-for-lot strategy. Aksen et al. [5] introduce a new dimension to this trade-off where a demand may not be satisfied to avoid high production and setup costs. Nevertheless, unsatisfied demands mean lost sales. Sandbothe and Thompson [51] address the same problem with and without capacity constraints. Unsatisfied demands are called stockouts.

Lost demands in lot-sizing problems will lead to the deterioration of customer service. Introducing lost sales in JIT lot-sizing models helps to evaluate, if it is not possible to avoid lost sales due to tight resource capacities, to what extend the quality service will be deteriorated. As mentioned by Aksen [6], lost demands in lot-sizing problems due to the deterioration of customer goodwill has not received as much attention in the operations research literature as other challenging characteristics of lot-sizing. There is an increasing interest in studying the customer goodwill loss, and we quote below the following recent papers that explicitly integrate shortages.

Different authors integrate the idea of lost sales in various lot-sizing models. Absi and Kedad-Sidhoum [1] introduce lost sales in the multi-item capacitated lot-sizing problem. They also use the idea of a maximum anticipation which is equivalent to the non customer specific (NCS) version of demand time windows introduced by Dauzère-Pérès et al. [20] (see Sect. 9.5). This maximum anticipation avoids producing a given demand before a given period. Liu et al. [34] and Hwang and van den Heuvel [28] address the ELS problem with lost sales and bounded inventories. Loparic et al. [35] address the ELS problem with lower bounds on stocks where the objective is to maximize sales instead of minimizing lost sales. In some recent works, the lost sales are also considered with capacity reservation aspects such as warm/cold processes [9]. The main idea of the ELS problem with lost sales is to add new decision variables and logical constraints to the standard model (9.1)–(9.5) while modifying the balance constraints for inventory flows and the objective function. Originally, the ELS problem with lost sales was addressed by Sandbothe and Thompson [51], the authors proposed an $O(T^3)$ dynamic programming algorithm to solve the problem optimally. Aksen et al. [5] improve this complexity to $O(T^2)$.

One of the main objectives of a JIT system is to avoid lost sales. By definition, lot-sizing problems that allow lost sales cannot be really considered as JIT models. The right modeling approach is to introduce overtimes. These overtimes represents the need of additional resources to satisfy total demand. Dixon et al. [23] deal with lack of capacity by considering overtimes but their model is restricted to a single item. The multi-item capacitated lot-sizing problem with setup times (MCLS) and overtime decisions is investigated by Diaby et al. [22], Özdamar and Birbil [40], and Özdamar and Bozyel [41]. The main idea of modeling overtimes in a capacitated lot-sizing problem is to add a new continuous decision variable to model the available capacity of each capacity constraint and penalize this variable in the objective function.

9.5 Time Windows Constraints and Early Productions

Delivery or production can be constrained to take place in a given time window, that is between a release date and a due date, rather than in a given period. The time window constraints are used to model two JIT constraints. The first constraint prevents the production of a given demand before a release date. If the size of the time window is small, the production is done as late as possible. The second constraint represents the JIT contracts with suppliers. In fact, production time windows can be defined depending on due dates of demands and the availability of raw materials or semi-finished goods that can be fixed according to the partnership with suppliers. These constraints ensure that suppliers are satisfied by respecting a maximum amount of raw materials to deliver.

The time window constraints are studied in the lot-sizing literature. The release dates model situations such as the availability dates imposed by perishable products or the availability of raw materials or some semi-finished products involved in the production of an item. This is particularly important in a multi-level environment. In lot-sizing problems, these aspects are modeled using the concept of production time window [20]. The delivery time windows [32], also called grace period represents the interval where the products can be delivered to the customers with no penalty. However, all the demands can still be produced as early as the first period of the planning horizon and the holding cost is considered to be null if a demand is satisfied within its corresponding time window. Dauzère-Pérès et al. [20] were the first to study the problem where a time window is imposed for producing a demand. They define the general case where no restriction on the time window structure is imposed. Such a problem is called Customer Specific (CS) problem. They propose a pseudo-polynomial dynamic programming algorithm to solve the problem. More recently, Hwang [27] proposes a dynamic programming algorithm to solve the problem with concave production costs in $O(T^5)$.

An interesting special case of the problem is called the NCS case where two time windows cannot be strictly included. In other words, for any pair of time windows $[s_1,t_1]$, $[s_2,t_2]$, either $s_1 \leq s_2$ and $t_1 \leq t_2$ or $s_1 \geq s_2$ and $t_1 \geq t_2$. The NCS problem is widely encountered in practical situations. Moreover, it is solved in polynomial time. An $O(T^4)$ time dynamic programming algorithm was proposed in [20] and later the complexity was improved to $O(T^2)$ by Wolsey [59]. The uncapacitated single item lot-sizing problem with NCS production time windows is formulated as follows:

$$\text{Minimize} \sum_{t=1}^{T} (p_t X_t + s_t Y_t + h_t I_t), \tag{9.7}$$

Subject to:

$$I_{t-1} + X_t = d_t + I_t \qquad \forall t = 1,\cdots,T, \tag{9.8}$$

$$\sum_{k=1}^{t} X_k \leq \sum_{k=1}^{t} \sum_{l=k}^{T} d_{kl} \qquad \forall t = 1,\cdots,T, \tag{9.9}$$

$$X_t \leq \left(\sum_{l=1}^{T} d_l \right) Y_t \qquad \forall t = 1,\cdots,T, \tag{9.10}$$

$$Y_t \in \{0,1\} \qquad \forall t = 1,\cdots,T, \tag{9.11}$$

$$X_t,\ I_t \geq 0 \qquad \forall t = 1,\cdots,T. \tag{9.12}$$

The objective (9.7) is to minimize the total production, setup, and holding costs. Constraints (9.8) are the inventory balance equations. In Constraints (9.9), the cumulative quantity of production for the first t periods should be at most equal to the sum of the demands that must start before or at t and be delivered at or after t. Constraints (9.10) relate the binary setup variables to the continuous production variables. The domain definitions of the variables are defined in Constraints (9.11) and (9.12). For more details on how to model lot-sizing problems with CS time windows, the reader can refer to [12,14,20].

In some practical situations, production time windows can be violated with extra costs. This typically happens when semi-finished goods are outsourced or when the cost of producing the demand within its time window is prohibitive, and thus production can be anticipated at a given cost. This cost corresponds for instance to the cost of early delivering raw materials or semi-finished products to process the demand. The release period of the demand is violated and a cost is incurred. In this model, we can also consider backlogs or lost sales when early production cost is still prohibitive. This model was first introduced in [3]. Introducing early productions when time windows are considered in JIT lot-sizing models allows the evaluation of satisfying the customer demand with an extra cost that corresponds to the production anticipation cost and to evaluate, if they cannot be avoided, backlogs and lost sales.

The model tries to find the trade-off between satisfying customers by respecting due dates or satisfying suppliers by respecting release dates. The time window violations represent the strength of the JIT partnership with customers and/or suppliers.

In order to introduce early deliveries to the lot-sizing problem with time windows, we need to define a new decision variable E_t that represents the quantity started at period t or before and that is not available at t. The new model is obtained from Model (9.7)–(9.12) by introducing the non-negative variable E_t, adding the cumulative early delivery cost to the objective function and replacing Constraints (9.9) by the following constraints:

$$\sum_{k=1}^{t} X_k - E_t \leq \sum_{k=1}^{t}\sum_{l=k}^{T} d_{kl} \qquad \forall\, t = 1,\cdots,T. \tag{9.13}$$

Absi et al. [3] propose another variant of this model by allowing lost sales and backlogs. They develop $O(T^2)$ solving algorithms.

9.6 Bounded Production and Inventory

To model some JIT manufacturing environments, different authors introduce lower and upper bounds on inventory levels [2,26,33,35,36,52]. In practice, lower bounds on inventory levels are generally considered as safety stocks that are used to counter demand randomness. If the difference between the demand forecast and the actual customer demand is large, it can cause stockouts and consequently the loss of customers.

JIT systems must take into account a strategic service level. In fact, productions must take place as late as possible to avoid inventory costs, this can be done by setting an upper bound on the inventory level. But a service level must be ensured to avoid the loss of demands and clients, this can be done by introducing a lower bound on the inventory level. This service level depends on the accuracy of the demand forecast, production, shortages and inventory cost components and the available capacities of production and storage.

The use of safety stocks is a strategic decision that ensures a service level towards customers. If safety stock levels are large, inventory costs are high but they buffer against demand uncertainty and errors. On the other hand, low safety stock levels will lead to small inventory costs but also a small service level.

Upper bounds on inventory levels are generally used to model a maximum level of anticipation or a limited storage space. If this parameter is low, production is done as late as possible. Companies should define inventory corridor levels using upper bounds and lower bounds on stocks to better control production and inventory levels.

Love [36] is the first author who extends the classical ELS problem by imposing upper and lower bounds on the inventory at each period. The author proposes an $O(T^3)$ dynamic programming algorithm to solve the problem. Different authors addressed lot-sizing problems with bounded inventories by proposing new dynamic programming algorithms or generalizing multi-item capacitated lot-sizing problems by introducing safety stock or maximum stock levels (Sandbothe and Thompson [52], Loparic et al. [35], Gutiérrez et al. [26] and Liu [33]).

Absi and Kedad-Sidhoum [2] propose the first lot-sizing model that integrates safety stock levels as a target to reach rather than a strong constraint to satisfy. They also consider the idea of target demand levels by using a lost sale model. The authors propose a polynomial dynamic programming algorithm to solve the extended version of the ELS problem.

Recently, van den Heuvel and Wagelmans [54] show that there exists an equivalence between lot-sizing problems dealing with bounded inventory, remanufacturing option, production time windows or cumulative capacities. These four models can be used to model different JIT constraints.

Upper and lower bounds on production quantities can also be used to model some JIT aspects. In fact, the JIT production leveling approach tries to smooth and balance workloads; this can be done by fixing maximum and minimum production levels. Love [36] is also the first who extends the ELS problem by introducing lower and upper bounds on production.

9.7 Dealing with Capacity Constraints

Generally, to solve practical problems, ELS models are embedded in more complicated models involving multiple items, tight capacity constraints, setup times and specific industrial constraints. The classical MCLS has often been studied in the literature, for which obtaining optimal and sometimes even feasible solutions remains challenging.

The MCLS problem consists in finding a production plan that satisfies the demand of N items and capacity constraints, and that minimizes the cumulative production, setup and inventory holding costs. The items share the same available resources. The MCLS problem can be formulated using an ELS model for each item and adding aggregate capacity constraints that link, at each period, the resource consumption of all items. Capacity constraints ensure that the overall used resources (unitary and setup times) at each period does not exceed the available capacity.

Chen and Thizy [16] prove that the MCLS problem is strongly NP-hard. Florian et al. [25] and Bitran and Yanasse [10] show that the single-item capacitated lot-sizing problem is NP-hard in the ordinary sense. They propose a pseudo-polynomial dynamic programming algorithm to solve this problem optimally. Different solving approaches have been proposed in the literature, exact methods based on dynamic programming, branch-and-bound, branch-and-cut, and cutting plane or heuristic approaches such as Lagrangian heuristics, meta-heuristics, and MIP-based heuristics.

In JIT systems, when capacity constraints are tight, production must be spread over the horizon to obtain feasible solutions. In this case, the JIT principle of the zero-inventory property cannot be used. Most Advanced Planner and Scheduler Advanced Planning Systems (APS) software offer a constructive method called *JIT approach* to solve production planning problems [55]. These heuristics are not based on a cost model but on a priority based selection of items and periods. These heuristics generally contain backward and forward passes that try to smooth workload over the horizon to satisfy capacity constraints. The main drawback of these heuristics is the resulting large number of setups.

Trigeiro et al. [48] are among the first who try to solve the classical MCLS problem. They propose a Lagrangian relaxation approach to get good lower bounds and a smoothing heuristic to obtain feasible production plans. They relax capacity constraints and penalize their violations in the objective function.

The general idea behind a smoothing heuristic is to start from the Lagrangian solution, which is most often infeasible, to construct a feasible solution. Production quantities are moved from one period to another in order to avoid capacity violations. Trigeiro et al. [48] propose a smoothing heuristic called TTM. This myopic heuristic iteratively performs slight modifications of the Lagrangian solution so as to satisfy capacity constraints. TTM has a maximum of four passes and stops either when a feasible solution is obtained for the original MCLS problem, or when the maximum allowed number of passes is reached. In the latter case, the procedure may not manage to remove overcapacity.

TTM starts with a backward pass that moves production quantities from the end to the beginning of the horizon to eliminate capacity violations, these moves are based on Lagrangian costs. If the solution is still infeasible, a forward pass is used to move production from the beginning to the end of the horizon to remove overcapacity. The third and fourth passes are identical to the first and second passes. When a feasible solution is found, a fix-up pass is used to eliminated unnecessary inventory.

This heuristic is somehow a JIT heuristic based on smoothing workloads. In fact, this is the goal of Production Leveling in JIT when several products are produced by the same system in order to reduce production, setup and inventory holding costs. The advantage of the TTM heuristic is that it is based on cost criteria. The smoothing heuristic of Trigeiro et al. [48] was generalized by different authors to solve different variants of the MCLS problem [2, 14, 38, 47], or used to obtain feasible solutions for other approaches such as Branch-and-Price [21].

Yavuz and Akçali [61] provide a good review on models and solution approaches for production smoothing in JIT manufacturing systems. The authors notice that the majority of existing studies concentrates on assembly line systems which are, in part, due to the fact that the JIT philosophy started from the Toyota Production System. They mention that there is a significant amount of works in the literature that neglect setup times which may be difficult to justify in practice. They also point that the final stage of manufacturing operations cannot be considered as a synchronized assembly line. The authors conclude that one of the challenges is to consider setup times in JIT production smoothing problems.

9.8 Integrated Lot-Sizing and Scheduling

Even though this is a very challenging task, more and more researchers are studying the integration of detailed scheduling constraints in lot-sizing decisions. From a JIT perspective, this helps to more precisely model the impact of lot sizes on the capacity. Setups can be reduced by selecting the right items and production sequence in a period or across multiple periods. In classical planning and lot-sizing models, the lead time of an item is assumed to be fixed, usually an integer number of periods, and independent of the lot size of this item but also of other items that are produced simultaneously. Thus, lead times maybe be underestimated in some periods and overestimated in other periods. When lead times are underestimated, lots will start and thus be completed too early, leading to undesired inventory. When lead times are overestimated, lots will be completed too late, thus inducing costly delays. Motivations for integrated models are discussed in detail in [19].

When explicitly integrating scheduling constraints, a schedule of the lots is determined. This integration is done in an indirect way in lot-sizing problems with small time buckets such as the DLSP (see Sect. 9.2). Time periods in the horizon are so small that only one product can be processed on a resource in each period, and scheduling constraints have to be modeled. However, the resulting models become rapidly intractable with the horizon length or the number of products (see [63]). An explicit lot-sizing and job-shop scheduling model is studied and solved in [31] and [18]. In this model, aggregate capacity constraints are replaced with detailed job-shop scheduling constraints. As discussed in [19], if lead times are fixed, they are considered as scheduling constraints to satisfy, that is each lot must be started and completed within a specified time window. However, lead times can also be determined since an actual schedule of the lots is calculated. These lead times may vary over time depending on the quantities that are planned to be produced for all items. Actually, a combination of fixed and variable lead times can be used.

9.9 Conclusion and New Research Directions

In this chapter, we tried to cover the main characteristics of lot-sizing models and show the links with a JIT approach. Lot-sizing problems help to tackle JIT issues such as optimized Production Leveling, and formalize JIT-related constraints such as inventory limitations or constraints induced by JIT partnership.

Lot-sizing is a rich field of research that has seen a renewed interest in the past ten years. This is due to the numerous possible applications in Advanced Planning Systems (APS). Many research avenues related to JIT need to be explored. There is a need to study single-item models with additional constraints on inventories, such as dynamic safety stocks. Current lot-sizing safety stock strategies are fixed a priori. An interesting research direction is to create new lot-sizing models that calculate dynamic safety stock levels depending not only on the problem parameters, but also on production levels that are fixed through the lot-sizing model. In fact, there is a strong relationship between safety stock levels and production periods. It is also relevant to extend to the multi-item case some of the complex single-item models, in particular when several characteristics are combined such as demand backlog and production time windows. Moreover, a lot remains to be done to develop effective models and approaches for multi-stage problems where demands of some items depend on the production or order quantities of others. This is probably in these problems that controlling inventories will the most cost-effective. The resulting models will support a better coordination between production, replenishment and distribution, and thus ensure JIT decisions throughout the supply chain. Finally, there are numerous research perspectives on combining lot-sizing and scheduling, whether it is scheduling constraints in lot-sizing decisions, as discussed in Sect. 9.8, or lot-sizing possibilities in JIT scheduling decisions, usually called lot-streaming. As mentioned in the review paper of Chang and Chiu [15], lot streaming combined with lot splitting and scheduling is one of the most effective techniques to quickly produce and deliver products by compressing manufacturing lead times. Lot-streaming models could help to determine the right number and sizes of lots in Kanban systems.

References

1. Absi N. and Kedad-Sidhoum S.: The multi-item capacitated lot-sizing problem with setup times and shortage costs. European Jounal of Operational Research **185(3)**, 1351–1374 (2008)
2. Absi N. and Kedad-Sidhoum S.: The multi-item capacitated lot-sizing problem with safety stocks and demand shortage costs. Computers & Operations Research **36(11)**, 2926–2936 (2009)
3. Absi N., Kedad-Sidhoum S., and Dauzère-Pérès S.: "Uncapacitated lot-sizing problem with production time windows, early productions, backlogs and lost sales". International Journal of Production Research, **49(9)**, 2551–2566 (2011)
4. Aggarwal A. and Park J.K.: Improved algorithms for economic lot-size problems. Operations Research **41**, 549–571 (1993)
5. Aksen D., Altinkemer K., and Chand S.: The single-item lot-sizing problem with immediate lost sales. European Journal of Operational Research **147**, 558–566 (2003)
6. Aksen D.: Loss of customer goodwill in the uncapacitated lot-sizing term problem. Computers & Operations Research **34(9)**, 2805–2823 (2007)
7. Allahverdi A. and Soroush H.M.: The significance of reducing setup times/setup costs. European Journal of Operational Research **187**, 978–984 (2008)
8. Belvaux G. and Wolsey L.A.: Modelling Practical Lot-Sizing Problems as Mixed-Integer Programs. Management Science **47(7)**, 993–1007 (2001)
9. Berk E., Toy A.O., and Hazir O.: Single item lot-sizing problem for a warm/cold process with immediate lost sales. European Journal of Operational Research **187**, 1251–1267 (2008)
10. Bitran G., and Yanasse H.H.: Computational complexity of the capacitated lot size problem. Management Science **28**, 1174–1186 (1982)
11. Bonney M.C., Zhang Z., Head M.A., Tien C.C., and Barson R.J.: Are push and pull systems really so different? International Journal of Production Economics **59**, 53–64 (1999)
12. N. Brahimi: Production planning: models and algorithms for lot-sizing problems. PhD Thesis, Ecole des Mines de Nantes, France (2004)
13. N. Brahimi, Najid N.M., Dauzère-Pérès S., and Nordli A.: Single item lot sizing problems. European Journal of Operational Research **168(1)**, 1–16 (2006)
14. N. Brahimi, Dauzère-Pérès S., and Najid N.: Capacitated multi-item lot-sizing problems with time windows. Operations Research **54(5)**, 951–967 (2006)

15. Chang J. H. and Chiu J.H.: A comprehensive review of lot streaming. International Journal of Production Research **43(8)**, 1515–1536 (2005)
16. Chen W.H. and Thizy J.M.: Analysis of relaxations for the multi-item capacitated lot-sizing problem. Annals of Operations Research **26**, 29–72 (1990)
17. Christensen W., Germain R. and Birou L.: Build-to-order and just-in-time as predictors of applied supply chain knowledge and market performance. Journal of Operations Management **23**, 470–481 (2005)
18. Dauzère-Pérès S. and Lasserre J.B.: Integration of lotsizing and scheduling decisions in a job-shop. European Journal of Operational Research **75(2)**, 413–426 (1994)
19. Dauzère-Pérès S. and Lasserre J.B.: On the importance of sequencing decisions in production planning and scheduling. International Transactions in Operational Research **9(6)**, 779–793 (2002)
20. Dauzère-Pérès S., Brahimi N., Najid N. and Nordli A.: The single-item lot sizing problem with time windows. Technical report 02/4/AUTO, Ecole des Mines de Nantes, France (2002)
21. Degraeve Z. and Jans R.: A New Dantzig-Wolfe Reformulation and Branch-and-Price Algorithm for the Capacitated Lot-Sizing Problem with Setup Times. Operations Research **55(5)**, 909-920 (2007)
22. Diaby M., Bahl H.C., Karwan M.H., and Zionts S.: A Lagrangean Relaxation Approach for Very-Large-Scale Capacitated Lot-Sizing. Management Science **38(9)**, 1329–1340 (1992)
23. Dixon P.S., Elder M.D., Rand G.K., and E.A. Silver: A heuristic algorithm for determining lot sizes of an item subject to regular and overtime production capacities. Journal of Operations Management, **3**, 121–130 (1983)
24. Federgruen A. and Tzur M.: A simple forward algorithm to solve general dynamic lot-sizing models with n periods in $O(n \log n)$ or $O(n)$ time. Management Science **37**, 909-925 (1991)
25. Florian M., Lenstra J.K., and Rinnoy Kan A.H.G.: Deterministic production planning : algorithms and complexity. Management Science **26**, 669–679 (1980)
26. Gutié rrez J., Sedeno-Noda A., Colebrook M. and Sicilia J.: A new characterization for the dynamic lot size problem with bounded inventory. Computers & Operations Research **30**, 383–395 (2002)
27. Hwang H.: Dynamic lot-sizing model with production time windows. Naval Research Logistics **54(6)**, 692–701 (2007)
28. Hwang H. and van den Heuvel W.: Economic Lot-Sizing Problem with Bounded Inventory and Lost-Sales. EI report series (Int. rep. EI 2009-01), Erasmus University Rotterdam (2009)
29. Hsu V.N. and Lowe T.J.: Dynamic economic lot size models with period-pair-dependent backorder and inventory costs. Operations Research **49(2)**, 316–321 (2001)
30. Jaber M.Y., Bonney M., and Moualek I.: Lot sizing with learning, forgetting and entropy cost. International Journal of Production Economics **118**, 19–25 (2009)
31. Lasserre J.B.: An Integrated Model for Job-Shop Planning and Scheduling. Management Science **38(8)**, 1201–1211 (1992)
32. Lee C.-Y., Cetinkaya S. and Wagelmans A.P.M.: A dynamic lot-sizing model with demand time windows. Management Science **47**, 1384–1395 (2001)
33. Liu T.: Economic lot sizing problem with inventory bounds. European Journal of Operational Research **185**, 204–215 (2008)
34. Liu X., Chu F., Chu C., and Wang C.: Lot sizing with bounded inventory and lost sales. International Journal of Production Research **45(24)**, 5881–5894 (2007)
35. Loparic M., Pochet Y., and Wolsey L.A.: The uncapacitated lot-sizing problem with sales and safety stock. Mathematical Programming **89**, 487–504 (2001)
36. Love S.F.: Bounded production and inventory models with piecewise concave costs. Management Science **20(3)**, 313–318 (1973)
37. Luss H. and Rosenwein M.B.: A lot-sizing model for just-in-time manufacturing. Journal of the Operational Research Society **41(3)**, 201–209 (1990)
38. Millar H.H. and Yang M.: Lagrangian heuristics for the capacitated multi-item lot-sizing problem with backordering. International Journal of Production Economics **34**, 1–15 (1994)
39. Monden Y.: Toyota Production System. Institute of Industrial Engineers Press, (1983)

40. Özdamar L. and Birbil S.I.: Hybrid heuristics for the capacitated lot sizing and loading problem with setup times and overtime decisions. European Journal of Operational Research, **110(3)**, 525–547 (1998)
41. Özdamar L. and Bozyel M.A.: The capacitated lot sizing problem with overtime decisions and setup times. IIE Transactions, **32(11)**, 1043–1057 (2000)
42. Pochet Y. and L.A. Wolsey: Production planning by mixed integer programming. Springer, (2006)
43. Riezebos J., Klingenberg W., and Hicks C.: Lean Production and information technology: Connection or contradiction? Computers in Industry **60**, 237–247 (2009)
44. Salomon M. and Kroon Leo G., Kuik R. and van Wassenhove Luk N.: Some Extensions of the Discrete Lotsizing and Scheduling Problem. Management Science **37(7)**, 801–812 (1991)
45. Sox C.R. and Gao Y.: The capacitated lot sizing problem with setup carry-over. IIE Transactions **31(2)**, 173–181 (1999)
46. Suerie C. and Stadtler H.: The capacitated lot-sizing problem with linked lot sizes. Management Science **49(8)**, 1039–1054 (2003)
47. Süral H., Denizel M., and Van Wassenhove L.N.: Lagrangean relaxation based heuristics for lot sizing with setup times. European Journal of Operational Research **194(1)**, 51–63 (2009)
48. Trigeiro W. and Thomas L.J., and McLain J.O.: Capacitated lot-sizing with setup times. Management Science **35**, 353–366 (1989)
49. van den Heuvel W. and Wagelmans A.P.M.: A holding cost bound for the economic lot-sizing problem with time-invariant cost parameters. Operations Research Letters **37**, 102–106 (2009)
50. Van Hoesel S., Kuik R., Salomon M., and Van Wassenhove L.N.: The single item discrete lot-sizing and scheduling problem: Optimization by linear and dynamic programming. Discrete Mathematics **48**, 289–303 (1994)
51. Sandbothe R.A. and G.L. Thompson: A forward algorithm for the capacitated lot size model with stockouts. Operations Research **38(3)**, 474–486 (1990)
52. Sandbothe, R.A. and Thompson G.L.: Decision horizons for the capacitated lot size model with inventory bounds and stockouts. Computers & Operations Research **20(5)**, 455–465 (1993)
53. White R. E. and Prybutok V.: The relationship between JIT practices and type of production system. Omega **29**, 113–124 (2001)
54. W. van den Heuvel and A.P.M. Wagelmans: Four equivalent lot-sizing models. Operations Research Letters **36(4)**, 465–470 (2008)
55. Vollmann T.E., Berry W.L., Whybark D.C., and Jacobs F.R.: Manufacturing planning and control systems for supply chain management. McGraw-Hill, (2005)
56. Vörös J.: Lot sizing with quality improvement and setup time reduction. European Journal of Operational Research **113**, 568–574 (1999)
57. A. Wagelmans, Van Hoesel S., and Kolen A.: Economic lot sizing: an $O(nlogn)$ that runs in linear time in the Wagner-Whitin case. Operations Research **40(1)**, S145-S156 (1992)
58. Wagner H.M. and Whitin T.M.: Dynamic version of the economic lot size model. Management Science **5(1)**, 89–96 (1958)
59. Wolsey L. A.: Lot-sizing with production and delivery time windows. Mathematical Programming **107(3)**, 471–489 (2006)
60. Yang J. and Deane R. H.: A lotsize reduction model for just-in-time manufacturing systems. Integrated Manufacturing Systems **13(7)**, 471–488 (2002)
61. Yavuz M. and Akçali E.: Production smoothing in just-in-time manufacturing systems: a review of the models and solution approaches. International Journal of Production Research **45(16)**, 3579–3597 (2007)
62. Zangwill W.: A backlogging model and a multi-echelon model of a dynamic economic lot size production system–A network approach. Management Science **15**, 506–527 (1969)
63. Zhu X. and Wilhelm W. E. :Scheduling and lot sizing with sequence-dependent setup: A literature review. IIE Transactions **38(11)**, 987–1007 (2006)

Chapter 10
Strategic Scheduling Games: Equilibria and Efficiency

Laurent Gourvès, Jérôme Monnot, and Orestis A. Telelis

Abstract Motivated by today's decentralized operation of interconnected computing platforms, classical task scheduling models are revisited under a game theoretic perspective. Instead of being designed by a central entity which aims at optimizing an aggregate efficiency measure, task allocations emerge through aggregated localized decisions taken by a group of autonomous self-interested agents. The outcome is sought as an equilibrium whose overall social efficiency typically diverges from the optimal group's choice. This divergence, captured by a measure that came to be known as the Price of Anarchy, can be alleviated by local scheduling policies called Coordination Mechanisms. This chapter reviews standard task scheduling models, dedicated coordination mechanisms and their influence on the price of anarchy. It also exemplifies the design and analysis of coordination mechanisms on a particular scheduling model with setup times, and discusses open research questions.

10.1 Introduction

The growth and popularity of large-scale distributed systems and applications has spawned in recent years novel theoretical models for their design and performance optimization. Such systems include computational grids for computationally intensive scientific applications, peer-to-peer file sharing communities, distributed replication and content dissemination systems and, last but not least, the World Wide Web itself as an application deployed over the internet. The underlying infrastructure consists of a vast amount of interconnected resources, including processors,

Laurent Gourvès and Jérôme Monnot
LAMSADE, CNRS FRE 3234, Université de Paris-Dauphine, 75775 Paris, France,
e-mail: {laurent.gourves, monnot}@lamsade.dauphine.fr

Orestis A. Telelis
Center for Mathematics and Computer Science (CWI) Amsterdam, The Netherlands,
e-mail: telelis@cwi.nl

R.Z. Ríos-Mercado and Y.A. Ríos-Solís (eds.), *Just-in-Time Systems*, Springer Optimization and Its Applications 60, DOI 10.1007/978-1-4614-1123-9_10,

distributed storage, routers, network links, to name a few basic. Users of modern distributed environments are driven by individual constraints and are self-interested, in that they aim at optimizing their own objectives. They are completely autonomous and, to this end, their individual decisions and actions in utilizing the wealth of system resources may harm the global system performance. Taking into account the users' autonomous and self-interested behavior is, therefore, of crucial importance in designing an environment of distributed resources. This concern has greatly contributed in shaping the fast growing field of *Algorithmic Game Theory* (AGT) within the computer science community. AGT lies at the intersection of Theoretical Computer Science and Game Theory as a field concerned with computational considerations on Game Theoretical grounds. We refer the reader to [42] for the multiple facets of AGT. In this short contribution, we examine an algorithmic game-theoretic approach to optimizing the performance of distributed systems utilized by autonomous self-interested users.

Koutsoupias and Papadimitriou proposed in their seminal work [36] that, in contexts modeled as strategic games [43] with autonomous self-interested agents, performance evaluation is carried in terms of the overall efficiency of the system at *Nash equilibrium* configuration. In this text, although some preliminaries are provided, we assume a certain level of familiarity with strategic games and the concept of Nash equilibrium; for an excellent introduction the reader is referred to [42, 43]. The overall efficiency of a Nash equilibrium configuration is naturally measured as an aggregate function over the agents' (players') individual objectives. This overall cost is often also referred to as *social cost*. In [36], the authors introduce the *worst-case coordination ratio* as the ratio of the cost of the socially most expensive Nash equilibrium over the socially optimum cost. This ratio has come to be widely referred to as the *Price of Anarchy*. The Price of Anarchy was largely exemplified in [36] on games emerging from the context of multi-processor scheduling. Each agent owns a job of a certain load and chooses one out of m machines to assign his job to. The cost of each agent is the total load assigned on the machine that he chooses, as a measure of the congestion incurred on that machine. The social cost is defined as the *maximum* cost over all players. If we view the players' costs as completion times then the social cost coincides with the *makespan* objective function, widely used in the scheduling literature. The makespan is simply the completion time of the machine that finishes last. The first bounds on the Price of Anarchy in strategic games were hence exemplified in [36] on a selfish scheduling game.

Strategic multiprocessor scheduling games are of importance in the wider context of managing distributed resources. The vast number of resources and users in modern distributed environments renders centralized deployment of global resource management policies expensive and inefficient. It is impossible for a centralized authority to optimize the assignment of loads to resources across the system. Such loads may of course be jobs to be processed, traffic to be forwarded across network links, data to be stored, replicated and disseminated etc. One reason is the scale of the system which makes information gathering and processing impossible; another is the autonomic nature of the loads' owners; they are expected to behave strategically anyway, even if they agree to comply with a suggestion, because they may misreport data associated with their loads. How can then a central authority manipulate the congestion caused on resources towards a cost efficient equilibrium? The idea is to specify local operational rules per resource, so as to *coordinate* users towards globally efficient system utilization. The deployment of such local rules was formalized in [19], under the notion of *coordination mechanisms* and demonstrated for scheduling and network congestion games. In scheduling, local policies are scheduling algorithms deployed on the machines; their set is referred to as a coordination mechanism. A policy decides the way that jobs are executed on each machine. This can be thought of as indirectly defining the completion time that agents experience on the machine they choose. As an example, in the game model studied in [36] the agents' completion times equal the total load assigned to the machines they choose. This cost can be interpreted as completion time under a scheduling policy that breaks down the jobs into infinitesimally small pieces and executes them in a round robin fashion, allocating each job execution time (processing power) proportionally to its size. This way all jobs complete at the same time, which coincides with the total load assigned to the machine.

The policies may be preemptive or non-preemptive, deterministic, or randomized; a policy decides the order of execution of assigned jobs on each machine and may also introduce delays in a systematic manner. A coordination mechanism induces a strategic game, by affecting the players' completion times. Apart from achievement of cost-efficient Nash equilibria, there are some other natural requirements imposed on the choice of appropriate coordination mechanisms. In particular, we are mostly interested for games where Nash equilibria can be found efficiently (i.e. in polynomial time) or can be reached efficiently by the agents, when they follow a simple natural protocol of assigning their loads to resources. In their most general form, Nash equilibria are configurations under which agents may play a probability distribution over their assignment choices (machines), so as to minimize the expectation of their cost (completion time). In practice though, it is unlikely that any such agent will randomize over his strategy space, as his regret for a choice is directly observable on the actual immediate outcome. Furthermore, we are not aware of some simple protocol under which agents may reach a mixed Nash equilibrium. Thus we restrict ourselves to *pure* Nash equilibria, where every agent plays deterministically one action from his strategy space. The drawback with pure Nash equilibria is that they do not always exist in strategic games. Therefore, in choosing a coordination mechanism, we always impose the constraint that the induced game has pure Nash equilibria. A protocol by which these equilibria may be reached in particular classes of games is the *iterative best response* procedure, where in each iteration a player is chosen to improve his strategy given the strategy of all other players; when no player may perform an improvement any more, the configuration reached is a pure Nash equilibrium. Monderer and Shapley showed in their influential work [38] that this procedure converges when the underlying strategic game belongs to the class of *potential games*. We are interested in coordination mechanisms that have pure Nash equilibria that can be found efficiently by algorithms/protocols that resemble closely the iterative best response algorithm.

10.1.1 Contents

In what follows we provide a brief account of selfish scheduling games, that have emerged by application of the notion of coordination mechanisms. We start by providing a few basics on strategic games in Sect. 10.2 and definitions with respect to cost-efficiency of stable configurations (Sect. 10.2.2). Our preliminary discussion is concluded with definitions and general remarks on scheduling games and coordination mechanisms in Sect. 10.2.3. In Sect. 10.3, we discuss several results regarding the most studied coordination mechanisms from the literature. These include the scheduling model (preemptive mechanism) introduced in [36] and further studied in [9, 20, 23, 24, 28, 46] (Sect. 10.3.1), non-preemptive ordering mechanisms studied in [5, 19, 22, 34] (Sect. 10.3.2) and a brief account of mechanisms for the most general scheduling environment of unrelated machines (Sect. 10.3.5) [10, 14, 34]. In Sect. 10.4, we revisit non-preemptive ordering mechanisms in the context of truthfulness [18, 19], where agents may misreport their processing loads to the mechanism so as to minimize their completion time in the outcome of the game.

Subsequently, we illustrate the design of coordination mechanisms and concerns raised by the introduction of *setup overheads*, additionally to the loads placed by autonomous users on the resources (machines) of the system [30]. The model is introduced in Sect. 10.5. In Sect. 10.5.1, we provide an analysis of the mechanism of [36] on this model. Then we introduce a class of mechanisms in Sect. 10.6, for this model, for which we prove existence of pure Nash equilibria and a lower bound on their performance on identical machines. In Sect. 10.6.2, we identify and analyze an optimum mechanism out of this class. We conclude our presentation by mentioning important open research problems for coordination mechanisms.

10.2 Preliminaries and Notation

10.2.1 Strategic Games and Equilibria

A *strategic game* Γ is a tuple $\langle N, (\Sigma_j)_{j\in N}, (c_j)_{j\in N}\rangle$, where N is the set of players and Σ_j is the *set of actions* for player j. In general, the *strategy space* of player j consists of all possible probability distributions over his set of actions Σ_j [43]. A probability distribution is then referred to as a *mixed strategy*. In this text, we will be studying deterministic players, which play *pure* strategies, that is an element of Σ_j with probability 1. Then, Σ_i coincides with the strategy space of each player i and will be referred to as such. A strategy for player j will be an element of Σ_j, denoted by s_j. A *strategy profile* or *configuration* – denoted by $s = (s_1, \dots, s_n)$ – is a vector of strategies, one for each player. The set of all strategy profiles is $\Sigma = \times_{j\in N}\Sigma_j$. Every player j has an individual *cost function* $c_j : \Sigma \to \mathbb{R}$ which depends on the strategy profile. We use the standard notation s_{-j} for $(s_1, \dots, s_{j-1}, s_{j+1}, \dots, s_n)$. For any subset of players $J \subseteq N$ we generalize this notation to s_{-J}, to refer to the part of s restricted to strategies of players in $N \setminus J$. We use s_J to refer to the part of s restricted to strategies of players in J. For any subset of players $J \subseteq N$, a strategy profile s, and s'_J such that $s'_j \neq s_j$ for all $j \in J$, let $s' = (s_{-J}, s'_J)$ refer to a new strategy profile under which $s'_j = s_j$ for all $j \in N \setminus J$.

The most prominent notion for describing the configurations that are *outcomes* (solutions) of a strategic game is the *Nash* equilibrium [40, 43]. In its generality it concerns players that may randomize over their action spaces Σ_i; we will restrict here to configuration outcomes that are *pure* Nash equilibria. These concern players that play deterministically a single element out of Σ_j. The Nash equilibrium expresses rationality of players, in that they select the strategy (action), that induces the lowest individual cost for each of them, given the strategies of the rest. Formally:

Definition 10.1. A strategy profile s is a pure Nash equilibrium if for every player $j \in N$ and for every strategy $s'_j \in \Sigma_j \setminus \{s_j\}$, $c_j(s) \leq c_j(s_{-j}, s'_j)$.

Thus, s is stable since no unilateral deviation from s is rational. We will use the abbreviation PNE for Pure Nash Equilibria. Let us note that PNE are not guaranteed to exist in strategic games; in contrast, John Nash's celebrated theorem proves that *mixed* strategy Nash equilibria always exist [40]. The notion of *Strong Equilibrium* (SE), due to Aumann [7] extends the notion of PNE and prescribes stability of configurations against pure joint/coalitional deviation of any subset of agents:

Definition 10.2. [1,7] A strategy profile s is a strong equilibrium if for every subset of players $J \subseteq N$ and every configuration s', where $s'_j \in \Sigma_j \setminus \{s_j\}$ for all $j \in J$, there is at least one player $j_0 \in J$ with $c_{j_0}(s) \leq c_{j_0}(s_{-J}, s'_J)$.

Every SE is also a PNE, because it is by definition resilient to deviations of trivial *singleton* coalitions, that is unilateral pure deviations; to this end it is also a Nash equilibrium, since every mixed (randomized) deviation of any single player is a linear combination of pure deviations. No such unilateral deviation may be profitable for the player under a PNE or SE, since there is no pure unilateral deviation that is profitable for him. For the sake of completeness however, we note that a strong equilibrium is *not* generally resilient to mixed deviations of non-trivial coalitions [1]; the concept only extends PNE with respect to *pure* deviations of coalitions. See [45] for further extensions that may handle also mixed deviations.

10.2.2 Social Cost-Efficiency of Equilibria

We will be interested in quantifying the overall cost efficiency of stable configurations for strategic games, that is configurations that are *Pure Nash equilibria* or *Strong Equilibria*. The *social cost* of a configuration s for a strategic game Γ characterizes how costly s is, aggregatively over the set of players. We use $C(s)$ to denote the social cost. Most commonly $C(s)$ is defined as $\sum_j c_j(s)$ or by

$\max_j c_j(s)$. The definition of the social cost is not fixed because it depends crucially on the strategic situation that the game captures. The social cost is minimized for some configurations that are referred to as *social optima*. These are not necessarily PNE or SE.

Let us give two famous examples which motivate the need for quantification of the social cost-efficiency of equilibria. Consider a network $G = (V, E)$ (e.g. a road network), where V contains a source A and a destination B, and E is the set of arcs (roads). A large *even* number of n players (e.g. cars), each incurring *unit* traffic volume, need to travel from A to B. Every arc e has a latency function ℓ_e whose argument is the total traffic volume traversing e, and which expresses how long this volume takes to pass through arc e. The strategy space of each player is the set of all paths from A to B and his unique goal is to minimize his total transit time. Given a configuration s, let $x_e(s)$ denote the number of players using arc e, that is $x_e(s) = |\{j | e \in s_j\}|$. A player's cost for a source to destination path P is $c_j(s) = \sum_{e \in P} \ell_e(x_e(s))$. Define the social cost to be $C(s) = \sum_j c_j(s) = \sum_{e \in E} x_e(s)\ell_e(x_e(s))$.

10.2.2.1 The Pigou Example

Pigou's example [44] shows how the players, guided by their individual cost functions, can ruin the social cost. The network is such that $V = \{A, B\}$ and two parallel arcs link A and B (see Fig. 10.1a for an illustration). The top link is a narrow – hence sensitive to traffic variation – road with latency function $\ell(x) = x$, where x denotes amount of traffic on this link. The bottom link is a large enough road that can accommodate any traffic volume in n time units. Observe that the social cost is minimized when $\frac{n}{2}$ players play the bottom link and $\frac{n}{2}$ players play the upper link. Let s^* denote this configuration and then, $C(s^*) = (n/2)^2 + n^2/2 = 3n^2/4$. Notice that s^* is not a PNE; any player playing the bottom link has individual cost n. Hence, the player prefers to switch to playing the upper link at an individual cost $\frac{n}{2} + 1$. Now consider a configuration s where all players play the upper link. In this case, the individual cost of every player n and the social cost $C(s)$ is equal to n^2, that is, $\frac{4}{3}$ times greater than $C(s^*)$. Configuration s is a PNE, because any player that may deviate to the bottom link, will not decrease his individual latency below n.

10.2.2.2 Braess' Paradox

In an effort to mitigate the impact of selfish behavior on the social cost, one may be tempted to add some efficient facilities. Braess' paradox [12, 13] goes against intuition since it shows that adding some facilities may damage the social welfare. The network is as depicted on Fig. 10.1b, without the dashed arc. There are two paths and the configuration which splits evenly the traffic on those paths is the only equilibrium, inducing a social cost of $3n^2/2$. Now suppose that a very fast road is built from node D to C, such that every player using that arc at D joins immediately C, ignoring the congestion on the arc (see the dashed arc on Fig. 10.1b). Hence, the previous

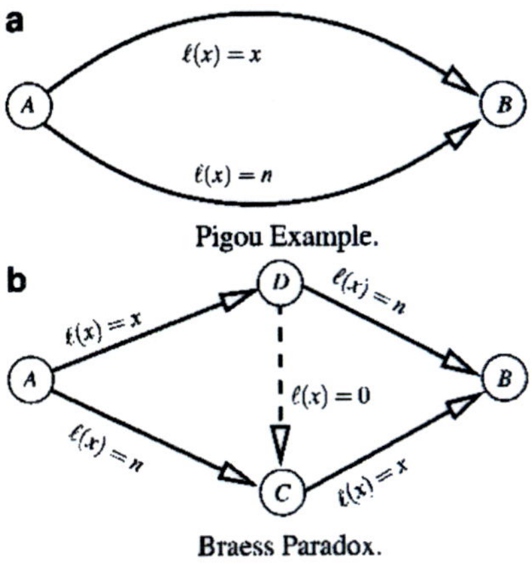

Fig. 10.1: Two classical examples that motivate quantification of efficiency of equilibria (**a**) Pigou example (**b**) Braess paradox

solution is not an equilibrium anymore since the path (A,D,C,B) is more profitable than (A,D,B) or (A,C,B). However, the only equilibrium in the augmented network is when all players adopt (A,D,C,B), inducing a social cost of $2n^2$.

To quantify the inefficiency of stable configurations in comparison to socially optimum ones, Koutsoupias and Papadimitriou introduced in [36] the notion of the *worst-case coordination ratio*, that became widely known as the *Price of Anarchy*; this is the worst-case ratio of the social cost of the most expensive Nash equilibrium over the socially optimum cost.

Definition 10.3. Let Γ denote the set of all game instances for particular strategic game, as defined by its parameters. Let $I \in \Gamma$ refer to any particular instance of the game, when the parameters are instantiated. Let $\mathscr{E}(I)$ denote the set of all stable configurations and for every configuration $s \in \mathscr{E}(I)$ its social cost is denoted by $C(s)$. Finally, let s^* be any configuration that minimizes the social cost C, that is $C(s^*) \leq C(s)$ for every configuration s. The Price of Anarchy of Γ, referred to as $PoA(\Gamma)$ is:

$$PoA(\Gamma) = \sup_{I \in \Gamma} \max_{s \in \mathscr{E}(I)} \frac{C(s)}{C(s^*)}.$$

When the stability concept used is PNE, we simply refer to the Price of Anarchy. When we are interested in SE, we refer to the *Strong* Price of anarchy, abbreviated as *SPoA*. We will drop the argument Γ, since the set of game instances we refer to will be clear from the context. The *Price of Stability* (PoS) was introduced in [4] as the worst-case ratio of the social cost of the least expensive Nash equilibrium over the socially optimum cost:

$$PoS(\Gamma) = \sup_{I \in \Gamma} \min_{s \in \mathscr{E}(I)} \frac{C(s)}{C(s^*)}.$$

This quantifies the efficiency that selfish players may achieve if they coordinate. Coordination is expressed also in terms of strong equilibria, where instability is caused by *joint* coordinated deviations of agents. Ultimately, because every strong equilibrium is a PNE, we have that $PoS(\Gamma) \leq SPoA(\Gamma) \leq PoA(\Gamma)$.

10.2.3 Selfish Scheduling and Coordination Mechanisms

We consider an environment of m resources that we refer to as machines, indexed by $i \in \mathcal{M} = \{1,\ldots,m\}$ and n jobs $j \in \mathcal{J} = \{1,\ldots n\}$. Each job is owned by a self-interested player, thus our set of players will be $N = \mathcal{J}$. We use interchangeably the terms *job* and *player*. Every job j is associated with a processing load $\ell_j > 0$. Each player chooses a machine to assign his job to; thus, the strategy space of any player j will be $\Sigma_j = \mathcal{M}$. Execution of a job $j \in \mathcal{J}$ on a machine $i \in \mathcal{M}$ incurs *processing time* p_{ji} (e.g. processing cycles, or total CPU time). The processing time is decided by the characteristics of the machines environment as follows:

- *Identical Machines:* For every pair of distinct machines $i_1, i_2 \in \mathcal{M}$ we have $p_{j,i_1} = p_{j,i_2} = \ell_j$.
- *Uniformly Related Machines:* Each machine $i \in \mathcal{M}$ is associated with a speed factor $v_i \geq 1$ and $p_{ji} = \ell_j / v_i$.
- *Restricted Assignment Machines:* Every job $j \in \mathcal{J}$ is associated with a subset of machines $M_j \subseteq \mathcal{M}$ such that $p_{ji} = \ell_j$ for every machine $i \in M_j$ and $p_{ji} = \infty$ for $i \in \mathcal{M} \setminus M_j$.
- *Unrelated Machines:* The processing time of job $j \in \mathcal{J}$ on machine $i \in \mathcal{M}$ is p_{ji}.

For any player $j \in \mathcal{J}$ denote by s_j the assignment of job j to machine $s_j \in \mathcal{M}$. We refer to $s = (s_1,\ldots,s_n)$ as an *assignment, configuration* or *strategy profile*. Given an assignment s, player $j \in \mathcal{J}$ experiences an individual cost $c_j(s)$, which denotes the *completion time* of his job on his chosen machine $s_j \in \mathcal{M}$. We denote by $C_i(s)$ the completion time of machine $i \in \mathcal{M}$ under assignment s. Next, we define the social cost function $C(s)$ that will quantify the cost efficiency of an assignment s; this is the maximum completion time of any player and coincides with the maximum completion time of any machine;

$$C(s) = \max_{j \in \mathcal{J}} c_j(s) = \max_{i \in \mathcal{M}} C_i(s).$$

$C(s)$ is commonly referred to in the scheduling literature as the *makespan* of the assignment s. Following the notation of Graham [31,32] we will refer to the makespan minimization problems that correspond to the mentioned machine environments of identical, uniformly related, restricted assignment and unrelated machines as $P,Q,B,R||C_{\max}$ respectively. For the rest of our discussion, the socially optimum cost $C(s^*)$ will be the optimum value of the optimization problems $P,Q,B,R||C_{\max}$.

Following [19] we define a coordination mechanism as the set of scheduling policies (algorithms), that are deployed on the machines; these indirectly define the individual completion times functions $c_j : \mathcal{M}^{|\mathcal{J}|} \rightarrow \Re_+$ of the players. Thus, choosing a coordination mechanism is equivalent to choosing completion time functions for players in $\mathcal{J}$. A coordination mechanism induces a strategic game for the players set $\mathcal{J}$. The design and study of coordination mechanisms concerns indirectly *choosing* a strategic game, that has cost-efficient – preferably pure – Nash equilibria (or even strong equilibria) which can be found efficiently.

A coordination mechanism can be characterized as *strongly local* or simply *local*, depending on the amount of information its constituting policies use to schedule the jobs assigned on the machines that the policies operate on. Under any assignment configuration s, if $J_i = \{j \in \mathcal{J} | s_j = i\}$ denotes the subset of jobs assigned to machine $i \in \mathcal{M}$, a policy deployed on i is *strongly local* if it uses only data of jobs in $j \in J_i$ with respect to machine i. This is better illustrated in the case of unrelated machines; a strongly local policy may use for example the index of any job $j \in J_i$ and its processing time p_{ji} on i. A policy that uses information of jobs in J_i that is relevant to machines other than i is referred to as simply *local*. Such a policy may for example use the processing time $p_{ji'}$ of $j \in J_i$, for any $i' \neq i$. A mechanism is local if it involves a local policy on at least one machine. It is strongly local if policies deployed on all machines are strongly local. Strongly, local mechanisms are desirable of course, as their policies are applicable independently of the characteristics of the machines environment they are deployed upon. Few simply local mechanisms have been studied in the literature [10, 14].

10.3 An Overview of Coordination Mechanisms

We describe below and survey the known results with respect to the most studied mechanisms.

10.3.1 Load Balancing: The `Makespan` *Mechanism*

The `Makespan` mechanism applies on each machine a scheduling policy first studied in a game-theoretic setting in [36]. This policy breaks a job's time requirement into infinitesimally small pieces and executes pieces of all jobs assigned to the machine interchangeably, in a round-robin fashion. Each job is allocated an amount of processing cycles proportionally to its total processing time. All jobs finish at the same time on each machine i, that is $c_j(s) = C_i(s)$ for all $j \in \mathcal{J}$ such that $s_j = i$. Because every player's completion time is the completion time of the machine that he plays, scheduling games under the `Makespan` mechanism are widely mentioned also as *load balancing* games. Selfish choices in this context are motivated by and affect directly the balancing of the total load across the machines.

10.3.1.1 Pure Nash Equilibria: Existence and Convergence

`Makespan` induces a game that has PNE, because it belongs to the class of *potential games*, as shown in [23]. This class was introduced by Monderer and Shapley in [38]. Under any configuration of a potential game, if there is a player that has incentive to deviate so as to improve his individual utility (or cost), then such an improvement also incurs an improvement to a potential function that is associated with the game. As this function may only receive values out of a finite set (one per configuration of the game), its indefinite improvement is impossible. Eventually, a configuration is reached where no agent has an incentive to deviate unilaterally. This configuration is a local optimum for the potential function and a PNE for the game. Using a similar potential function argument, Andelman et al. [1] also proved existence of *strong* equilibria under the `Makespan` mechanism for the most general environment of unrelated machines. We further generalize their proof in Sect. 10.5.1 to prove the same result for even more general scheduling games (Theorem 10.3).

The existence of a potential function that is improved along with every selfish improvement performed by each player suggests a simple procedure for finding PNE: initialize an arbitrary configuration and have iteratively an arbitrary player optimize his strategy against the fixed strategies of all other players. This is known as *Iterative Best Response*. Rate of convergence of Iterative Best Response under the `Maskespan` mechanism was studied extensively by Even-Dar, Kesselman, and Mansour [23]. They considered $R||C_{\max}$ scheduling with m machines and n agents of at most K different processing times on the machines. Apart from iterative best response they studied also *iterative improvements* algorithms, where each agent is not restricted to *best* responses, but merely to strictly improving his completion time. The authors show that even these less restricted iterative algorithms converge to PNE, by introducing a total ordering of all possible configurations of the $R||C_{\max}$ game. Any improvement is shown to lead to a configuration lower in the global ordering, hence the game is an *ordinal potential game*. For $R||C_{\max}$ the space of all possible configurations is shown to have cardinality $\min\{[O((n/Km)+1)]^{Km}, m^n\}$ which is a general upper bound on the of iterative improvements needed for a PNE to be reached. An $O(4^P)$ bound is shown when the processing times on all machines are integers and P denotes the *worst-case* sum of processing times of jobs. Even-Dar et al. proved several more bounds for the $P,Q||C_{\max}$ cases, by using specific rules for choosing the agent that will perform an improvement next [23].

Let us note that computation of PNE for `Makespan` can be carried out efficiently, in polynomial time, for the $P||C_{\max}$ and $Q||C_{\max}$ models. The simple algorithm which assigns the longest unassigned job next to the machine on which it will finish first, given the current partial assignment, finds a PNE. We will study this algorithm for $P||C_{\max}$ under the more general model of *setup times* in Sect. 10.6.1 (Theorem 10.6). This algorithm is also referred to as `LPT` scheduling; the reader is referred to [42] (Chap. 20) for the case of $Q||C_{\max}$ which has an essentially similar proof as for the case of $P||C_{\max}$, even under the model of setup times examined in Sect. 10.6.1. Gairing et al. devised in [27] algorithms for computing cost-efficient PNE for the case of restricted assignment, even when the machines are also uniformly related (a combination of $B||C_{\max}$ and $Q||C_{\max}$). Their algorithms are based on techniques of flow computation. For the case of $B||C_{\max}$ alone, they provide the first polynomial-time algorithm for computing a PNE with social cost within strictly less than 2 of the socially optimum cost.

10.3.1.2 Efficiency of Equilibria

A considerable amount of literature on scheduling games and coordination mechanisms has been concerned with the study of efficiency of the `Makespan` mechanism. The results are summarized in Table 10.1. For $P||C_{\max}$, results of Schuurman and Vredeveld [46] (and, earlier, of Finn and Horowitz [25]) imply that the *PoA* is $2m/(m+1)$. We provide a proof of this result, along with a tight lower bound for strong equilibria.

Theorem 10.1. *The Price of Anarchy of the* `Makespan` *mechanism for selfish scheduling on identical machines is* $2-2/(m+1)$ *and this bound is tight even for strong equilibria.*

Proof. Under an equilibrium assignment s, let i denote the machine that finishes last, that is $C_i(s) \geq C_{i'}(s)$ for all $i' \in \mathscr{M}$. Now consider j to be the job of the smallest processing load among the jobs assigned to i under s, that is $\ell_j = \min\{\ell_k : k \in \mathscr{J} : s_k = i\}$. Then, without loss of generality, there are at least two jobs assigned to machine i, that is there is $j' \neq j$ with $s_{j'} = i$. If this is not the case we obtain $C_i(s) = \ell_j \leq C(s^*)$, because $C(s^*) \geq \ell_j$ for every job $j \in \mathscr{J}$. Now if $s_{j'} = s_j = i$ and $\ell_j \leq \ell_{j'}$, we have $C(s) = C_i(s) \geq \ell_j + \ell_{j'} \geq 2\ell_j$. Because s is a PNE, j does not have incentive to switch to any machine $i' \neq i$, that is $c_j(s) = C_i(s) \leq C_{i'}(s) + \ell_j$ hence $C_{i'}(s) \geq C_i(s) - \ell_j \geq C(s) - \frac{1}{2}C(s) = \frac{1}{2}C(s)$. For the socially optimum configuration s^* we have:

$$C(s^*) \geq \frac{1}{m}\sum_j \ell_j = \frac{1}{m}\sum_{i'} C_{i'}(s) = \frac{1}{m}C_i(s) + \frac{1}{m}\sum_{i' \neq i} C_{i'}(s) \geq \frac{1}{m}C(s) + \frac{(m-1)}{2m}C(s).$$

The latter yields $C(s) \leq 2m/(m+1)C(s^*)$.

Now let us illustrate a lower bound that is also a strong equilibrium. For any instance with m machines, define $n = 2m$ jobs, with $\ell_1 = \cdots = \ell_m = m$ and $\ell_j = 1$, for $j = m+1, \ldots, 2m$. In the socially optimum assignment, we have, $s^*_j = j$ and $s_{m+j} = j$ for $j = 1, \ldots, m$. The makespan of this assignment is $C(s^*) = m+1$. Consider now an equilibrium assignment s, under which $s_j = j$ for $j = 1, \ldots, m-1$, $s_m = 1$ and the m unit-length jobs are executed on machine m, see Fig. 10.2a for an illustration. Then $C(s) = \ell_1 + \ell_m = 2m$. We claim that s is a *strong* equilibrium. By contradiction, assume that there are a coalition $\mathscr{J}' \subseteq \{1, \ldots, 2m\}$ and a state $s' = (s_{-\mathscr{J}'}, s'_{\mathscr{J}'})$ such that $\forall j \in \mathscr{J}'$, $c_j(s') < c_j(s)$. Notice that $c_j(s) = m = \ell_j$ for jobs $j = 2, \ldots, m-1$. Then jobs in $\{2, \ldots, m-1\}$ cannot belong to $\mathscr{J}'$. Also, jobs of $\mathscr{J}'$ must deviate to

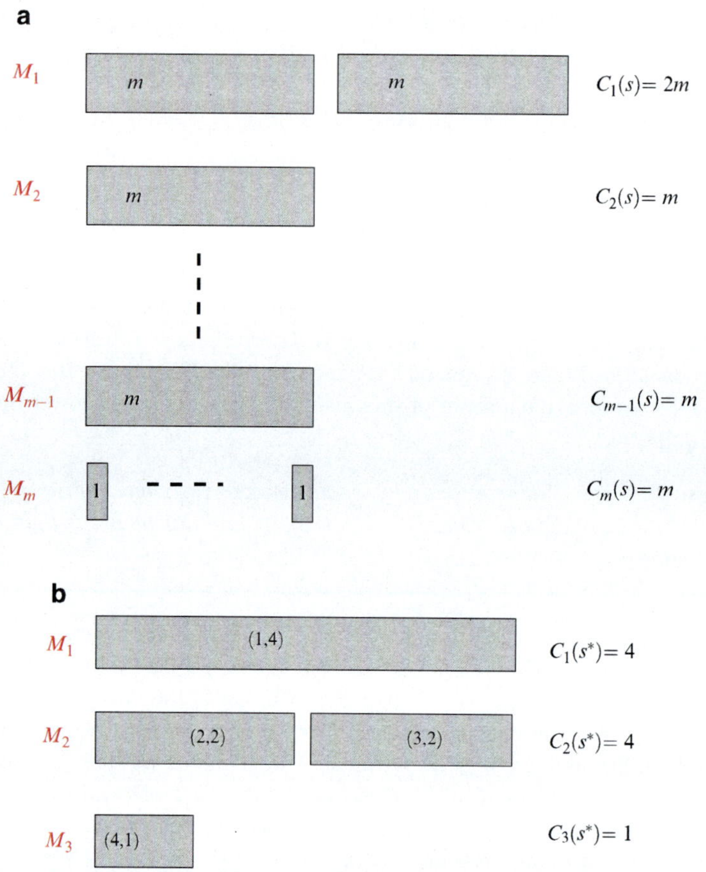

Fig. 10.2: Worst-case SE and unstable social optimum for $P||C_{\max}$ under Makespan (**a**) A strong equilibrium s for $P||C_{\max}$ under Makespan (**b**) A social optimum s^* which is not a Nash equilibrium. Every couple on the picture designates a job's index and load (e.g. $(1,4)$ for job 1 with load 4)

machines 1 or m since $c_j(s) \geq m$ for all jobs and jobs in $\{2,\ldots,m-1\}$ cannot move. Obviously, at least one job of 1 or m must deviate to machine m since otherwise none jobs of $\{m+1,\ldots,2m\}$ can deviate and also the two jobs 1 and m cannot deviate both (otherwise, from previous argument they move together to machine m and then $c_1(s') \geq \ell_1 + \ell_m = 2m = c_1(s)$). Thus, we deduce that exactly one job, either job 1 or job m must deviate to machine m and also that some jobs of $\{m+1,\ldots,m\}$ must deviate to machine 1 (since $c_j(s) = m$ for $j \in \{m+1,\ldots,2m\}$). These latter jobs of $\{m+1,\ldots,2m\} \cap \mathscr{J}'$ increase their completion times, which is impossible. □

Andelman, Feldman, and Mansour showed in [1] that there is always a socially optimum configuration that is a strong equilibrium under Makespan, even for $R||C_{\max}$. However, there may exist socially optimum configurations that are not even PNE. Consider the following instance with three identical machines and four jobs $\mathscr{J} = \{1,\ldots,4\}$ where the processing times satisfy $\ell_1 = 4$, $\ell_2 = \ell_3 = 2$ and $\ell_4 = 1$. State $s^* = (1,2,2,3)$ is a social optimum since $C(s^*) = C_1(s^*) = \ell_1$, see Fig. 10.2 for an illustration. Now, if job 2 moves to machine 3 (corresponding to state $s' = (s^*_{-2}, 3)$), its completion time drops to $3 = c_2(s') < 4 = c_2(s^*)$. Hence, s^* is not a Nash equilibrium. When $m = 2$, the situation is more optimistic:

Proposition 10.1. *For the* `Makespan` *mechanism for selfish scheduling on* $m = 2$ *identical machines ($P||C_{\max}$), any social optimum is a strong equilibrium.*

Proof. Let s^* be an social optimum, and assume that machine 1 is the most loaded machine, that is, $C(s^*) = C_1(s^*) \geq C_2(s^*)$. Moreover, let $\mathscr{J}_i = \{j \in \mathscr{J} \,|\, s^*_j = i\}$ for $i = 1,2$. Finally, let $\mathscr{J}'$ be a coalition and consider the state $s' = (s^*_{-\mathscr{J}'}, s'_{\mathscr{J}'})$ where $s'_j = 1$ for $\mathscr{J}' \cap \mathscr{J}_2$ and $s'_j = 2$ for $\mathscr{J}' \cap \mathscr{J}_1$. We affirm that there is $j \in \mathscr{J}'$ such that $c_j(s') \geq c_j(s^*)$ leading to the conclusion that s^* is a strong equilibrium.

- If $\mathscr{J}' \subseteq \mathscr{J}_2$, then let $j_2 \in \mathscr{J}'$. We have: $c_{j_2}(s') = C_1(s^*) + \sum_{j \in \mathscr{J}'} \ell_j > C_1(s^*) \geq C_2(s^*) = c_{j_2}(s^*)$.
- If $\mathscr{J}' \subseteq \mathscr{J}_1$, then let $j_1 \in \mathscr{J}'$. By contradiction, suppose that: $\forall j \in \mathscr{J}', c_j(s') < c_j(s^*)$. In particular, we obtain: $C_2(s') = c_{j_1}(s') < c_{j_1}(s^*) = C_1(s^*)$. Also we get: $C_1(s') = C_1(s^*) - \sum_{j \in \mathscr{J}'} \ell_j < C_1(s^*)$. Thus, $C(s') = \max\{C_1(s'); C_2(s')\} < C_1(s^*) = C(s^*)$, which is a contradiction since s^* is supposed to be a social optimum.
- If $\mathscr{J}' \cap \mathscr{J}_1 \neq \emptyset$ and $\mathscr{J}' \cap \mathscr{J}_2 \neq \emptyset$, then let $j_i \in \mathscr{J}' \cap \mathscr{J}_i$ for $i = 1,2$. Finally, let $\delta = \sum_{j \in \mathscr{J}' \cap \mathscr{J}_1} \ell_j - \sum_{j \in \mathscr{J}' \cap \mathscr{J}_2} \ell_j$. We get that $c_{j_1}(s') = c_{j_2}(s^*) + \delta = C_2(s^*) + \delta$ and $c_{j_2}(s') = c_{j_1}(s^*) - \delta = C_1(s^*) - \delta$. Hence, $c_{j_1}(s') + c_{j_2}(s') = C_1(s^*) + C_2(s^*)$. Now, if $\forall j \in \mathscr{J}', c_j(s') < c_j(s^*)$, we deduce that $c_{j_1}(s') < c_{j_1}(s^*) = C_1(s^*)$ and $c_{j_2}(s') < c_{j_2}(s^*) = C_2(s^*)$. Thus, $c_{j_1}(s') + c_{j_2}(s') < C_1(s^*) + C_2(s^*)$, which is a contradiction with the previous equality.

In conclusion, for any $\mathscr{J}' \subseteq \mathscr{J}$, there exist $j \in \mathscr{J}'$ such that $c_j(s') \geq c_j(s^*)$ or equivalently s^* is a strong equilibrium. □

For uniformly related machines ($Q||C_{\max}$), Czumaj and Vöcking proved in [20] that `Makespan` achieves $PoA = \Theta(\log m / \log\log m)$. The same bound was shown

Table 10.1: Performance of the `Makespan` mechanism for multiprocessor scheduling

`Makespan` mechanism [36]		
Model	PoA	SPoA
$P\|\|C_{\max}$	$\frac{2m}{m+1}$ [25,46]	$\frac{2m}{m+1}$ [25,46]
$Q\|\|C_{\max}$	$\Theta\left(\frac{\log m}{\log\log m}\right)$ [20]	$\Theta\left(\frac{\log m}{(\log\log m)^2}\right)$ [24]
$B\|\|C_{\max}$	$\Theta\left(\frac{\log m}{\log\log m}\right)$ [9,28]	$\Omega\left(\frac{\log m}{\log\log m}\right)$ [9,28]
$R\|\|C_{\max}$	∞	m [1,24]

independently by Gairing et al. [28] and Awerbuch et al. [9] for restricted assignment machines. We discuss the lower bound for both cases; for the case $B||C_{\max}$ we show that the lower bound is true for strong equilibria as well. In both cases we will use the *Gamma Function* Γ, which is defined by $\Gamma(k+1) = k!$ for any integer $k \geq 1$. Both Γ and its inverse, Γ^{-1} are monotone increasing. It is known that $\Gamma^{-1}(k) = \Theta(\log k / \log\log k)$ as k grows.

Proposition 10.2. *The Price of Anarchy of* `Makespan` *for uniformly related machines is* $\Omega(\log m / \log\log m)$.

Proof. Consider $k+1$ disjoint groups of machines, G_r, $r = 0, \ldots, k$, so that $\mathscr{M} = \cup_{r=0}^{k} G_r$. For $r = 0, \ldots, k$, group G_r contains $|G_r| = \frac{k!}{r!}$ machines, where $0! = 1$. Each machine in group G_r has speed 2^r. Furthermore, We make r jobs of processing length 2^r, for every machine in group G_r. Consider the assignment s under which r jobs of processing length 2^r each are assigned to each of the machines in group G_r, for $r = 1, \ldots, k$. Notice that no jobs are assigned to any of the machines in group G_0. We claim that this assignment is a PNE. Notice that for every machine $i \in G_r$ we have $C_i(s) = r$, for $r = 0, \ldots, k$. Hence, the only potentially profitable deviation for a job j with $s_j \in G_r$ for $r \geq 1$, is to play $s'_j \in G_{r'}$ with $r' < r$. But then

$$c_j(s') = r' + \frac{2^r}{v_{s'_j}} = r' + 2^{r-r'} \geq r' + (r - r' + 1) = r + 1 > c_j(s),$$

since $2^x \geq x + 1$ when $x \geq 1$.
Now the total number of machines is $m = \sum_{r=0}^{k} \frac{k!}{r!} = k! \sum_{r=0}^{k} \frac{1}{r!} \leq 3\Gamma(k+1)$. Then we obtain $k \geq \Gamma^{-1}(\frac{m}{3}) - 1 = \Omega(\log m / \log\log m)$ since Γ^{-1} is an increasing mapping. We show that the socially optimum configuration s^* has makespan $C(s^*) = 2$. Define s^* as the assignment of $\frac{k!}{r!} \times r$ jobs of processing length 2^r on machines of group G_{r-1}. Because $|G_{r-1}| = \frac{k!}{(r-1)!} = \frac{k!}{r!} \times r$, one job is assigned on each machine of G_{r-1} and the makespan is $2^r / 2^{r-1} = 2$. Thus, $PoA \geq C(s)/C(s^*) = \frac{k}{2} = \Omega(\log m / \log\log m)$. □

Proposition 10.3. *The Strong Price of Anarchy of* `Makespan` *for restricted assignment machines is* $\Omega(\log m / \log\log m)$.

Proof. Now let us utilize the same grouping of machines for the case of restricted assignments, with identical machines; assume they all have unit speed. For $r = 1, \ldots k$, make up r jobs per machine in group G_r of unit processing length each. Call J_r the set of these jobs. These jobs may only be assigned to machines of groups G_{r-1} or G_r. Consider the assignment s where jobs J_r are assigned to group G_r, r jobs per machine. This assignment has makespan $C(s) = k$ (actually, for every machine $i \in G_r$ we have $C_i(s) = r$). We claim that s is a strong equilibrium. Consider any joint deviation of any coalition $J \subseteq \mathscr{J}$ of jobs, under s. We show two simple claims.

Claim. If a job $j_1 \in J$ with $s_j \in G_r$ deviates to any machine $s'_{j_1} \in \mathscr{M} \setminus \{s_{j_1}\}$, at least one job $j_2 \neq j_1$, $j_2 \in J$ with $s_{j_2} = s'_{j_1}$ leaves machine $s_{j_2} = s'_{j_1}$ for some machine $s'_{j_2} \neq s_{j_2}$.

This follows by the fact that j_1 may only switch to a machine within G_r, or to a machine within G_{r-1}. If no job leaves j_2 we have, in the first case, $c_{j_1}(s') \geq r+1 > c_{j_1}(s)$ and in the second case it is $c_{j_1}(s') \geq (r-1)+1 = r = c_{j_1}(s)$. Thus, in either case, j_1 would not participate in the joint deviation of J.

Claim. If there is a deviating job $j \in J$ with $s_j \in G_r$, then there must exist at least one job $j' \in J$ with $s_{j'} \in G_r$, that deviates to some machine in G_{r-1}.

This follows because jobs are optimally distributed to each group of machines under s; hence, simply swapping machines among them will not lessen their completion time. Using this fact recursively down to group G_1, there is at least one job $j \in J$ with $s_j \in G_1$, that deviates to some machine of G_0. But j may not decrease its completion time this way because it was $c_j(s) = 1$ and since no jobs are assigned to machines of G_0, it will still be $c_j(s') = 1$. Thus, s is strong. For the optimum configuration s^*, we assign jobs of J_r to machines G_{r-1}, for $r = 1,\ldots,k$. Thus, $C_i(s^*) = 1$ for every machine $i \in G_r$ when $r = 0,\ldots,k$ and then $C(s^*) = 1$. In conclusion, Thus, $PoA \geq C(s)/C(s^*) = k = \Omega(\log m/\log\log m)$. □

A trivial example, however, shows that `Makespan` has unbounded *PoA* for $R||C_{\max}$; take two machines and two jobs with $p_{11} = p_{22} = \varepsilon$ and $p_{12} = p_{21} = W$. Then the configuration $(2,1)$ is a PNE of makespan W while $(1,2)$ is the socially optimum configuration (also a PNE). Thus, $PoA \geq W/\varepsilon$, which may be arbitrarily large for big values of W and arbitrarily small values of ε. Andelman et al. [1] first studied strong equilibria for selfish scheduling under the `Makespan` mechanism. They show that the *SPoA* for $R||C_{\max}$ lies between m and $2m-1$. Fiat et al. subsequently showed that $SPoA = \Theta\left(\log m/(\log\log m)^2\right)$ for $Q||C_{\max}$ and tightened the *SPoA* for $R||C_{\max}$ to m.

10.3.2 Ordering Policies

An ordering policy is intuitively one that orders execution of jobs non-preemptively on a machine. The `Makespan` policy discussed above is a preemptive policy (hence, not an ordering one). To order the jobs on their assigned machine, an *ordering* policy may use the jobs' processing time on the machine, or even their distinct index, to resolve ties in a definitive manner. A characterization of *strongly local* deterministic ordering policies proposed in [10] is through the *Independence of Irrelevant Alternatives* (IIA) property; this roughly dictates that the relative ordering of any two jobs on a machine under a policy satisfying the IIA is independent of the presence of another job. More formally, let $J_i(s) \subseteq \mathscr{J}$ denote the subset of jobs that are assigned on machine i under configuration s. Suppose that for two distinct jobs $j_1, j_2 \in J_i(s)$, j_1 precedes j_2, that is $j_1 \prec_\Pi j_2$, according to an ordering $\prec_\Pi$ induced under s, by the strongly local policy Π deployed on $i \in \mathscr{M}$. Then under any configuration s' with $j_1, j_2 \in J_i(s) \cap J_i(s')$, j_1 we still have $j_1 \prec_{\Pi(s')} j_2$. Notice that under ordering policies the completion time of a job j is dependent on the ordering policy deployed on the machine that j is assigned and on the rest of the jobs assigned to this machine. In particular, if $\Pi(i)$ denotes the ordering policy deployed on machine $i \in \mathcal{M}$, we have:

$$c_j(s) = p_{j,s_j} + \sum_{j' \prec_{\Pi(s_j)} j : s_{j'} = s_j} p_{j',s_j}.$$

Below we describe known results with respect to two well studied ordering policies satisfying the IIA property. Bounds on the *PoA* of these policies and on the randomized ordering policy discussed subsequently are summarized in Table 10.2.

Shortest Processing Time (SPT) Under this mechanism jobs on all machines are executed by shortest processing time first. Ties are resolved consistently across all machines on the basis of jobs' indices (i.e. smaller or larger index first), hence jobs are totally ordered under SPT. This is an ordering non-preemptive policy, in contrast to Makespan which is preemptive. It was shown in the classical work of Graham [32] that SPT achieves $PoA = 2 - \frac{1}{m}$ for $P||C_{\max}$. The related *shortest-first greedy algorithm* was shown to exhibit approximation ratio of $O(\log m)$ for $Q||C_{\max}$ and $B||C_{\max}$, by Aspnes et al. [5] and Azar et al. [11], respectively. Cho and Sahni showed a lower bound of $\Omega(\log m)$ for $Q||C_{\max}$ [17]. Immorlica et al. used these results in [34] to prove a *PoA* of $\Theta(\log m)$ for SPT in $Q,B||C_{\max}$. In particular, they showed that the set of pure Nash equilibria under SPT, even for $R||C_{\max}$, coincides with the set solutions that can be returned by the shortest-first greedy algorithm. They also proved a general upper bound of $O(\log m)$ on the *PoA* of any deterministic ordering policy for $Q||C_{\max}$ and for $B||C_{\max}$. We provide the proof for the $Q||C_{\max}$ below in Theorem 10.2. For $R||C_{\max}$ a *PoA* of m for SPT emerges by the analysis of the shortest-first greedy algorithm, a result due to Ibarra and Kim [33]. An asymptotically matching lower bound was shown by Azar et al. in [10] for any deterministic ordering policy (we survey their results later in Sect. 10.3.5). Interestingly, SPT induces a potential game, even on unrelated machines as shown in [34]. Furthermore, an iterative best response performed by the players converges in exactly n rounds, when each job is given the chance for best response in every round, in arbitrary order.

Longest Processing Time (LPT) Under LPT jobs are scheduled on all machines by LPT first. Ties are resolved consistently across all machines on the basis of jobs' indices (i.e. smaller or larger index first), hence jobs are totally ordered under LPT. It is a non-preemptive ordering policy that achieves $PoA = \frac{4}{3} - \frac{1}{3m}$ on identical machines and at most $O(\log m)$ on uniformly related and restricted assignment machines. The first bound emerges by the analysis of the *longest-first greedy algorithm* by Graham [32] (upper bound) and by the analysis of Christodoulou et al. [19] (lower bound). The second comes from the analysis of [34]. Immorlica et al. show in [34] that the set of pure Nash equilibria obtained under LPT for $P,Q,B||C_{\max}$ is exactly the set of solutions that can be returned by the *longest-first greedy algorithm*. This yields $PoA \in [1.52, 1.59]$ for $Q||C_{\max}$, by the works of Dobson [21] and Friesen [26] on the longest-first greedy algorithm. Immorlica et al. proved a general $PoA = \Theta(\log m)$ for any deterministic ordering policy for $B||C_{\max}$. They also showed that under LPT players converge in PNE after n rounds of iterative best responses but only for the $P,Q,B||C_{\max}$ scheduling models. It is unknown whether iterative best response converges for $R||C_{\max}$.

We provide the proof of a general $O(\log m)$ upper bound for the *PoA* of any deterministic ordering policy (including SPT, LPT) for $P,Q||C_{\max}$. This appeared originally in [34]. Notice, however, that both SPT and LPT perform much better for $P||C_{\max}$ (and LPT performs much better for $Q||C_{\max}$). The performance of SPT and LPT for $P||C_{\max}$ is given in a separate subsection below.

Theorem 10.2. *The Price of Anarchy of any deterministic ordering policy for uniformly related machines is* $O(\log m)$.

Proof. Assume that machines are ordered in such a way that $v_1 \geq v_2 \geq \cdots \geq v_m$. Let α be an integer so that $\alpha C(s^*) \leq C(s) \leq (\alpha+1)C(s^*)$, where s is a PNE and s^* is the socially optimum configuration. For $i \in \{1,\ldots,\alpha-1\}$, let μ_i be the minimum index of a machine so that $C_{\mu_i+1}(s) < (\alpha-i)C(s^*)$ and $C_\mu(s) \geq (\alpha-i)C(s^*)$ for all machines $\mu \leq \mu_i$. We show first that $C_1(s) \geq (\alpha-1)C(s^*)$, which means that $\mu_1 \geq 1$. Subsequently, we will prove that $\mu_i \geq (i-r-1)\mu_r$, for any $r \in \{1,\ldots,i-1\}$ and for $i \leq \alpha-1$. This will establish that $\mu_{\alpha-1} \geq 2\mu_{\alpha-4} \geq 2^q \cdot \mu_{\alpha-3q-1} \geq 2^{\frac{\alpha-2}{3}} \cdot \mu_1$. Because $m \geq \mu_{\alpha-1}$ and $\mu_1 \geq 1$, we will obtain $\alpha = O(\log m)$.

For proving $\mu_1 \geq 1$ and $C_1(s) \geq (\alpha-1)C(s^*)$, let j denote the job scheduled last on the machine of maximum makespan under s; thus, we deduce $c_j(s) \geq \alpha C(s^*)$. Then, if $C_1(s) < (\alpha-1)C(s^*)$, setting $s'_j = 1$ yields $c_j(s_{-j},s'_j) = C_1(s) + \frac{\ell_j}{v_1} < (\alpha-1)C(s^*) + C(s^*) < \alpha C(s^*) \leq c_j(s)$ since $C(s^*) \geq C_i(s^*) \geq \frac{\ell_j}{v_i} \geq \frac{\ell_j}{v_1}$ where i is the machine containing job j in assignment s^*, which contradicts s being a PNE. Therefore, $C_1(s) \geq (\alpha-1)C(s^*)$ and $\mu_1 \geq 1$.

Now we prove that $\mu_i \geq (i-r-1)\mu_r$ for any $r \in \{1,\ldots,i-1\}$ and $i \leq \alpha-1$. Let $J \subseteq \mathscr{J}$ denote the set of jobs that are scheduled on machines $1,2,\ldots,\mu_r$ in s and have completion times $c_j(s) \geq (\alpha-i+1)C(s^*)$, for all $j \in J$. Take any job $j \in J$; we claim that in s^* job j is assigned to one of the machines $1,2,\ldots,\mu_i$. If j is assigned to a machine $\mu > \mu_i$ in s^*, then $\frac{\ell_j}{v_{\mu_i+1}} \leq \frac{\ell_j}{v_\mu} \leq C(s^*)$, which means that j has an incentive to move to $s'_j = \mu_i+1$ under s, because

$$c_j(s_{-j},s'_j) = C_{\mu_i+1}(s) + \frac{\ell_j}{v_{\mu_i+1}} < (\alpha-i) + C(s^*) < (\alpha-i+1)C(s^*) \leq c_j(s).$$

Thus, all jobs in J are assigned to machines $1,2,\ldots,\mu_i$ in s^*. The sum of processing lengths of jobs in J on any machine $q \leq \mu_r$ is at least $((\alpha-r)-(\alpha-i+1))C(s^*)v_q = (i-r-1)C(s^*)v_q$. The sum of processing lengths of all jobs in J is then: $\sum_{j\in J}\ell_j \geq (i-r-1)C(s^*)\sum_{q=1}^{\mu_r} v_q$. This total processing time is assigned to machines $1,2,\ldots,\mu_i$ in the socially optimum configuration s^*; thus, $\sum_{j\in J}\ell_j \leq C(s^*)\sum_{q=1}^{\mu_i} v_q$. The latter two inequalities yield $C(s^*)\sum_{q=1}^{\mu_i} v_q \geq (i-r-1)$ $C(s^*)\sum_{q=1}^{\mu_r} v_q$, which becomes $\mu_i \geq (i-r-1)\mu_r$, because machines are indexed by non-increasing order of speeds. This completes the proof. □

Table 10.2: Performance of ordering mechanisms for multiprocessor scheduling

Ordering mechanisms			
Model	SPT	LPT	RANDOM
$P\|\|C_{\max}$	$2-\frac{1}{m}$ [31,34]	$\frac{4}{3}-\frac{1}{3m}$ [19,32]	$2-\frac{2}{1+m}$ [25,46]
$Q\|\|C_{\max}$	$\Theta(\log m)$ [5,34]	$[1.52, 1.59]$ [21,26,34]	$\Theta\left(\frac{\log m}{\log\log m}\right)$ [20]
$B\|\|C_{\max}$	$\Theta(\log m)$ [5,34]	$\Theta\left(\frac{\log m}{\log\log m}\right)$ [11,34]	$\Theta\left(\frac{\log m}{\log\log m}\right)$ [9,28]
$R\|\|C_{\max}$	m [10,34]	∞	$\Theta(m)$ [34]

10.3.3 Efficiency of SPT and LPT on Identical Machines

On the basis of a classical work due to R. L. Graham [32], we show that the PoA for SPT and LPT is $2-1/m$ and $(4/3-\frac{1}{3m})$, respectively, for the case of $P||C_{\max}$.

Mechanism SPT Suppose w.l.o.g. that $\ell_1 \leq \ell_2 \leq \cdots \leq \ell_n$. Let s be a PNE whereas s^* denotes the socially optimum configuration. First remark that in s, the completion time of job n coincides with the makespan, that is $C(s) = c_n(s)$. If it were not the case then the job whose completion time coincides with the makespan is not on machine s_n, and it can move to machine s_n and benefit. Denote by T the time at which the execution of job n starts in solution s:

$$C(s) = T + \ell_n. \tag{10.1}$$

Now observe that every machine is busy between times 0 and T, otherwise job n could move to another machine and complete earlier:

$$T \leq \frac{1}{m}\sum_{i=1}^{n-1} \ell_i. \tag{10.2}$$

As usual, we have that

$$C(s^*) \geq \frac{1}{m}\sum_{i=1}^{n} \ell_i \text{ and } C(s^*) \geq \ell_n. \tag{10.3}$$

Using (10.1)–(10.3), we get that $C(s) = T + \ell_n \leq \frac{1}{m}\sum_{i=1}^{n-1}\ell_i + \frac{\ell_n}{m} + (1-\frac{1}{m})\ell_n \leq (2-\frac{1}{m})C(s^*)$.

A tight example for every $m \geq 2$ can be the following: there are $(m-1)^2$ job of load 1, one job of load m and one job of load $m-1$. In a strategy profile s, the jobs of load m and $m-1$ are on the first machine, and every other machine hosts $m-1$ jobs of length 1. It is not difficult to see that s is a PNE with $C(s) = 2m-1$. In a strategy profile s^*, the job of load m is alone on the first machine, the job of load $m-1$ shares the second machine with one job of load 1, and every machine out of the $m-2$ remaining ones hosts m jobs of length 1. We get that $C(s^*) = m$ so $C(s)/C(s^*) = 2-1/m$.

Mechanism LPT We suppose w.l.o.g. that $\ell_1 \geq \ell_2 \geq \cdots \geq \ell_n$. Again, s denotes a PNE whereas s^* denotes a social optimum. Let k be the job whose completion time in s coincides with the makespan, that is $C(s) = c_k(s)$. By a minimality argument on the number of jobs in an instance achieving a given PoA, one can suppose w.l.o.g. that $k = n$.[1] Denote by T the time at which the execution of job n starts in solution s, (10.1)–(10.3) still hold so we get that $C(s) \leq C(s^*) + (1 - \frac{1}{m})\ell_n$. Now suppose that $\ell_n \leq C(s^*)/3$, it would be $C(s) \leq (\frac{4}{3} - \frac{1}{3m})C(s^*)$. Now suppose that $\ell_n > C(s^*)/3$. By hypothesis $\ell_i > C(s^*)/3$ for all i. Hence at most two jobs share the same machine in s^*. In that case s must be optimal, that is up to a renaming of the machine indexes, s is such that job j is on machine j when $1 \leq j \leq m$, and job j is on machine $2m+1-j$ when $m+1 \leq j \leq 2m$.

A tight example for every $m \geq 2$ can be the following: there are three jobs of load m and for every $\ell \in [m+1..2m-1]$ there are two jobs of load ℓ. The $2m+1$ jobs are such that $\ell_1 \geq \ell_2 \geq \cdots \geq \ell_{2m+1}$. In a strategy profile s, the m largest jobs choose distinct machines, that is job j is on machine j for $1 \leq j \leq m$. For $m+1 \leq j \leq 2m$, job j is on machine $2m+1-j$. Job $2m+1$ is on machine 1. It is not difficult to see that s is a PNE with $C(s) = 4m-1$. In a strategy profile s^*, the three jobs of load m are on the first machine, the second and third machine both host one job of load $2m-1$ and one job of load $m+1$, the fourth and fifth machine both host one job of load $2m-2$ and one job of load $m+2$, and so on. We get that $C(s^*) = 3m$ since the total load on every machine is $3m$. Hence $C(s)/C(s^*) = \frac{4}{3} - \frac{1}{3m}$.

10.3.4 RANDOM*: Randomized Ordering*

Let us close this section by mentioning a randomized non-preemptive (ordering) mechanism. Under the RANDOM mechanism each machine executes the jobs by order chosen uniformly at random (random permutation). RANDOM simulates in some sense Makespan in a randomized manner and yields a similar expression to the individual completion time of jobs. It can be verified that the completion time of a job under RANDOM is:

$$c_j(s) = p_{j,s_j} + \frac{1}{2} \sum_{j' \neq j : s_{j'} = s_j} p_{j',s_j}.$$

Because of the technical similarity of the players' cost under RANDOM with that under Makespan, the *PoA* bounds of $(2 - 2/(1+m))$ for $P||C_{\max}$ and $O(\log m / \log\log m)$ for $Q,B||C_{\max}$ under Makespan hold also for the case of RANDOM. Immorlica et al. showed $PoA \leq 2m-1$ and $PoA \geq m-1$ for RANDOM in $R||C_{\max}$ [34]. It is not known whether RANDOM does induce games with PNE on environments other than identical machines ($P||C_{\max}$). For uniformly related machines, Dürr and Nguyen Kim showed that it induces PNE if the speeds of the

---

[1] If $k < n$, then remove every job of index $i > k$. It is not difficult to see that s remains a Nash equilibrium with the same makespan. At the same time, $C(s^*)$ cannot increase.

machines do not differ much [22]. They also proved that a potential game is induced by the policy on two unrelated machines. Perhaps the most interesting feature of RANDOM is that it is the only known non-preemptive *non-clairvoyant* policy; it does not need to know the processing lengths of the jobs and it is an ordering policy. However, it cannot handle anonymous jobs (since it needs to order them). Completing the study of its performance remains an important open question.

10.3.5 Unrelated Machines Scheduling

Scheduling unrelated machines has received considerable attention as the most general machines environment. Under the framework of coordination mechanisms it has also attracted interest from the perspective of the performance of the iterative best response procedure; this is a simple practical local search procedure compared to the polynomial-time 2-approximation method that was developed in the seminal paper of Lenstra et al. [37]. The most recent developments in this context are due to Azar et al. [10] and Caragiannis [14].

Azar et al. [10] study coordination mechanisms consisting of local policies. For any set of strongly local ordering policies deployed upon a set of m unrelated machines, they prove that the *PoA* of the mechanism is at least $\frac{m}{2}$. This matches asymptotically the upper bound of m shown by Ibarra and Kim [33] for the performance of SPT scheduling; it is known that the output of SPT scheduling corresponds to PNE of the SPT mechanism. In an effort to improve over the performance of strongly local ordering policies, Azar et al. study a simply local ordering policy that orders the jobs on each machine according to their *inefficiency* ratio. The inefficiency of a job $j \in \mathscr{J}$ on machine $i \in \mathscr{M}$ is defined as $e_{ji} = \frac{p_{ji}}{p_j}$, where $p_j = \min_i p_{ji}$ is the smallest processing time of j over all machines. The inefficiency-based policy schedules jobs on a machine in order of non-decreasing inefficiency e_{ji}. Clearly, this is not a strongly local policy, because it uses information about the jobs assigned to a machine, that is relevant to all other machines. In [10], it is shown that the inefficiency based policy achieves a *PoA* of $O(\log m)$; its drawback is that it does not always induce PNE and, even when it does, an iterative best response procedure may not converge to any of them. The upper bound on the performance of this policy is shown to be tight for the whole class of simply local ordering policies. The authors then propose a *preemptive* mechanism that has slightly worse $PoA = O(\log^2 m)$, but guarantees fast convergence of iterative best response to PNE in at most $O(n)$ rounds for all players, that is at most $O(n)$ best responses per player.

Caragiannis [14] continued the study of coordination mechanisms in the unrelated machines model, to uncover the potential of *preemptive* scheduling policies. He designed three mechanisms with different properties. They are all deterministic, preemptive and simply local. The first one induces a potential game with $PoA = O(\log m)$, in which players converge to PNE in at most n rounds of iterative best response. The second breaks through the $\Omega(\log m)$ lower bound of Azar et al. for simply local non-preemptive mechanisms, by exhibiting $PoA = O(\log m / \log\log m)$;

however, iterative best response may not converge under this mechanism. The third proposed mechanism induces a potential game with $PoS = O(\log m)$, $PoA = O(\log^2 m)$ and players can converge to PNE within $O(n \log n)$ rounds of iterative best response. The latter mechanism is the first one with the properties of inducing a potential game with bounded PoA, that can also handle *anonymous* jobs, that is it does not need to know a specific index for each job.

10.4 Selfish Scheduling and Truthfulness

In this section, we suppose that a job's load is only known to his owner. In this case, every player must communicate this information. For feasibility reasons, every machine should allocate a time window to every job it has to execute and this slot must be at least as large as the processing time of the job that it accommodates. However, the computation of these slots are based on the declaration of the agents. This is an opportunity for some agents to report a false load and decrease their individual cost. Then the agents are definitely not trustworthy. With false information, unreasonable solutions can emerge if the system is not designed to cope with manipulation. Therefore, the problem is at the intersection of scheduling theory and *mechanism design*, a field recently explored by Nisan and Ronen in [41], in which the social cost is minimized while misreporting a private value should not be profitable. A mechanism is called *truthful* (or *strategy-proof*) if misreporting a private value is not profitable to a single player.

Nisan and Ronen's seminal article follows the well-established formalism of mechanism design in which an output function and a payment vector is computed on the basis of the agents' declarations. Nevertheless, many models were investigated in the literature (see e.g. [2, 3, 6, 18, 19, 39, 41] for a non exhaustive list of results), depending on several parameters: the players are either the jobs or the machines, payments are allowed or not, the strategy of an agents is vector or a singleton, preemption is allowed or not, etc. Here, we present some results which apply for a model close to the one mainly discussed in this chapter.

10.4.1 The Model

We consider the model introduced in Sect. 10.2 and we enrich it as follows. Every player's action, in addition to a machine number, contains a real (denoted by b_i for player i) which represents the declared length of his job. The machine environment is $P||C_{\max}$ (i.e. the machines are identical and the social cost coincides with the makespan) and the mechanism must be deterministic and non-preemptive. Every machine must compute a feasible schedule, that is a time window for every job that chooses to be executed on it. Then a job j starts being executed when its corresponding time window is opened and it completes ℓ_j time units after. A player would not be satisfied if his job were not fully executed so we assume that he always reports a value $b_i \geq \ell_i$ and the machine allocates him a time slot of at least b_i time units. We say that a mechanism (a set of m algorithms, one per machine, which allocates time windows to the players, on the basis of their declarations) is *truthful* (without money) if a player cannot unilaterally report a false length and decrease his completion time. The next two subsections are based on [18, 19].

10.4.2 LPT *Is Not Truthful*

Suppose that the mechanism is LPT, that is every machine proceeds as follows for the subset of jobs that it has to execute: it takes the player who declared the largest length (ties are broken indexwise) and allocates him a time window of size equal to the bidded length; next the player who declared the second largest length is allocated a time window of size equal to the bidded length, and so on. No idle time is inserted between two consecutive time slots. As we have seen before, the PoA is $\frac{4}{3} - \frac{1}{3m}$ if the players report their true lengths.

Consider an instance with two machines and three jobs 1, 2 and 3 with loads 1, 2, and 3, respectively. The state where job 3 is alone on the first machine while the others are on the second machine is optimal but it is not a Nash equilibrium. Indeed job 1 completes at time 3 (it is executed just after job 2) but if it declares $2+\varepsilon$ (with $1 > \varepsilon > 0$) then it ends at time $2+\varepsilon$ because the mechanism executes it before job 2. Suppose that job 1 does so, the second job completes at time $4+\varepsilon$. Then, player 2 can move to machine 1 and declare $3+\varepsilon$ and complete at time $3+\varepsilon$, etc. Clearly, LPT is not truthful.

10.4.3 SPT *and a Matching Lower Bound*

Suppose that the mechanism is SPT, that is every machine proceeds as follows for the subset of jobs that it has to execute: it takes the player who declared the smallest length (ties are broken indexwise) and allocates him a time window of size equal to the bidded length; next, the player who declared the second smallest length is allocated a time window of size equal to the bidded length, and so on. No idle time is inserted between two consecutive time slots. As we have seen before, the PoA is $2 - \frac{1}{m}$ if the players report their true lengths.

It is not difficult to see that SPT is truthful. If player i declares a length b_i satisfying $b_i > \ell_i$ then the time window that the mechanism allocates him can only begin after the one the player gets when he reports his true length. As mentioned above, no player can rationally report a value below his true length.

Now suppose that there is a deterministic and non preemptive mechanism, say MEC, which is truthful and its PoA is strictly less than $2 - \frac{1}{m}$. Consider a first instance $\mathscr{I}_1$ made of $m(m-1)+1$ jobs of length 1. In this case, the optimal makespan is m.

Since MEC is deterministic and non preemptive, there must be a job of $\mathscr{I}_1$, say j', which starts to be executed at time $t \geq m-1$. Consider a second instance $\mathscr{I}_2$ made of $m(m-1)$ jobs of length 1 plus one job of length m. For this second instance, the optimal makespan is still m (the large job is alone on one machine). Since MEC has PoA strictly less than $2-\frac{1}{m}$, any job of $\mathscr{I}_2$ completes before time $(2-\frac{1}{m})m = 2m-1$. In particular, MEC gives to the job of length m a time slot which starts before time $2m-1-m=m-1$. Combining instance $\mathscr{I}_1$ and $\mathscr{I}_2$ we reach a contradiction: job j' in $\mathscr{I}_1$ starts at time $t \geq m-1$ but if it declares m instead of 1 then MEC allocates him a time window which starts strictly before $m-1$, as it is shown for instance $\mathscr{I}_2$. MEC cannot be truthful. We deduce that SPT is optimal among truthful mechanisms.

10.5 Setup Times: A Case Study

Let us exemplify the study of coordination mechanisms for a slightly more involved class of multiprocessor job scheduling games; we will introduce setup times, where each job is of a certain *type*. On any machine, jobs of a given type may be executed *only* after a type-dependent preprocessing (performed once for all same-type jobs) called *setup*. Each machine schedules its assigned jobs according to a *scheduling policy* (algorithm). Given the deployed scheduling policies, players assign their jobs selfishly, to minimize their individual completion times. We examine the impact of selfish behavior on the *overall* (social) cost of *stable* assignments (PNE and SE) and how this can be alleviated, by deployment of appropriate scheduling policies on the machines.

This study is motivated by concerns in performance optimization of large-scale distributed systems (computational grids, P2P file sharing systems, etc.). Setup overheads in these systems may well dominate the net processing load of tasks assigned by users; consider for example loading application environments, booting operating systems, establishing QoS for network connections. Scheduling with setup times also provides for modeling another very natural situation; the case where many autonomous users may benefit from the output of an identical process. In this case, users may only care for the output of the setup (which takes non-negligible time) but their jobs may have virtually zero load for the machine. As an example, consider the setup corresponding to computation and writing of output to a remote file which users simply read. Then the machine may also have to deal *obliviously* of users' presence; only one of them may actually declare his presence by requesting execution of the setup, whereas the rest simply benefit from the request.

Preemptive multi-processor scheduling *with setup times* [8] requires a minimum makespan preemptive schedule, such that the setup is executed by a machine between execution of two job portions of different type. The problem is known to be **NP**-hard, see for example [29] [SS6]. The best known approximation algorithm has a performance guarantee of $\frac{4}{3}$ [47] (see [15] for a previous $\frac{3}{2}$ factor). For equal setup times a PTAS is given in [47] and an FPTAS for two machines in [48]. See [16] for a slightly different version. In our context, we will only maintain the concept of setup times, to investigate the congestion effects that it introduces in a game-theoretic setting. We will study a straightforward adaptation of the `Makespan` mechanism and then introduce a class of ordering mechanisms and characterize their performance.

Definition 10.4. We mainly consider an environment of *identical* machines. Each job $j \in \mathscr{J}$ is associated with a *type* $t_j \in \mathscr{U}$, where $\mathscr{U}$ is the universe of all possible types.[2] The subset of $\mathscr{U}$ corresponding to the set of jobs $\mathscr{J}$ is denoted by $\mathscr{T} = \{t_j | j \in \mathscr{J}\}$ and define $k = |\mathscr{T}|$. We refer to any specific type by θ. Each job $j \in \mathscr{J}$ and each type $\theta \in \mathscr{U}$ are respectively associated to processing length $\ell_j \geq 0$ and setup time $w(\theta) \geq 0$. If $w(\theta) = 0$, then $\ell_j > 0$ for all j with $t_j = \theta$. Otherwise, we allow $\ell_j = 0$. In this model, the makespan of a socially optimum assignment s^* can be lower bounded as

$$\begin{aligned} \textbf{(a)} \quad mC(s^*) &\geq \sum_{\theta \in \mathscr{T}} w(\theta) + \sum_{j \in \mathscr{J}} \ell_j \qquad (10.4) \\ \textbf{(b)} \quad C(s^*) &\geq w(t_j) + \ell_j \qquad \text{for any } j \in \mathscr{J} \\ \textbf{(c)} \quad (k-1)C(s^*) &\geq \sum_{\xi \in \mathscr{T} \setminus \{\theta\}} w(\xi) \qquad \text{for any } \theta \in \mathscr{T} \end{aligned}$$

The only restriction that we impose on the scheduling policies is that the setup of any type θ on any machine i is executed before execution of type θ jobs on i.

10.5.1 On the `Makespan` *Mechanism*

We study an adaptation of the preemptive `Makespan` mechanism introduced in [36] and discussed in Sect. 10.3.1; under `Makespan` completion time of j equals completion time of the machine that j is assigned to. The proof of the following result is a generalization of the proof of [1]. Their proof depends crucially on all jobs having non-zero processing length (see Lemma *A*.1 in the full version [1]). To overcome this, we used a more detailed vector potential function.

Theorem 10.3. *Strong Equilibria exist in the scheduling game with setup times, under the* `Makespan` *mechanism.*

Proof. We suppose that $\mathscr{J} = \{1, \ldots, n\}$. To every state s we associate a permutation π_s of the set of players such that $\pi_s(j) \leq \pi_s(j')$ if and only if $c_j(s) \geq c_{j'}(s)$. Obviously, such a permutation always exists (for instance we sort the jobs by non increasing completion time). In particular, $\pi_s^{-1}(1)$ is a job whose completion time is the makespan under s, that is, $C(s) = c_{\pi_s^{-1}(1)}(s)$. Associate the vector $v_s = \langle c_{\pi_s^{-1}(1)}(s), c_{\pi_s^{-1}(2)}(s), \cdots, c_{\pi_s^{-1}(n)}(s) \rangle$ to every state s. In the following, v_s^i denotes the i-th coordinate of v_s. Given two assignments s and r, we say that v_r is

[2] For example the set of application environments installed on each machine.

lexicographically smaller than v_s iff $v_r^1 < v_s^1$ or there is an index $i^* > 1$ such that $v_r^i = v_s^i$ for $i \in \{1, \cdots, i^* - 1\}$ and $v_r^{i^*} < v_s^{i^*}$. We show by contradiction that, if s is the lexicographically smallest assignment, then it is a strong equilibrium. Let r be an assignment and $J \subseteq \mathscr{J}$ be a nonempty coalition such that $\forall j \in J\ s_j \neq r_j$ and $\forall j \notin J$ $s_j = r_j$. Moreover, by construction of a coalition, we must get: $\forall j \in J\ c_j(s) > c_j(r)$. Let $c_{\max} = \max_{j \in J} c_j(s)$, that is $c_{\max}$ is the maximum completion time of the jobs of the coalition J under state s. We prove the following properties:

(i) For any $j \in \mathscr{J}$ such that $c_j(s) > c_{\max}$, $c_j(r) = c_j(s)$.
(ii) For any $j \in J$, $c_j(r) < c_{\max}$.
(iii) If $j \in \mathscr{J} \setminus J$ and $c_j(s) = c_{\max}$, then $c_j(r) \leq c_{\max}$.
(iv) For $j \in \mathscr{J} \setminus J$ with $c_j(s) < c_{\max}$, $c_j(r) < c_{\max}$.

For (i). Let j be a player such that $c_j(s) > c_{\max}$. Since $c_{\max} \geq c_p(s)$ for all $p \in J$, we deduce that $j \in \mathscr{J} \setminus J$ and $s_j = r_j$. If at least one job of J, say p, moves to machine s_j then $c_p(s) \leq c_{\max} < c_j(s) \leq c_p(r)$, contradiction with $c_p(r) < c_p(s)$. Then no job quit or moves to s_j, the completion time on that machine remains unchanged.

For (ii). If $j \in J$ then $c_j(r) < c_j(s)$ and $c_j(s) \leq c_{\max}$.
For (iii). Since $j \in \mathscr{J} \setminus J$, it is $s_j = r_j$. If no job of J moves to machine s_j then $c_j(r) \leq c_j(s) = c_{\max}$ (the completion of s_j cannot increase). If at least one job of J, say p, moves to s_j then $c_j(r) = c_p(r)$ and $c_p(r) < c_{\max}$ by item (ii).

For (iv). The proof is quite similar to item (iii). By definition we have $s_j = r_j$, that is j stays on machine s_j. If no job of J moves to s_j, then $c_j(r) \leq c_j(s)$ (the completion of machine s_j cannot increase). If at least one job of J, say p, moves to s_j then $c_j(r) = c_p(r)$ and $c_p(r) < c_{\max}$ by item (ii).

Items (i) – (iv) lead to $|\{i : c_i(r) = c_{\max}\}| \leq |\{i : c_i(s) = c_{\max}\}|$. Moreover, using item (ii), we deduce that $|\{i : c_i(r) = c_{\max}\}| \neq |\{i : c_i(s) = c_{\max}\}|$ since by construction there exists at least one job $j \in J$ with $c_j(s) = c_{\max}$ (and $c_j(r) < c_{\max}$ by item (ii)). Thus globally, items (i) to (iv) imply that v_r is lexicographically smaller than v_s (contradiction with the minimality of v_s) because $v_r^i = v_s^i$ when $v_r^i > c_{\max}$ and $|\{i : v_r^i = c_{\max}\}| < |\{i : v_s^i = c_{\max}\}|$. □

Theorem 10.4. *The PoA of the* `Makespan` *mechanism for the scheduling game with setup times is m when $m \leq k$, at most $k + 1 - k/m$ when $m > k$ and at least $\frac{k+1}{1+\varepsilon}$ for $m \geq 3k - 2$ and $\varepsilon = \frac{2k-1}{m-k+1}$.*

Proof. **Case** $m \leq k$: The most expensive PNE s has makespan $C(s) \leq \sum_\theta w(\theta) + \sum_j \ell_j$ (all jobs on one machine). By (10.4a), it is $PoA \leq m$. For the lower bound take $k = m$ types of jobs, each type θ having $w(\theta) = 1$ and containing m jobs of zero processing length. A job of each type is assigned to each machine in s . Then, $C(s) = m$ and s is clearly a PNE. In the social optimum s^* each type is assigned to a dedicated machine, thus $C(s^*) = 1$.

Case $m > k$: Assume that for the most expensive PNE s it is $C(s) = C_1(s)$. Let x be a job of type $t_x = \theta$ executed on machine 1. x cannot decrease its cost $c_x(s) = C_1(s)$ by switching to any machine $i \neq 1$. Then $C_1(s) \leq C_i(s) + \ell_x$ if $\theta \in T_i(s)$ or

$C_1(s) \leq C_i(s) + w(\theta) + \ell_x$ otherwise. We sum up the inequalities over all machines, assuming that θ does not appear on α machines, and add $C_1(s)$ to both sides to obtain $mC_1(s) \leq \sum_{i=1}^{m} C_i(s) + \alpha w(\theta) + (m-1)\ell_x \leq m\sum_{\xi\in\mathcal{T}} w(\xi) + (m-1)\ell_x + \sum_{j\in\mathcal{J}} \ell_j$. Divide by m and rewrite it as:

$$C_1(s) \leq \frac{m-1}{m}\Big(\sum_{\xi\in\mathcal{T}\setminus\{\theta\}} w(\xi) + w(\theta) + \ell_x\Big) + \frac{1}{m}\Big(\sum_{\xi\in\mathcal{T}} w(\xi) + \sum_{j\in\mathcal{J}} \ell_j\Big).$$

Using (10.4a,b,c), $C(s) \leq \frac{m-1}{m}((k-1)C(s^*) + C(s^*)) + C(s^*) = (k+1-\frac{k}{m})C(s^*)$.

For the lower bound take k types, $m \geq 3k-2$, and let $w(1) = 0$ and $w(\theta) = 1$ for $\theta \in \{2,\ldots,k\}$. There are $k+1$ jobs of type 1 and length 1 and $\frac{m-1}{\varepsilon}$ jobs of type 1 and length $\varepsilon = \frac{2k-1}{m-k+1}$. Types $\theta \in \{2,\ldots,k\}$ have $m-1$ jobs each, of processing length 0. A PNE s is as follows. $k+1$ jobs of type 1 and length 1 are assigned to machine 1. One job from each type $\theta \geq 2$ is assigned to each machine $i = 2,\ldots,m$. $\frac{1}{\varepsilon}$ jobs of type 1 and length ε are also assigned to each machine $i \geq 2$. Thus $C_i(S) = k$ for $i \geq 2$ and $C_1(s) = k+1$. No job may decrease its completion time (equal to the makespan of the machine it is assigned to) by switching machine. In the optimum assignment s^*, assign two jobs of type 1 – with lengths 1 and ε – to each machine $i = 1\ldots k+1$. Every machine $i = k+2\ldots 2k$, has $m-1$ jobs of type $i-k$, each of length 0. Every machine $i = 2k+1\ldots m$, has $1/\varepsilon + 1$ jobs of type 1, of length ε. The makespan of s^* is $1+\varepsilon$. See Fig. 10.3 for an illustration of the example. □

Theorem 10.5. *The Price of Anarchy of strong equilibria under* `Makespan` *for the scheduling game with setup times is 2 for $m \geq 3$, and $\frac{3}{2}$ for $m = 2$ machines.*

Proof. We give the proof for the case $m \geq 3$. The reader is referred to [30] for the case $m = 2$. Let s be a SE, s^* the socially optimum assignment, and

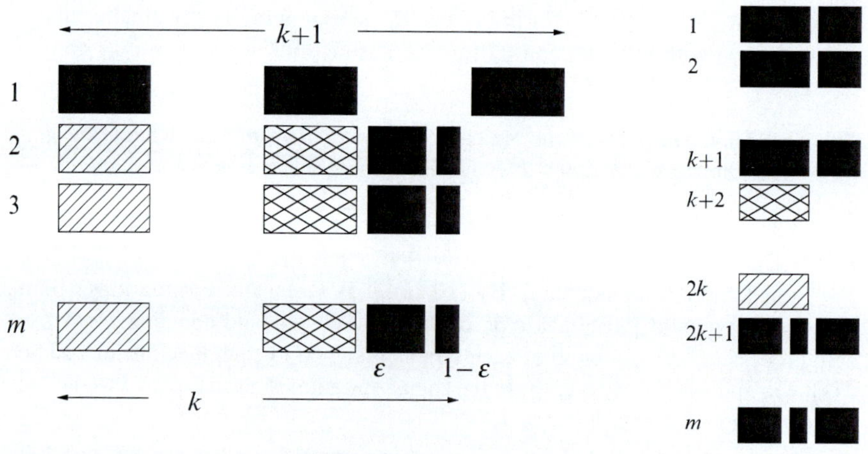

Fig. 10.3: A lower bound of the Price of Anarchy for `Makespan`, on identical machines

$C(s) = C_1(s)$. If $C_1(s) \leq C(s^*)$ we get $SPoA = 1$. If $C_1(s) > C(s^*)$, there is machine $i \neq 1$ with $C_i(s) \leq C(s^*)$, because otherwise s would not be a SE; all jobs would reduce their completion time by switching from s to s^*. For any job x with $s_x = 1$, it is $c_x(s) \leq c_x(s_{-x}, i)$. Thus, $C_1(s) = c_x(s) \leq C_i(s) + w(t_x) + \ell_x$. Thus, $C(s) = c_x(s) \leq 2C(s^*)$ because $C_i(s) \leq C(s^*)$ and (10.4b). For the lower bound, take three machines and four jobs, with $t_1 = t_2 = \theta_1$ and $t_3 = t_4 = \theta_2$. Set $w(\theta_1) = \varepsilon$, $\ell_1 = \ell_2 = 1$ and $w(\theta_2) = 1$, $\ell_3 = \ell_4 = \varepsilon$. An assignment where jobs 1,2 play machine 1 and jobs 3,4 play machines 2,3, respectively, is a strong equilibrium of makespan $2+\varepsilon$. In the social optimum jobs 3,4 are assigned to the same machine and 1 and 2 on dedicated machines; the makespan becomes then $1+2\varepsilon$. Thus, $SPoA \geq \frac{2+\varepsilon}{1+2\varepsilon} \to 2$, as $\varepsilon \to 0$. □

10.6 Type Ordering Mechanisms

We describe a class of (deterministic) *type ordering* mechanisms, for *batch scheduling* of same-type jobs. Each machine i groups together jobs of the same type θ, into a *batch* of type θ. A type batch is executed by the machine as a whole; the setup is executed first, followed by *preemptive* execution of all jobs in the batch, in a `Makespan` fashion. Jobs within the same batch have equal completion times and are scheduled preemptively in parallel. Type batches are executed serially by each machine.

Policies in type ordering mechanisms satisfy a version of the property of IIA [10]. Under the IIA property, for any set of jobs $J_i \subseteq \mathscr{J}$ assigned to machine $i \in \mathscr{M}$ and for any pair of types $\theta, \theta' \in \mathscr{U}$ with jobs in J_i if the θ-type batch has smaller completion time than the θ'-type batch, then the θ batch has a smaller completion time than the θ' batch in any set $J_i \cup \{j\}$, $j \in \mathscr{J} \setminus J_i$. Presence of j does not affect the relative order of execution of θ and θ' batches. The IIA property was used in [10] for proving a lower bound on the *PoA* of a class of job ordering mechanisms in the context of unrelated machines scheduling. Type ordering policies do not introduce delays in the execution of batches, but only decide their relative order of their execution, based on a batch's type index and setup time. They do not use the number of jobs within each batch; otherwise the IIA property may not be satisfied. Job lengths are used only for `Makespan`-wise scheduling within batches. Hence, type ordering mechanisms function obliviously of "hidden" players with zero job lengths.

10.6.1 Pure Nash Equilibria

We prove next existence of PNE for any number of machines, and SE for $m = 2$ under type ordering mechanisms. An algorithm for finding PNE follows. Let $o(i)$ be the ordering of types on machine i, and $O = \{o(i) | i \in \mathscr{M}\}$ be the set of all orderings of the mechanism. By $\prec_o$ denote the precedence relation of types, prescribed by $o \in O$. Let M_o be the set of machines that schedule according to $o \in O$. Initialize $o \in O$ arbitrarily, and *repeat* until all jobs are assigned:

1. Find the earliest type θ according to $\prec_o$, with at least one unassigned job.
2. Let j be the largest length unassigned job with $t_j = \theta$.
3. Pick $i \in \mathscr{M}$ minimizing completion time of j[3] (break ties in favor of $i \in M_o$).
4. *If* $i \in M_o$ set $s_j = i$ *else* switch ordering o to $o(i)$.

Theorem 10.6. *The scheduling game with setup times has pure Nash equilibria, under type ordering mechanisms.*

Proof. The algorithm terminates in polynomial time; once a job is assigned, it is never considered again and within every $O(m+n)$ iterations some job is always assigned. For any type θ, denote by $\hat{s}_\theta$ the partial assignment up to the time after the last job of type θ has been assigned. We show by contradiction that no job j has incentive to deviate under an assignment s returned by the algorithm.

Assume that j does have incentive to deviate from s_j, and let s' be the resulting assignment after deviation of j. At the time corresponding to the partial assignment $\hat{s}_{t_j}$, there is no type $\theta \neq t_j$ and machine i such that $\theta \in T_i(\hat{s}_{t_j})$ and $t_j \prec_{o(i)} \theta$. If it was the case, the first job of type $\theta \neq t_j$ assigned to i would have been chosen before jobs of type t_j were exhausted, which contradicts step 1. of the algorithm. Thus, batches of type t_j are scheduled – under $\hat{s}_{t_j}$ – last on all machines with $t_j \in T_i(\hat{s}_{t_j})$. Furthermore, if j wishes to deviate to a machine $i \neq s_j$, then $c_j(s) = c_j(\hat{s}_{t_j}) > C_i(\hat{s}_{t_j}) + \ell_j = c_j(s')$, if $t_j \in T_i(\hat{s}_{t_j})$, and $c_j(s) = c_j(\hat{s}_{t_j}) > C_i(\hat{s}_{t_j}) + w(t_j) + \ell_j = c_j(s')$, if $t_j \notin T_i(\hat{s}_{t_j})$. Let j' be the last job of type t_j assigned to machine s_j (it may be $j' = j$). Because $\ell_{j'} \leq \ell_j$, it is also $c_{j'}(\hat{s}_{t_j}) = c_j(\hat{s}_{t_j}) > C_i(\hat{s}_{t_j}) + \ell_{j'}$ or $c_{j'}(\hat{s}_{t_j}) = c_j(\hat{s}_{t_j}) > C_i(\hat{s}_{t_j}) + w(t_{j'}) + \ell_{j'}$ accordingly. By the time j' was assigned, the completion time of i was at most $C_i(\hat{s}_{t_j})$. This contradicts step 3. of the algorithm with respect to j'. □

The following result identifies performance limitations of type ordering mechanisms, due to lack of a priori knowledge of $\mathscr{T} \subseteq \mathscr{U}$.

Theorem 10.7. *The Price of Anarchy of the scheduling game with setup times is* $\frac{m+1}{2}$ *for every deterministic type ordering mechanism.*

Proof. For any deterministic type ordering mechanism, assume there is a subset $\mathscr{T} \subseteq \mathscr{U}$ of $k = 2m-1$ types, say $\mathscr{T} = \{1, \cdots, 2m-1\}$, such that: all types of $\mathscr{T}$ are scheduled in order of ascending index in a machines and in order of descending index in $d = m - a$ machines. Then, there is a family of instances with $PoA \geq \frac{m+1}{2}$. Next, we prove existence of $\mathscr{T}$. Set $w(\theta) = 1$ for all $\theta \in \mathscr{U}$. When $a = m$ or $d = m$, take an instance of m zero length jobs for each type $\theta \in \{1, \cdots, m\}$. Placing one job

[3] j incurs processing load $w(t_j) + \ell_j$ if a t_j-type job is not already assigned to i.

of each type on every machine yields a PNE with makespan m. An assignment of makespan 1 has all same-type jobs assigned to a dedicated machine, thus $PoA \geq m$. When $a \geq 1$ and $d \geq 1$, the instance has:

- a jobs of zero length for each type $\theta \in \{1, \cdots, m-1\}$
- d jobs of zero length for each type $\theta \in \{m+1, \cdots, 2m-1\}$
- $m-1$ jobs of zero length and type m
- One job of length 1 and type m
- No jobs for $\theta \in \mathscr{U} \setminus \mathscr{T}$

Assign one job of type $\theta \in \{1, \cdots, m-1\}$ on each of the a *ascending* type index machines, and one job of type $\theta \in \{m+1, \cdots, 2m-1\}$ on each of the d *descending* type index machines. Put one job of type m on every machine. This is a PNE of makespan $m+1$. Placing all jobs of type $\theta \in \{i, 2m-i\}$ on machine i yields makespan 2. Thus, it is $PoA \geq \frac{m+1}{2}$.

We show existence of $\mathscr{T}$ for sufficiently large universe $\mathscr{U}$. We use the fact that any sequence of n different real numbers has a monotone (not necessarily contiguous) subsequence of $\sqrt{n}$ terms (a corollary of Theorem 4.4, page 39 in [35]). By renaming types in $\mathscr{U}$ we can assume w.l.o.g. that $\mathscr{U}$ is ordered monotonically (index-wise) on machine 1, and set $T_1 = \mathscr{U}$. Then, there is $T_2 \subseteq T_1$ such that $|T_2| \geq \sqrt{|T_1|}$ and all the types of T_2 are ordered monotonically according to index, on machines 1 and 2. After $m-1$ applications of the corollary, we obtain a set $T_m \subseteq T_{m-1} \subseteq \cdots \subseteq T_1 = \mathscr{U}$ with $|T_m| \geq |\mathscr{U}|^{2^{1-m}}$ and all its types are scheduled monotonically to their index on every machine. We set $\mathscr{T} = T_m$, and take a universe $\mathscr{U}$ of types with $|\mathscr{U}| = (2m-1)^{2^{m-1}}$, to ensure existence of $\mathscr{T}$ with $k = |\mathscr{T}| = 2m-1$ types. □

Let us note that "longest" or "shortest batch first" policies are no more powerful than type ordering mechanisms; they reduce to them for zero length jobs.

10.6.2 An Optimal Type Ordering Mechanism

We analyze the *PoA* of a type ordering mechanism termed AD (short for *Ascending–Descending*), that schedules type batches by ascending type index on half of the machines, and by descending type index on the rest. If m is odd, one of the policies in applied to one machine more. First, we prove the following lemma.

Lemma 10.1. *Let $\mathscr{T}' \subseteq \mathscr{T}$ include types with non-zero setup times. If two jobs of the same type in $\mathscr{T}'$ play an ascending and a descending index machine, respectively, under the* AD *mechanism, their type batches are scheduled last on the respective machines.*

Proof. We show the result by contradiction. Let jobs x_1, x_2 with $t_{x_1} = t_{x_2} = \theta$ be assigned on the ascending and descending machines 1, 2, respectively. Assume that a job y, $t_y = \theta' \neq \theta$, is scheduled on 1 after type θ. Because s is a PNE, job x_2 does not decrease its completion time if it moves to machine 1; because y is scheduled after x_1 on 1:

$$c_{x_1}(s) \geq c_{x_2}(s) - \ell_{x_2}, \text{ and } c_y(s) \geq c_{x_1}(s) + w(\theta') + \ell_y. \tag{10.5}$$

If y switches to processor M_2, then it will be scheduled before type θ; thus, its completion time will be at most $c_{x_2}(s) - w(\theta) - \ell_{x_2} + w(\theta') + \ell_y$ if $\theta' \notin T_2(s)$ (and at most $c_{x_2}(s) - w(\theta) - \ell_{x_2} + \ell_y$, otherwise). In the worst case, we obtain

$$c_y(s) \leq c_{x_2}(s) - w(\theta) - \ell_{x_2} + w(\theta') + \ell_y. \tag{10.6}$$

By (10.5) and (10.6), $c_y(s) \leq c_y(s) - w(\theta) < c_y(s)$, a contradiction, because $\theta \in \mathscr{T}'$. □

The next result identifies upper bounds on the *PoA* of AD. A proposition that follows proves tightness, through lower bounds on the Price of Stability, the ratio of the *least* expensive PNE makespan over the optimum makespan. We take $k \geq 2$; AD is identical to `Makespan` for $k = 1$.

Theorem 10.8. *The Price of Anarchy of the* AD *mechanism for the scheduling game with setup times is at most* $\frac{m+1}{2}$ *when* $m \leq k$ *and at most* $\frac{k+3}{2} - \rho$ *(*$\rho = \frac{k}{m}$ *when* m *is even and* $\rho = \frac{k-1}{m-1}$ *otherwise), when* $m > k$.

Proof. Let s be the most expensive PNE assignment and $C(s) = C_1(s) = \max_i C_i(s)$. Let θ_0 be the type scheduled last on machine 1 and x a job with $t_x = \theta_0$. Define $\mathscr{T}'_C \subseteq \mathscr{T}'$ to be types with jobs assigned to both ascending and descending machines under s. Let $T'_A \subseteq T' \setminus T'_C$ and $T'_D \subseteq T' \setminus T'_C$ contain types exclusively assigned to ascending and descending machines respectively. Notice that at most one type $\theta_1 \in \mathscr{T}'_C$ may appear in at least $\frac{m}{2} + 1$ machines (when m even) and $\frac{m+1}{2}$ machines (when m odd); thus any type in $\mathscr{T}'_C \setminus \{\theta_1\}$ appears on at most $\frac{m}{2}$ machines (actually, $\frac{m-1}{2}$ machines when m is odd). We study two cases depending on whether θ_1 exists and whether it coincides with θ_0 or not.

CASE 1: $\theta_0 = \theta_1$ or θ_1 does not exist. Job x will not decrease its completion time by moving to machine p for $p = 2, \ldots, m$. If $M_{\theta_0}(s)$ are the indices of machines which contain type θ_0, then

$$\forall p \in M_{\theta_0}(s),\ c_x(s) \leq C_p(s) + \ell_x \text{ and } \forall p \notin M_{\theta_0}(s),\ c_x(s) \leq C_p(s) + w(\theta_0) + \ell_x. \tag{10.7}$$

To obtain the upper bound we sum up (10.7) for $p \in \{2, \ldots, m\}$, add $C_1(s)$ in the left and right hand part, and take $\sum_{\theta \notin \mathscr{T}'} w(\theta) = 0$. We will do this analysis below, collectively for cases 1 and 2.

CASE 2: $\theta_0 \neq \theta_1$ and θ_1 exists. Assume $\theta_0 < \theta_1$ and let R contain the indices of ascending machines which have at least one job of type θ_1 assigned (if $\theta_0 > \theta_1$, we consider the indices of descending machines). Let R be the indices of these machines and $R' \subseteq R$ be the indices of machines that are also assigned type θ_0 jobs (note that

$\theta_0 \notin \mathcal{T}'_C$ if $R' \neq \emptyset$). If job x moves to a machine with index in $p \in R'$, the completion time of x becomes at most $C_p(s) - w(\theta_1) + \ell_x$ and $C_p(s) - w(\theta_1) + w(\theta_0) + \ell_x$ if $p \in R'' = R \setminus R'$. Since s is a PNE:

$$\forall p \in R',\ c_x(s) \leq C_p(s) - w(\theta_1) + \ell_x,\ \forall p \in R'',\ c_x(s) \leq C_p(s) - w(\theta_1) + w(\theta_0) + \ell_x. \tag{10.8}$$

We will sum up inequalities (10.7) or (10.8) for $p \in \{2,\dots,m\}$ depending on whether $p \in R$ or not. As in case 1 we add $C_1(s)$ to left and right hand parts and consider $\sum_{\theta \notin \mathcal{T}'} w(\theta) = 0$. Before summing note that when m is even, each type in $\mathcal{T}'_A \cup \mathcal{T}'_D$ has jobs assigned to at most $\frac{r}{2}$ machines, for $r = m$. When m is odd assume w.l.o.g. that there are $\frac{m+1}{2}$ descending machines. *We ignore one of them* – different than M_1 – in the summation (we assume $m \geq 3$; otherwise $m = 1$ and $C(s) = C(s^*)$). Then, in case 2, type θ_1 appears at most $\frac{r}{2}$ times, $r = m-1$, in the remaining $m-1$ machines

$$rC_1(s) \leq \frac{r}{2}\left(\sum_{\theta \in \mathcal{T}'_A \cup \mathcal{T}'_D \setminus \{\theta_0\}} w(\theta) + \sum_{\theta \in \mathcal{T}'_C \setminus \{\theta_0\}} w(\theta)\right) + rw(\theta_0) + \sum_{j \in \mathcal{J}} \ell_j + (r-1)\ell_x$$

$$= \frac{r}{2}\left(\sum_{\theta \in \mathcal{T}} w(\theta) + \sum_{j \in \mathcal{J}} \ell_j\right) + \frac{r}{2} w(\theta_0) + (r-1)\ell_x - \frac{r-2}{2}\sum_{j \in \mathcal{J}} \ell_j \tag{10.9}$$

$$\leq \frac{r}{2}\left(\sum_{\theta \in \mathcal{T}} w(\theta) + \sum_{j \in \mathcal{J}} \ell_j\right) + \frac{r}{2}(w(\theta_0) + \ell_x) \text{ since } \textstyle\sum_{j \in \mathcal{J}} \ell_j \geq \ell_x. \tag{10.10}$$

When $k \geq m$ we use (10.4a,b) with (10.10) to obtain $C_1(s) \leq \frac{m+1}{2} OPT$. When $k < m$ we rewrite (10.9) as

$$C(s) \leq \frac{1}{r}\left(\sum_{\theta \in \mathcal{T}} w(\theta) + \sum_{j \in \mathcal{J}} \ell_j\right) + \left(\frac{1}{2} - \frac{1}{r}\right)\left(\sum_{\theta \in \mathcal{T}} w(\theta) + \ell_x\right) + \frac{1}{2}(w(\theta_0) + \ell_x). \tag{10.11}$$

Using (10.4b,c), we get $kC(s^*) \geq \sum_{\theta \in \mathcal{T}} w(\theta) + \ell_x$ and replacing $r = m$ and $r = m-1$ for even and odd m, respectively, yields the stated bounds with respect to k. □

Theorem 10.9. *The Price of Stability of the scheduling game with setup times under the* AD *mechanism is* $\frac{m+1}{2}$ *when* $k > m$ *and* $\frac{k+3}{2} - \rho$ *(*$\rho = \frac{k}{m}$ *when* m *is even and* $\rho = \frac{k-1}{m-1}$ *otherwise) when* $k \leq m$.

Proof. For $k > m$ we use the same example as in the proof of Theorem 10.7, but replace the zero length jobs with very small $\varepsilon > 0$ length. For AD the described assignment for $a, d \geq 1$ applies, and it is a PNE with makespan $m + 1 + (m-1)\varepsilon$; the socially optimum makespan has length $2 + m\varepsilon$. In any PNE, all jobs of types 1 and $2m-1$ will play exactly the strategies specified in the described PNE assignment, because a lower completion time is not achievable for them in any assignment. Inductively, jobs of types i and $2m-i$, $i = 2 \dots m-1$, follow the same practice, given the strategies of jobs of types $i-1$, $2m-i+1$. For the jobs of type m, the strategies

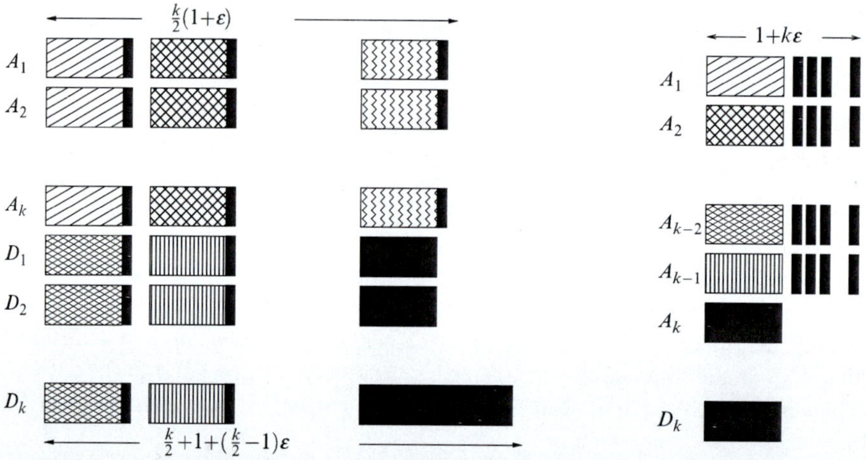

Fig. 10.4: A lower bound for the performance of AD for even m and when $k = \frac{m}{2}$

described in the aforementioned assignment are best possible, given the strategies of all other jobs. Therefore, the described PNE is unique, hence $PoS \to \frac{m+1}{2}$ for $\varepsilon \to 0$. The same uniqueness argument holds when $k \le m$, for the instances given below.

For $k < m$, take k even and $m \ge 2k$. Assume that machines $1,\dots,\lceil\frac{m}{2}\rceil$ are ascending index and machines $\lceil\frac{m}{2}\rceil+1,\dots,m$ are descending index ($\lfloor\frac{m}{2}\rfloor$ in total). Let $\hat{\theta}$ denote the *median* type index $\frac{k}{2}+1$. For all types $\theta \neq \hat{\theta}$, that is $\theta \in \{1,2,\dots,k\}\setminus\{\hat{\theta}\}$, set $w(\theta) = 1$. For each type $\theta < \hat{\theta}$ assume $\lceil\frac{m}{2}\rceil$ jobs of small processing length $\varepsilon > 0$. For each type $\theta > \hat{\theta}$ assume $\lfloor\frac{m}{2}\rfloor$ jobs of small processing length ε. Finally, let $w(\hat{\theta}) = 0$ and let there be $m-k+1$ jobs of type $\hat{\theta}$ and processing length 1 each. Let us describe first the socially optimum configuration s^*. For each type $\theta \in \{1,\dots,k\}\setminus\{\hat{\theta}\}$ assign its jobs to a dedicated machine for this type. Assign each of the jobs of type $\hat{\theta}$ to a separate machine each, out of the remaining $m-k+1$ ones. The makespan of s^* is $C(s^*) = 1+\lceil\frac{m}{2}\rceil\varepsilon$. See the right of Fig. 10.4 for an example when $m = 2k$.

Now let us describe a PNE configuration s. Each of the first $\lceil\frac{m}{2}\rceil$ ascending index machines has one job of each type $\theta \in \{1,\dots,\hat{\theta}-1\}$. Each of the last $\lfloor\frac{m}{2}\rfloor$ descending index machines has one job of each type $\theta \in \{\hat{\theta}+1,\dots,k\}$. See the left of Fig. 10.4 for an example when $m = 2k$. Finally, we assign appropriately the remaining length-1 jobs of type $\hat{\theta}$. In doing so, notice that right before this final assignment, the ascending index machines have currently makespan $\frac{k}{2}(1+\varepsilon)$ and the descending index ones $(\frac{k}{2}-1)(1+\varepsilon)$. Under this current partial assignment no unilateral deviation may occur, because same-type jobs have equal completion times (see left of Fig. 10.4). No job of a type scheduled last on an ascending index machine may decrease its completion time, by switching to a decreasing index machine with makespan, because its batch of total length $(1+\varepsilon)$ will be scheduled last on this machine as well.

Now assign first one job of type $\hat{\theta}$ to each of the $\lceil \frac{m}{2} \rceil$ descending index machines, so that their makespan becomes $\frac{k}{2} + \varepsilon(\frac{k}{2} - 1)$. Now continue assigning the remaining $m - k + 1 - \lfloor \frac{m}{2} \rfloor > 0$ (because $m \geq 2k$) type $\hat{\theta}$ jobs on descending index machines and increase their makespan to $\frac{k}{2} + 1 + \varepsilon(\frac{k}{2} - 1)$. In the left of Fig. 10.4 we have added one last such job on the last descending index machine. Now because k is even and $m \geq 2k$, we have that $\rho = \frac{k}{m} \leq \frac{1}{2}$ when m is even and $\rho = \frac{k-1}{m-1} \leq \frac{1}{2}$ when m is odd. Then, we obtain $\lim_{\varepsilon \to 0} \frac{C(s)}{C(s^*)} \geq \frac{k}{2} + 1 \geq \frac{k+3}{2} - \rho$. Configuration s is a unique PNE, by a similar argument to the one used for the case $k > m$, which we omit for brevity. Therefore, the lower bound applies to the *PoS* as stated. □

10.7 Research Directions

Let us summarize here some of the important current research directions that emerge from the study of recent literature. One of the most important questions emerging from the work of Caragiannis [14], concerns resolving the performance of *preemptvie* scheduling policies for unrelated machines scheduling. Caragiannis managed to break through a $\Theta(\log m)$ barrier of *PoA* by devising a mechanism with $PoA = \Theta(\log m / \log\log m)$ that does not always induce PNE. However, there is no general lower bound at the time of this writing, with respect to the performance of preemptive policies.

Two related questions concern non-preemptive mechanisms and stems from the results of Azar et al. [10]; they proved an $\Omega(\log m)$ lower bound for the *PoA* of the class of deterministic ordering mechanisms, and devised a mechanism with $PoA = O(\log m)$, that does not always induce PNE though. The first question concerns devising a mechanism that achieves the same upper bound and always induces PNE that can be computed efficiently. The second concerns breaking through the $\Omega(\log m)$ lower bound using randomization; what is the performance of randomized non-preemptive mechanisms?

Concerning randomized mechanisms, the performance of RANDOM is yet to be fully determined. It is not known whether RANDOM induces PNE for $Q||C_{\max}$, $B||C_{\max}$ or $R||C_{\max}$ in general; the results of Dürr and Nguyen Kim [22] show that RANDOM does induce PNE for uniformly related machines, only when the speeds do not differ too much. This mechanism is the only known non-preemptive non-clairvoyant mechanism, which also achieves a bounded *PoA* even for unrelated machines scheduling. Therefore, settling its theoretical merits remains an important open question.

We believe that designing and analyzing of coordination mechanisms has reached a maturity, that should allow it to serve as a model for decentralized congestion manipulation in large scale systems of distributed resources. To this end, more involved congestion effects should be studied, motivated by the practice of these systems. The introduction of *setup times* is such an attempt, to bring more involved considerations in the scene, that require the design of novel machinery for dealing with them. Some open questions naturally arise from this attempt. Notice that the universe of types $\mathscr{U}$ is required to be huge in the proof of Theorem 10.7 (double exponential). This size is non-realistic for most interesting practical settings. Is there a lower size of $\mathscr{U}$ that also yields $PoA \geq \frac{m+1}{2}$ for type ordering mechanisms? For example, the proof of Theorem 10.7 requires that $|\mathscr{U}| \geq 9$ when $m = 2$, although $|\mathscr{U}| \geq 3$ is enough. The performance of type ordering mechanisms is not fully characterized by Theorem 10.7; there may be certain sizes of $|\mathscr{U}|$ below which these mechanisms may perform better. Another interesting issue to be examined, is when type ordering mechanisms are a priori aware of the subset of types $\mathscr{T}$ that corresponds to players $\mathscr{J}$. What is the impact of such an a priori knowledge to the achievable *PoA* by type ordering mechanisms? Finally, we have not considered in the context of setup times any simply local mechanisms, or more challenging machine environments (uniformly related or unrelated machines). All these constitute very interesting aspects for future developments on the subject.

Acknowledgements This work is supported by French National Agency (ANR), project COCA ANR-09-JCJC-0066-01. This work was carried out during the tenure of an ERCIM "Alain Benssousan" Fellowship Program of Orestis A. Telelis.

References

1. Andelman, N., Feldman, M., Mansour, Y.: Strong price of anarchy. In: Proceedings of the ACM-SIAM Symposium on Discrete Algorithms (SODA), pp. 189–198 (2007)
2. Angel, E., Bampis, E., Pascual, F.: Truthful algorithms for scheduling selfish tasks on parallel machines. Theor. Comput. Sci. **369**(1-3), 157–168 (2006)
3. Angel, E., Bampis, E., Thibault, N.: Randomized truthful algorithms for scheduling selfish tasks on parallel machines. In: Proceedings of the 9th Latin American Theoretical Informatics Symposium (LATIN), Springer LNCS 6034, pp. 38–48 (2010)
4. Anshelevich, E., Dasgupta, A., Kleinberg, J.M., Tardos, E., Wexler, T., Roughgarden, T.: The Price of Stability for Network Design with Fair Cost Allocation. SIAM Journal on Computing **38**(4), 1602–1623 (2008)
5. Aspnes, J., Azar, Y., Fiat, A., Plotkin, S.A., Waarts, O.: On-line routing of virtual circuits with applications to load balancing and machine scheduling. Journal of the ACM **44**(3), 486–504 (1997)
6. Auletta, V., Prisco, R.D., Penna, P., Persiano, G.: Deterministic truthful approximation mechanisms for scheduling related machines. In: Proceedings of the Symposium on Theoretical Aspects of Computer Science (STACS), Springer LNCS 2996, pp. 608–619 (2004)
7. Aumann, R.J.: Acceptable points in games of perfect information. Pacific Journal of Mathematics **10**, 381–417 (1960)
8. Ausiello, G., Crescenzi, P., Gambosi, G., Kann, V., Marchetti-Spaccamela, A., Protasi, M.: Complexity and Approximation: Combinatorial Optimization Problems and Their Approximability Properties. Springer (1999)
9. Awerbuch, B., Azar, Y., Richter, Y., Tsur, D.: Tradeoffs in worst-case equilibria. Theoretical Computer Science **361**(2–3), 200–209 (2006)
10. Azar, Y., Jain, K., Mirrokni, V.S.: (almost) optimal coordination mechanisms for unrelated machine scheduling. In: Proceedings of ACM-SIAM Symposium on Discrete Algorithms (SODA), pp. 323–332 (2008)

11. Azar, Y., Naor, J., Rom, R.: The competitiveness of on-line assignments. Journal of Algorithms **18**, 221–237 (1995)
12. Braess, D.: Über ein paradoxon aus der Verkehrsplanung. Unternehmensforschung **12**, 258–268 (1968)
13. Braess, D.: On a Paradox of Traffic Planning. Transportation Science **39**(4), 446–450 (2005)
14. Caragiannis, I.: Efficient coordination mechanisms for unrelated machine scheduling. In: Proceedings of ACM-SIAM Symposium on Discrete Algorithms (SODA), pp. 815–824 (2009)
15. Chen, B.: A better heuristic for preemptive parallel machine scheduling with batch setup times. SIAM Journal on Computing **22**, 1303–1318 (1993)
16. Chen, B., Ye, Y., Zhang, J.: Lot-sizing scheduling with batch setup times. Journal of Scheduling **9**(3), 299–310 (2006)
17. Cho, Y., Sahni, S.: Bounds for list schedules on uniform processors. SIAM Journal on Computing **9**, 91–103 (1980)
18. Christodoulou, G., Gourvès, L., Pascual, F.: Scheduling Selfish Tasks: About the Performance of Truthful Algorithms. In: Proceedings of the International Computing and Combinatorics Conference (COCOON), Springer LNCS 4598, pp. 187–197 (2007)
19. Christodoulou, G., Koutsoupias, E., Nanavati, A.: Coordination Mechanisms. In: Proceedings of the International Colloquium on Automata,Languages and Programming (ICALP), Springer LNCS 3142, pp. 345–357 (2004)
20. Czumaj, A., Vöcking, B.: Tight bounds for worst-case equilibria. ACM Transactions on Algorithms **3**(1) (2007)
21. Dobson, G.: Scheduling independent tasks on uniform processors. SIAM Journal on Computing **13**, 705–716 (1984)
22. Dürr, C., Nguyen, T.K.: Non-clairvoyant Scheduling Games. In: Proceedings of the International Symposium on Algorithmic Game Theory (SAGT), Springer LNCS 5814, pp. 135–146 (2009)
23. Even-Dar, E., Kesselman, A., Mansour, Y.: Convergence time to Nash equilibrium in load balancing. ACM Transactions on Algorithms **3**(3) (2007)
24. Fiat, A., Kaplan, H., Levy, M., Olonetsky, S.: Strong Price of Anarchy for Machine Load Balancing. In: Proceedings of the International Colloquium on Automata, Languages and Programming (ICALP), Springer LNCS 4596, pp. 583–594 (2007)
25. Finn, G., Horowitz, E.: A linear time approximation algorithm for multiprocessor scheduling. BIT **19**, 312–320 (1979)
26. Friesen, D.K.: Tighter bounds for LPT scheduling on uniform processors. SIAM Journal on Computing **16**, 554–560 (1987)
27. Gairing, M., Lücking, T., Mavronicolas, M., Monien, B.: Computing Nash equilibria for scheduling on restricted parallel links. In: Proceedings of the ACM Symposium on Theory of Computing (STOC), pp. 613–622 (2004)
28. Gairing, M., Lücking, T., Mavronicolas, M., Monien, B.: The Price of Anarchy for Restricted Parallel Links. Parallel Processing Letters **16**(1), 117–132 (2006)
29. Garey, M.R., Johnson, D.S.: Computers and Intractability: A Guide to the Theory of NP-completeness. W.H.Freeman & Co Ltd (1979)
30. Gourvès, L., Monnot, J., Telelis, O.: Selfish Scheduling with Setup Times. In: Proceedings of the International Workshop on Internet and Network Economics (WINE), Springer LNCS 5929, pp. 292–303 (2009)
31. Graham, R.L.: Bounds for certain multiprocessing anomalies. Bell System Technical Journal **45**, 1563–1581 (1966)
32. Graham, R.L.: Bounds for certain multiprocessing anomalies. SIAM Journal of Applied Mathematics **17**, 416–429 (1969)
33. Ibarra, O.H., Kim, C.E.: Heuristic Algorithms for Scheduling Independent Tasks on Nonidentical Processors. Journal of the ACM **24**(2), 280–289 (1977)
34. Immorlica, N., Li, L., Mirrokni, V.S., Schulz, A.: Coordination Mechanisms for Selfish Scheduling. Theoretical Computer Science **410**, 1589–1598 (2009)
35. Jukna, S.: Extremal Combinatorics with Applications in Computer Science. Springer-Verlag (2001)

36. Koutsoupias, E., Papadimitriou, C.H.: Worst-case Equilibria. In: Proceedings of the International Symposium on Theoretical Aspects of Computer Science (STACS), Springer LNCS 1543, pp. 404–413 (1999)
37. Lenstra, J.K., Shmoys, D.B., Tardos, E.: Approximation Algorithms for Scheduling Unrelated Parallel Machines. Mathematical Programmming **46**, 259–271 (1990)
38. Monderer, D., Shapley, L.: Potential Games. Games and Economic Behanior **14**, 124–143 (1996)
39. Mu'alem, A., Schapira, M.: Setting lower bounds on truthfulness: extended abstract. In: Proceedings of the ACM-SIAM Symposium on Discrete Algorithms (SODA), pp. 1143–1152 (2007)
40. Nash, J.: Noncooperative Games. Annals of Mathematics **54**, 289–295 (1951)
41. Nisan, N., Ronen, A.: Algorithmic mechanism design (extended abstract). In: Proceedings of the ACM Symposium on Theory of Computing (STOC), pp. 129–140 (1999)
42. Nisan, N., Roughgarden, T., Tardos, E., Vazirani, V. (eds.): Algorithmic Game Theory. Cambridge University Press (2007)
43. Osborne, M.J., Rubinstein, A.: A Course in Game Theory. MIT Press (1994)
44. Pigou, A.C.: The Economics of Welfare. Macmillan (1920)
45. Rozenfeld, O., Tennenholtz, M.: Strong and Correlated Strong Equilibria in Monotone Congestion Games. In: Proceedings of the International Workshop on Internet & Network Economics (WINE), Springer LNCS 4286, pp. 74–86 (2006)
46. Schuurman, P., Vredeveld, T.: Performance Guarantees of Local Search for Multiprocessor Scheduling. INFORMS Journal on Computing **19**(1), 52–63 (2007)
47. Schuurman, P., Wöginger, G.J.: Preemptive scheduling with job-dependent setup times. In: Proceedings of the ACM-SIAM Symposium on Discrete Algorithms (SODA), pp. 759–767 (1999)
48. Wöginger, G.J., Yu, Z.: A heuristic for preemptive scheduling with set-up times. Computing **49**(2), 151–158 (1992)

Chapter 11
Artificial Neural Networks-Based Forecasting: An Attractive Option for Just-in-Time Systems

Mauricio Cabrera-Ríos, María Angélica Salazar-Aguilar, María Guadalupe Villarreal-Marroquín, and Ángela Patricia Anaya Salazar

Abstract Just-in-time (JIT) systems focus on lead-time reduction and equalization to make them respond rapidly to changes in demand. Lead-time variability in real life production, however, does affect the performance of JIT systems. This makes demand forecasting an important task to ponder. In this chapter, the use of artificial neural networks (ANNs) is advocated as an attractive approach to forecast demand for JIT systems. ANNs' capabilities to accommodate nonlinear dependencies and to generate forecasts for multiple periods ahead are among the most important reasons to consider for their adoption. A general method to build ANNs for time series prediction is presented aiming to circumvent some of the perceived difficulties associated to these models. Two case studies are also provided to illustrate the intended use.

11.1 Introduction

Just-in-time (JIT), as a philosophy and praxis, encompasses the coordinated flow of information and material. Usually in a JIT manufacturing system, information flows backwards while the material moves forward in tight synchronization. Ideally,

M. Cabrera-Ríos
Industrial Engineering Department, University of Puerto Rico at Mayagüez, Puerto Rico
e-mail: Mauricio.cabrera1@upr.edu

M. A. Salazar-Aguilar
CIRRELT–HEC, Université de Montréal, Quebec, Canada
e-mail: angelica.salazar@cirrelt.ca

M. G. Villarreal-Marroquín
Integrated Systems Engineering Department, The Ohio State University, Columbus, OH, USA
e-mail: lupita.villarreal@gmail.com

A. P. Anaya Salazar
Industrial Engineering Department, Universidad de San Buenaventura-Cali, Cali, Valle, Colombia
e-mail: angelitanaya@gmail.com

R.Z. Ríos-Mercado and Y.A. Ríos-Solís (eds.), *Just-in-Time Systems*, Springer Optimization and Its Applications 60, DOI 10.1007/978-1-4614-1123-9_11,

a JIT system should be able to react immediately to sudden changes in demand due to its emphasis on lead-time reduction and equalization. In real life, however, lead-time variation is ever present and it does have an effect on the performance of JIT systems, therefore, changes in demand must be anticipated to insure a proper response. Forecasting – the ability to generate estimates of future demand – then plays an important role on keeping JIT systems feasible. More precise forecasts lead to better planning and control as well as the reduction of inventories, a major focus of JIT systems.

Although the number of forecasting techniques is large in the literature, the competitive use of these techniques is often hampered by the lack of proper statistical training of forecasters, which precludes both an effective model selection and an adequate determination of the chosen model's parameters.

The objective of this chapter is to advocate the use of artificial neural networks (ANNs) as an attractive forecasting option for JIT systems. Some of the advantages of adopting ANN-based systems include (1) the flexibility of the models to represent linear and nonlinear dependencies on time series and (2) the ability to generate a single forecast at a time or multiple forecasts simultaneously. Because the origin of ANNs is not in the field of statistics, their application typically does not require major statistical training. This last fact can be certainly seen as an advantage or a disadvantage for a particular reader; however, it is important to reiterate that many people involved in forecasting might not have a strong statistical background.

The use of ANNs, however, is not without caveats as it is necessary to determine an adequate number of parameters to use in the model, the values of these parameters, and the number of significant historical periods (lags) that must be taken into account to make a forecast. It is precisely with the idea of circumventing these difficulties that a general strategy is presented here to facilitate the adoption of ANNs in JIT systems.

Two case studies are also included in this chapter, the first one illustrates the use of ANNs to generate a single forecast and the second one does the same for the simultaneous forecast of multiple periods.

11.2 Artificial Neural Networks as Forecasting Tools

Concisely described, an ANN is a series of nodes connected by directed arcs with weights on them. The nodes are called neurons and the arcs are called connections. The weights in the connections are usually called the same – weights. The way an ANN is organized on its neurons and connections is called its architecture, and the algorithm used to determine its weights is called its training algorithm.

The dynamic of an ANN is rather simple: connections apply their weights to whatever information they are transmitting, neurons concentrate the weighted information from incoming connections, apply a mathematical transformation, and pass their output forward through outgoing connections. The mathematical transformation is commonly referred to as transfer function. An ANN is a small-scale model of the human brain looking to mimic the brain's ability to carry out very complex tasks from the simple process of activating different networks of neurons.

Figure 11.1 shows the general architecture of an ANN used for forecasting, called the three-layer feedforward backpropagation ANN. This means that it has three layers of neurons, with forward connections linking them, and that its weights are found through error backpropagation, a common squared-error minimization task.

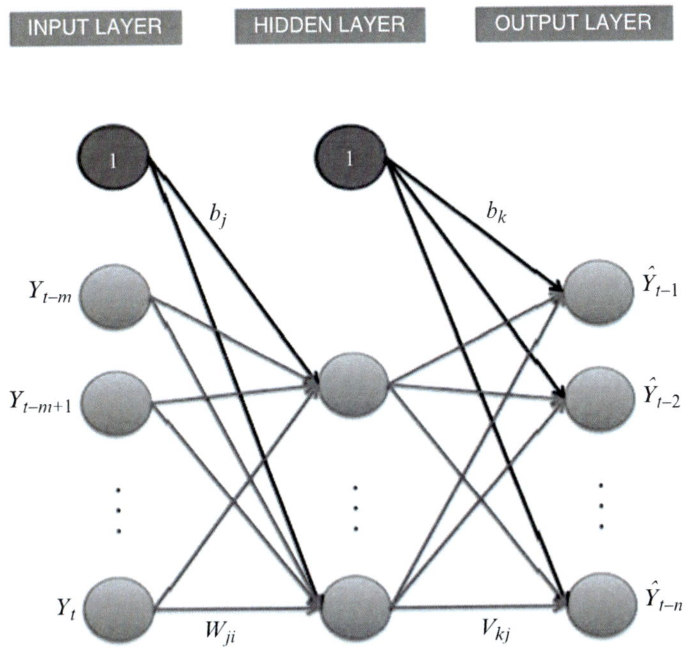

Fig. 11.1: Three-layer feed forward back propagation ANN

The ANN's first layer of neurons (input layer) receives historic data in a time series, which are then sent to the intermediate layer (called the hidden layer) for processing. The number of neurons in the hidden layer determines the degree of flexibility in the forecasting model. Once processed by the hidden layer, the information is finally sent to the output layer, which can contain either a single output neuron for a single-period forecast, or multiple output neurons for the forecast of multiple periods. When the transfer function of the neurons of the hidden layer is the hyperbolic tangent, and the transfer function of the neurons of the output layer is a linear function, the resulting ANN architecture has well studied and demonstrated universal approximation capabilities (White [16]; Hornik [11]); as well as well-documented good performance in forecasting (Zhang et al. [19]; Zhang [17]; Liao et al. [13]; Hansen et al. [9]).

Referring to Fig. 11.1, ANN inputs $Y_i, i = t-m, t-m+1, ..., t$ correspond to the m lags previous to period $t+1$, which is to be forecasted. Determining m is crucial because it indicates the extent in historic data to which a particular data point is correlated. As explained before, the number of neurons in the hidden layer determine the level of nonlinearity of the model, thus it is also an important parameter to determine. A number of hidden neurons larger than necessary would result in the loss of the ANN's prediction capability. If too few neurons are included, however, the model might not be flexible enough to be a good approximation. Finally, the neuron in the output layer computes the forecast for the n future periods $t+1, t+2, ..., t+n$, represented by $\hat{Y}_{t+1}, \hat{Y}_{t+2}, ..., \hat{Y}_{t+n}$ respectively. Accordingly, the transfer function for the hidden neurons is the hyperbolic tangent, and the identity function for the output neurons.

Additionally in Fig. 11.1, W_{ji} is the weight applied in the connection incoming to the jth hidden neuron from the ith input neuron, V_{kj} is the weight applied in the connection incoming to the kth output neuron from the jth hidden neuron. These weights modify the information that passes through their respective connections and can be understood as fitting parameters. Special weights b_j and b_k are known as biases.

There is a large collection of algorithms in the literature to train an ANN, that is, finding all ANN weight values. Two of the most utilized are the Levenberg–Marquardt's (lm) algorithm, which is valued for its convergence speed, and Bayesian Regularization (br), valued for its capability to preserve the model's generalization capability. Both of these algorithms have been shown to be quite competitive, although for different reasons as explained before (Bishop [1]; Hagan et al. [8]).

When working with time series, several options arise on how to handle the data. One option is to either leave the data as demand points on each period or used them as the difference between two adjacent periods. Also, one must decide to either normalize the data to fall within the range $[-1, 1]$, or use their original scale.

Also, in order to measure the ANN forecasting error, the mean square error (MSE), the mean absolute error (MAE), the largest-in-magnitude positive error or overprediction (S_Pred), and the largest-in-magnitude negative error or underprediction (B_Pred) can be used. Both MSE and MAE have been widely utilized in the literature, while S_Pred and B_Pred are important measures to show the worst cases in a particular analysis.

11.3 A General Strategy for Artificial Neural-Networks-Based Forecasting

In this section, a general method to set the parameters of the ANN in a simultaneous and systematic manner is described. The method is presented here as applicable to generic feedforward backpropagation ANNs (Salazar et al. [15]).

The basic idea behind the method is to represent the ANN as a system with controllable variables that exert influence over certain ANN's outputs. In this way, by considering the ANN parameters as controllable variables, we can explore their respective ranges and characterize their effect on key performance measures (outputs) through regression techniques. This characterization, in turn, will allow the use of analysis of variance (ANOVA) to screen significant controllable variables as well as determine if two or more performance measures (PMs) depend on disjoint subsets of these variables. In this last case, independent optimization problems can be created, thereby simplifying this task.

Because the ultimate goal is to set these parameters to obtain the best possible PM values, the problem indeed falls in the optimization realm. The steps of the method are described next:

1. *Description of the ANN as a system.* The objective in this step is to specify the characteristics of the ANN to be used. It is necessary to identify the ANN controllable parameters as well as to define the ANN's PMs to be included in the study.
2. *Design and Analysis of Experiments.* In this step, the focus is on planning, executing and interpreting the results of a statistically designed experiment that includes the previously identified parameters and PMs.
3. *Metamodeling.* A regression model must be obtained through proper statistical techniques (including residual analysis) to describe the response surface corresponding to each PM as a function of the controllable parameters.
4. *Optimization Problem.* An optimization problem is built with the metamodels obtained in the previous steps as objective functions.
5. *Solution.* Solve the optimization problem through proper techniques, that is, multiple criteria optimization if necessary and multiple starting points.

The rationale for the proposed method obeys the point of view of an experiment, where planned changes are introduced in a system with the objective to analyze the variation induced in PMs of interest. Because the size of the experiment to be run depends on the number of controllable parameters, the use of an adequate experimental design is critical. If there are but few parameters, then a full factorial design will suffice. A factorial design contains as many experimental runs as the complete enumeration of the combinations of values to be sampled per parameter (factor). For example, if four parameters are to be studied, each varied at three, four, three and five values or levels, then the factorial design will have $3 \times 4 \times 3 \times 5 = 180$ experimental runs.

For our purposes, then, under each combination of controllable parameters (run) an ANN will be built and trained to obtain a quantification of the prediction quality through the chosen PMs as the responses of the experiment.

It is advisable to consider at least three different values for each parameter where possible to allow the determination of response curvature right from the beginning.

Once the experiment is complete, it is followed by its analysis aiming to characterize the variation induced in the PMs. In order to achieve this, an ANOVA is run based on an underlying full quadratic regression model of each PM as a function of the controllable parameters. The regression model proposed is shown in (11.1), where the residuals, ε, are assumed to be identically, independently and normally distributed with a mean of 0 and an unknown, but constant variance.

$$y = \beta_0 + \sum_{i=1}^{k} \beta_i x_i + \sum_{i=1}^{k} \beta_{ii} x_i^2 + \sum_{i=1}^{k-1} \sum_{j=i+1}^{k} \beta_{ij} x_i x_j + \varepsilon. \qquad (11.1)$$

In (11.1), dependent variable y represents the value of the PM of interest, x_i corresponds to the ith parameter, β_0 is the intercept of the model, β_i is the regression coefficient associated to the variation of x_i, β_{ii} is the regression coefficient for x_i^2, and β_{ij} is the regression coefficient for the interaction between x_i and x_j; and k is the total number of controllable parameters in the experiment.

The regression coefficients are typically computed through a squares reduction technique, available in most commercial statistical software packages. Checking the residuals' assumptions to verify the adequacy of the model is also conveniently carried out through these same packages.

Finally, each of the resulting regression models (one per PM) is included as an objective function in an optimization problem, where finding the parameter settings to achieve the best possible value for the objective function is seek.

At the end, the optimization problem will most likely be nonlinear and nonconvex. This makes the optimization problem hard to solve, and therefore the highest aspiration tends to be finding an attractive local optimal solution.

11.4 Case Study 1: Forecasting One Period Ahead

The strategy previously described was put in practice in a telecommunications company for testing purposes. The company's ultimate objective, as the vast majority of companies in the world, was – and still is – to be profitable through providing a high service level to its customers. Its major resource was a transmission network, loosely defined as a set of interconnected devices with a finite transmission capacity.

Customers demanded different services at different levels through such network in a stochastic fashion and challenged the installed network capacity every so often. Because decisions on capacity expansion take time and equipment purchase and installation are not immediate, it was critical for the company to be able to estimate the demand behavior at least for the next period of time. Reliable forecasts would allow the decision-maker to plan the required capacity expansions to meet the expected demand without having too much nonutilized installed capacity. The company provided historic network utilization data organized in monthly periods, that is, a time series. Nonlinearity seemed evident from the beginning. This observation was further confirmed with the poor forecasts from a previously tried linear regression. However, as none of the previous was completely conclusive, we relied on the fact that ANNs have been shown to be competitive even in cases of linearity (Hwarng [12]; Medeiros et al. [14]; Zhang [18]). Based on these pieces of information, it was decided that an ANN would be a sensible choice. The parameter selection was approached through the strategy described in the previous section.

As part of the first step of the strategy, the following parameters or controllable factors were selected: (1) number of lags, (2) number of hidden neurons, (3) use of a transformation in the data, (4) scale of the data, and (5) training algorithm to be used. The number of lags was varied in the levels $\{2,3,6\}$; the number of hidden neurons in the levels $\{2,5,7\}$; the transformation of the data consisted in keeping the data as timed data points or taking the data as differences between two consecutive periods; the scale of the data referred to keeping the data in their original scale or standardize them to fall within -1 and 1; finally, two training algorithms were tried, the Levenberg–Marquardt and the Bayesian regularization. The PMs were MSE, MAE, B_Pred and S_Pred. Specific values for the ANN parameters are, then, as follows: lags = $\{2,3,6\}$, neurons = $\{2,5,7\}$, transformation = {none, differences}, scale = {original, $[-1,1]$} and algorithm = {lm, br}. The labels "1", "2" and "3" were used to denote the levels of these parameters and correspond to the order of the listed values.

When conducting the experiment, it was deemed adequate to use a factorial design $3^2 \times 2^3$, which resulted in a total of 72 combinations. Table 11.1 shows the experimental results when the training algorithm was Levenberg–Marquardt's.

For the metamodelling part of the strategy, the resulting experiment was used to characterize the results through the use of regression analysis and ANOVA. This analysis, along with the factorial structure and the regression metamodel, allows the assessment of the statistical contribution of the parameters independently, as well as in second-order interactions. Second order regression models were obtained for each of the PMs. The resulting coefficients are shown in Table 11.2.

Because using regression metamodels and ANOVA implies assumptions regarding the residuals, namely normality, independence and constant variance, it is important that these assumptions be verified through residual analysis to make sure that the subsequent conclusions are statistically valid.

Figure 11.2 shows an example on how residual analysis can be carried out graphically as a first approximation. This figure shows the analysis pertaining to B_Pred. The one on the left allows verifying if the residuals are following a normal distribution. In the graph in the upper left corner, a straight-line pattern is seek and in the graph in the lower left corner a bell shape similar to the normal distribution should be approximately evident. The graphs in the right have the purpose of checking for independence of the residuals vs. the model predictions (fits) and vs. the run order. The behavior of the residuals in these graphs should be random. These two graphs also help to see if there was a dramatic change in the spread of the residuals (non-constant variance). A series of formal statistical tests usually follows the graphical analysis. In this case, after statistical testing, the conclusion was that there was no significant violation to the residual assumptions. Further details on these tests can be found in (Devore [7]). A similar analysis was carried out for each PM, with similar conclusions.

According to the proposed strategy, metamodeling must be followed by the creation of optimization problems. In these optimization problems, the objective function z is defined by the metamodel representing the PM of interest. In general, the resulting optimization problem, followed by Tables 11.1 and 11.2, to end with Figure 11.2.

Table 11.1: Results with Levenberg–Marquardt's training algorithm

Neurons	Lags	Scale	Transformation	Algorithm	MAE	MSE	B_Pred	S_Pred
2	2	1	1	1	15.38	717.49	69.89	125.52
2	2	1	2	1	14.83	822.03	73.26	130.74
2	2	2	1	1	14.99	609.14	66.36	105.76
2	2	2	2	1	14.41	808.62	71.63	132.36
2	3	1	1	1	15.39	725.56	70.07	125.99
2	3	1	2	1	15.25	724.62	68.28	135.72
2	3	2	1	1	14.89	613.74	66.36	105.42
2	3	2	2	1	12.53	523.33	69.62	135.46
Neurons	Lags	Scale	Transformation	Algorithm	MAE	MSE	B_Pred	S_Pred
2	6	1	1	1	15.31	566.63	69.56	116.49
2	6	1	2	1	15.02	578.48	69.35	98.61
2	6	2	1	1	13.43	501.43	69.18	112.18
2	6	2	2	1	10.60	248.54	65.77	41.95
5	2	1	1	1	15.09	698.58	70.89	125.17
5	2	1	2	1	11.80	448.97	65.35	128.56
5	2	2	1	1	12.47	407.59	43.30	114.36
5	2	2	2	1	9.93	440.55	69.97	134.07
5	3	1	1	1	15.01	715.72	69.37	126.64
5	3	1	2	1	14.69	759.51	69.42	142.16
5	3	2	1	1	10.21	355.73	65.85	108.67
5	3	2	2	1	9.78	251.53	70.43	55.35
5	6	1	1	1	15.46	553.00	59.24	101.95
5	6	1	2	1	12.71	545.57	68.51	120.65
5	6	2	1	1	7.04	169.25	30.79	74.33
5	6	2	2	1	3.29	32.99	21.04	13.86
7	2	1	1	1	14.81	700.10	70.41	125.41
7	2	1	2	1	9.09	178.11	65.23	26.45
7	2	2	1	1	8.89	218.43	63.52	57.49
7	2	2	2	1	7.30	147.09	65.69	32.18
7	3	1	1	1	15.28	722.19	69.49	126.43
7	3	1	2	1	11.85	443.52	68.94	114.13
7	3	2	1	1	9.20	310.22	48.55	108.67
7	3	2	2	1	5.49	68.57	17.46	29.64
7	6	1	1	1	14.45	607.09	64.92	121.57
7	6	1	2	1	13.90	461.54	58.44	90.33
7	6	2	1	1	3.59	19.41	13.88	10.82
7	6	2	2	1	0.80	1.68	5.04	4.68

$$\text{Find} \quad x_i \quad \forall_i \in I = \{1,2,\cdots,k\} \text{ to}$$

$$\text{Minimize } z = \beta_0 + \sum_{i=1}^{k} \beta_i x_i + \sum_{i=1}^{k} \beta_{ii} x_i^2 + \sum_{i=1}^{k-1} \sum_{j=i+1}^{k} \beta_{ij} x_i x_j$$

$$\text{Subject to} \quad x_{li} \le x_i \le x_{ui} \quad \forall i \in I$$

$$x_i \in Z^+ \quad \forall i \in I.$$

Table 11.2: Regression coefficients for each ANN performance measure

Regression Term	MAE	MSE	B_Pred	S_Pred
Constant	29.82	1726.89	102.61	219.74
x_1	−1.08	-84.40	0.13	4.36
x_2	0.55	-4.72	1.58	20.73
x_3	−1.92	−178.61	0.88	-23.03
x_4	−1.67	−100.62	9.10	-9.47
x_5	−14.15	-935.46	-61.57	−145.98
x^2_1	0.02	-0.15	0.07	−1.41
x^2_2	−0.08	-6.00	−0.13	-2.83
$x_1\,x_2$	0.01	7.41	-0.61	0.61
$x_1\,x_3$	−0.52	-25.08	-2.99	-4.35
$x_1\,x_4$	-0.21	-21.05	-1.08	-3.57
$x_1\,x_5$	1.00	68.57	3.94	9.57
$x_2\,x_3$	-0.74	-37.02	-3.41	-7.30
$x_2\,x_4$	0.06	4.64	-0.61	-2.19
$x_2\,x_5$	0.52	25.07	4.13	6.03
$x_3\,x_4$	-1.71	-101.01	-6.87	-7.87
$x_3\,x_5$	4.84	291.74	19.42	42.71
$x_4\,x_5$	2.91	237.47	7.45	27.82

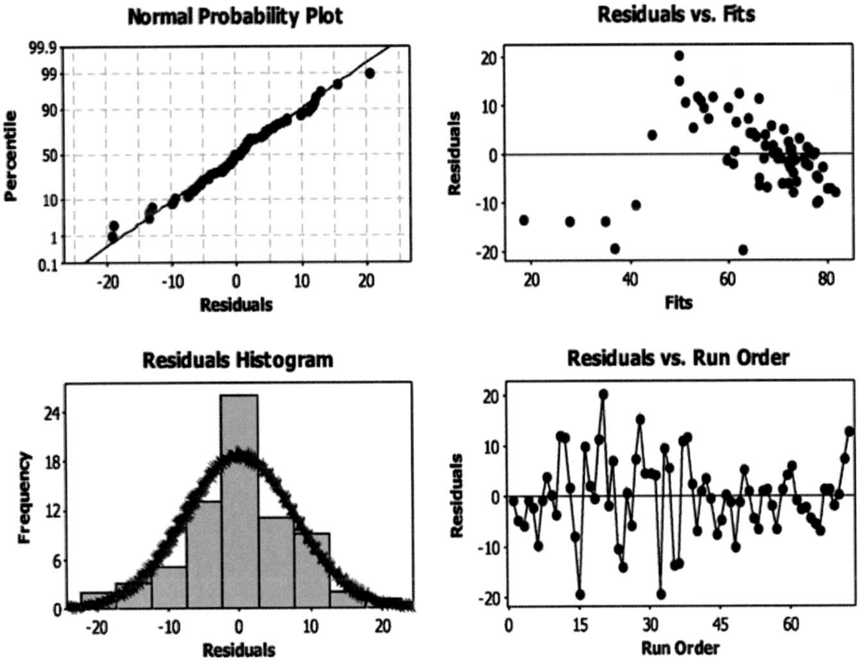

Fig. 11.2: Residual analysis for the regression metamodel associated to B_Pred

Where the decision variables x_i, represent the ith ANN parameter through the variation of which the objective function z is to be minimized. k stands for the number of controllable parameters whose values must fall within a preset lower bound x_{li} and a preset upper bound x_{ui}. These bounds usually correspond to the limits of the region explored in the experimental design. Finally, the decision variables must be positive integers as indicated by the last constraint. In this case study $k = 5$, corresponds to the parameters: neurons, lags, scale, transformation, and algorithm.

Solving each optimization problem independently and using multiple starting points to increase the probability of finding an attractive local optimum, it was found that the input data should be handled in differences and scaled to fall within the range $[-1,1]$. The number of lags (now in differences) was determined to be six periods, and the ANN should have seven hidden neurons. The training algorithm was prescribed to be Levenberg–Marquardt's. An ANN with these resulting characteristics provides forecasts with performance as shown in Fig. 11.3. It can be observed that the model is indeed a good approximation to the time series. In fact, this configuration coincided with the best experimental combination in Table 11.1. An adequate forecasting performance is highly probable after the application of the proposed method.

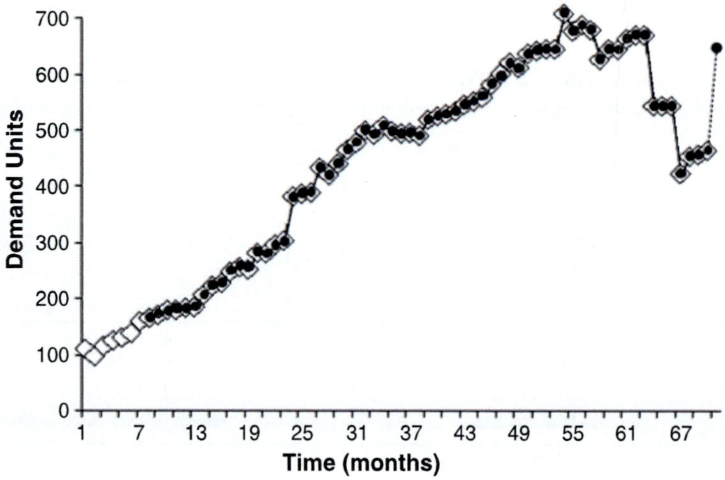

Fig. 11.3: ANN predictions (*solid circles*) vs. the real (*diamonds*) time series

The final ANN, whose performance is depicted in Fig. 11.3 was the best one found by the optimization procedure when minimizing each one of the PMs. This result is a clear indication of a large degree of correlation without conflict among the PMs. Had the optimization problems arrived to solutions different to each other, selecting a combination would have not been a trivial task. In such case, it would have been necessary to use multiple criteria optimization techniques. These techniques try to find the best compromises between all PMs, which are formally called Pareto-efficient solutions. Because treatment of multiple criteria optimization procedures is beyond the scope of this work, the interested reader is referred to Deb [6] and Hillermeier [10]. Also, some recent applications of multiple criteria optimization in manufacturing using Data Envelopment Analysis can be found in (Cabrera-Ríos et al. [2, 3], Castro et al. [5] and in Castro et al. [4].

In this case study, the computational tools were Matlab™, Minitab™ and MS Excel™. The first one was used to build the ANNs, the second one was used for statistical analysis, and the third one for optimization purposes through its XL Solver module.

11.5 Case Study 2: Forecasting Multiple Periods Simultaneously

Figure 11.4 shows two of the time series facilitated by the company. It was required to generate 12 periods ahead with an ANN similar to the one shown in Fig. 11.1. The information provided was scaled to fall within −1 and 1, and the training algorithm used was the Levenberg–Marquardt's based on our previous results.

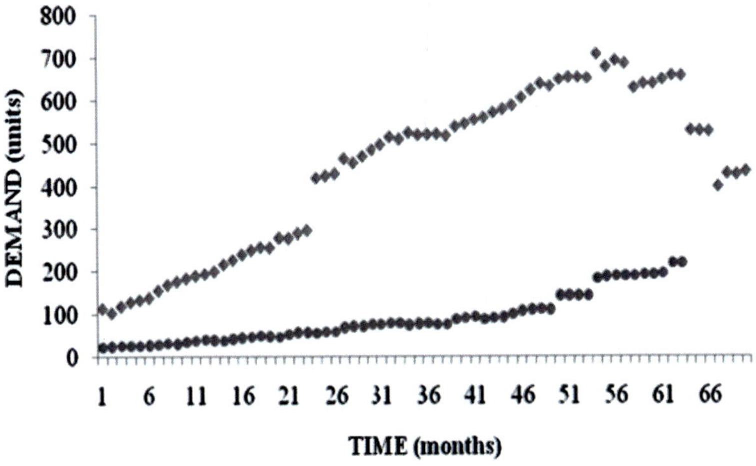

Fig. 11.4: Demand behavior (Series 1 in *diamonds*, Series 2 in *circles*)

Two controllable parameters were considered: (1) the number of neurons in the ANN's hidden layer, and (2) the number of lags to be used to generate the 12 forecasts.

In order to determine the weights for the ANN, two data sets were prepared: a training data set and a validation data set. Approximately 70% of the data were used for training and the rest for validation. The forecasting performance of the ANN was assessed through the use of the MSEs for training (MSE_T) and for validation (MSE_V). The training phase is itself a nonlinear optimization procedure aimed to find the best possible set of ANN weights. In this case, the resulting problem is also nonconvex, therefore special care must be given to try different starting points for the optimization procedure. For this reason, the weights were initialized all on -1, -0.5, 0, 0.5 and 1 as well as from random values. Thus, for each experimental run, six different ANNs were created. The ANN with the lowest value of MSE_V was kept as the experimental response.

Five levels were chosen to sample the controllable parameters for the Series 1 instance. The number of hidden neurons was varied on the values $\{2,4,6,8,10\}$ and the number of lags on $\{3,6,9,12,15\}$. The experimental design was a full factorial, where all possible combinations are tried, that is, in this case, $5 \times 5 = 25$ runs. The initial regression metamodels, however, did not result in a good fit. With this in mind, it was decided to focus the experimental region where low MSE values were observed. This reduced region included the grid formed by the combinations of the first three levels of both controllable parameters. After this adjustment, the metamodels did show a good fit. It is important to secure a good fit to the experimental data since providing a higher fidelity of the metamodels will increase the probability of finding attractive solutions in the optimization step.

Solving the optimization problems independently, one for minimizing MSE_T and another for minimizing MSE_V, it was clear that these converged to different solutions, that is, there was a conflict between them. The former – involving MSE_T – converged to an ANN with five hidden neurons and considering seven lags. The latter – involving MSE_V – converged to five hidden neurons and five lags.

In this case, since the conflict was only on the number of lags, it was decided to give a larger weight to MSE_V, thus keeping five hidden neurons and five lags. Figure 11.5 shows the forecasts obtained with the ANN chosen for forecasting Series 1 as prescribed by the method as well as the comparison against the use of linear regression. A comparison vs. linear regression is presented because this method showed the best forecasting performance, as measured through MSE, among eight linear statistical methods for this series: (1) naïve method, (2) average, (3) moving average, (4) single exponential smoothing, (5) ARIMA (0,1,1), (6) linear regression, (7) double exponential smoothing, and (8) ARIMA (0,2,2).

The proposed strategy was also applied to Series 2, resulting in a prescribed ANN with three hidden neurons and five lags. Figure 11.6 shows the forecasts with this ANN as well as a comparison against those from double exponential smoothing. Again the comparison is against the best linear statistical method from the eight previously listed.

The results in both cases showed a better forecasting performance when the ANNs were used. The implementation of the strategy outlined previously provided the company with a tool to handle a wide array of time series while having the ability to set the parameters of the ANNs to make them perform adequately. In the developed forecasting system, the task of experiments planning and execution was automated in order to facilitate the construction of the ANNs model.

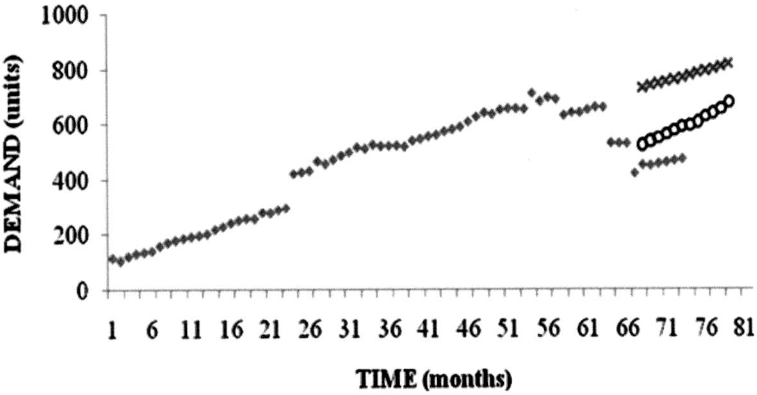

Fig. 11.5: Real (*diamonds*) vs. ANN (*circles*) vs. linear regression (*crosses*), Series 1

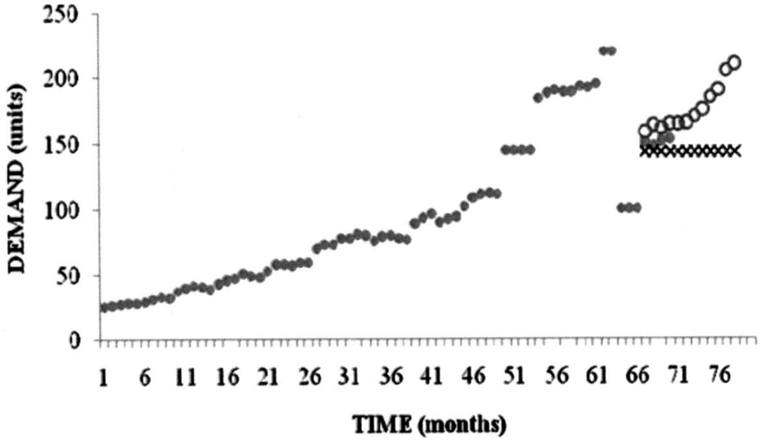

Fig. 11.6: Real (*solid circles*) vs. ANN (*empty circles*) vs. double exponential smoothing (*crosses*), Series 2

11.6 Final Thoughts

JIT systems are ideally capable to deal with demand fluctuations. In practice, however, lead time variability will lead to pay special attention to production planning and production control and, thus, to demand forecasting. This chapter is aimed for the interested reader to consider the adoption of ANNs for forecasting to capitalize in these tools' flexibility to accommodate nonlinearity as well as the simultaneous prediction of multiple demand periods. This practice will be particularly useful if the personnel in charge of forecasting does not have enough statistical background to make a competitive statistical model selection and the associated parameter setting. The strategy presented in this chapter to build ANNs focuses on circumventing some of the most common caveats when using this kind of model. The use of the strategy was also demonstrated through two case studies.

Acknowledgments

The support of Guillermo J. Moreno is gratefully acknowledged in this work. The authors would also like to thank the support of CONACYT-México and UANL PAICYT Program.

References

1. Bishop, C.M.: Neural Networks for Pattern Recognition. Oxford University Press, Oxford (1995)
2. Cabrera-Ríos, M., Castro, J.M., Mount-Campbell, C.A.: Multiple quality criteria optimization in reactive in-mold coating with a data envelopment analysis approach II: a case with more than three performance measures. Journal of Polymer Engineering **24**, 435–450 (2004)
3. Cabrera-Ríos, M., Castro, J.M., Mount-Campbell, C.A.: Multiple quality criteria optimization in in-mold coating (IMC) with a data envelopment analysis approach. Journal of Polymer Engineering **22**, 305–340 (2002)
4. Castro, C.E., Cabrera-Ríos, M., Lilly, B., Castro, J.M., Mount-Campbell, C.A.: Identifying the best compromise between multiple performance measures in injection holding (IM) using data envelopment analysis (DEA). Journal of Integrated Design and Process Science **7**, 77–87 (2003)
5. Castro, J.M., Cabrera-Ríos, M., Mount-Campbell, C.A.: Modelling and Simulation in reactive polymer processing. Modelling and Simulation in Materials Science and Engineering **3**, S121-S149 (2004)
6. Deb, K.: Multi-Objective Optimization Using Evolutionary Algorithms. Wiley, New York (2004)
7. Devore, J.L.: Probability and Statistics for Engineering the Sciences.4th Edition, Duxbury Press, California (1995)
8. Hagan, M.T., Demuth, H.B., Beale, M.: Neural Network Design. PWS Publishing Company, Boston (1996)
9. Hansen, J.V., Nelson, R.D.: Forecasting and recombining time-series components by using neural networks. Journal of the Operations Research Society **54**, 307–317 (2003)
10. Hillermeier, C.: Nonlinear Multiobjective Optimization: A Generalized Homotopy Approach. Birkhauser Verlag, Munich (2001)
11. Hornik, K., Stinchcombe, M., White, H.: Multilayer feedforward networks are universal approximators. Neural Networks **5**, 359–366 (1989)
12. Hwarng, H.B.: Insights into neural-network forecasting of time series corresponding to ARMA (p,q) structures. Omega: The International Journal of Management Science **29**, 273–289 (2001)
13. Liao, K.-P., Fildes, R.: The accuracy of a procedural approach to specifying feedforward neural networks for forecasting. Computers & Operations Research **32**, 151–2169 (2005)
14. Medeiros, M.C., Pedreira, C.E.: What are the effects of forecasting linear time series with neural networks? Logistic and Transportation Review **31**, 239–251 (2001)

15. Salazar-Aguilar, M.A., Moreno Rodríguez, G.M., Cabrera-Rios, M.: Statistical Characterization and Optimization of Artificial Neural Networks in Time Series Forecasting: The One Period Forecast Case. Computación y Sistemas **10**, 69–81 (2006)
16. White, H.: Connectionist nonparametric regression: Multilayer feedforward networks can learn arbitrary mappings. Neural Networks **3**, 535–549 (1990)
17. Zhang, G.P.: Neural Networks in Business Forecasting. Idea Group Publishing, Georgia (2004)
18. Zhang, G.P., Hu, M.: A simulation study of artificial neural networks for nonlinear time series forecasting. Computers & Operations Research **28**, 381–396 (2001)
19. Zhang, G., Patuwo, E., Hu, M.: Forecasting with artificial neural networks the state of the art. International Journal of Forecasting **14**, 35-62 (1998)

Chapter 12
Just-in-Time Management of a Building Project in the Middle-East

Low Sui Pheng and Thilini Shiromani Jayawickrama

Abstract The Just-in-Time (JIT) philosophy originated from and is widely used in, but not limited to the manufacturing industry. Many studies have been conducted to examine the possibility of adopting JIT principles to specific areas of the construction industry. This study aims to identify the application of the JIT principles for a building project in a region of the Middle-East from the perspective of the main contractor through a case study approach. The study provides anecdotal evidence of the project applying the seven principles of JIT successfully to improve productivity and quality standards, which helped the main contractor to achieve the target completion date. The study also reveals top management commitment as a major underpinning for implementing the JIT system as well as some of the hindrances which may be inevitable because of the corresponding barriers encountered. Some of these barriers are due to the inherent features of the construction industry such as the multidisciplinary nature of practices and complexity of projects. Other barriers are region-specific and include perceptibly lenient and less effective resource management techniques and philosophies of many local clients driven by their high liquidity and strong spending capacities.

12.1 Introduction

The Just-in-Time (JIT) philosophy, also commonly known as the Toyota Production System is a concept that originated in Japan to improve the productivity in the manufacturing industry. In the early 1950s, Mr Taiichi Ohno, former Executive Vice-President of Toyota Motor Corporation developed and promoted this concept and they were able to achieve improved productivity and were able to manufacture cars with higher quality and reliability through the application of this concept [1, 3].

Low Sui Pheng and Thilini Shiromani Jayawickrama
Department of Building, National University of Singapore, Singapore,
e-mail: {bdglowsp, a0066405}@nus.edu.sg

R.Z. Ríos-Mercado and Y.A. Ríos-Solís (eds.), *Just-in-Time Systems*, Springer Optimization and Its Applications 60, DOI 10.1007/978-1-4614-1123-9_12,

Although the JIT philosophy originated from and is widely used in the manufacturing industry, many studies have been conducted to examine the possibility of adopting JIT principles to specific areas of the construction industry [7]. Correspondingly, there were situations where the construction industry had successfully applied JIT principles to some extent to improve productivity and quality standards (e.g. [3,5]).

These studies have identified that the JIT philosophy is applicable in the construction industry with some modifications even though the nature of the construction industry is distinctively different in many aspects from the manufacturing industry. This was further discussed by Low [2] through a comparative study between these two industries.

Hence, based on these earlier works, the aim of this study is to further identify the application of the JIT concept by a contracting organization for a building project in the Middle-East and the implementation of JIT principles will be discussed from the perspective of the main contractor. It is hoped that the case study presented will help to encourage more contractors to consider the implementation of best practices based on the JIT principles in their daily operations to further raise productivity and quality standards in the construction industry.

12.2 Key JIT Management Concepts and Principles

One of the dominant thrusts behind the JIT philosophy is the provision of the right items with the right quality, in the right quantities, in the right place only at the right time [6]. This can be accomplished through seven broad principles which are very much intertwined and related to one another as explained by Low and Choong [4].

12.2.1 Elimination of Waste

The fundamental principle of the JIT philosophy is to eliminate waste [3]. It highlights the elimination of waste in any kind of resources. According to Low and Choong [4], under the JIT environment, waste is defined as anything that does not contribute to the value of the product such as inspection, rework, transportation, processing waste, waste of waiting time, waste of motion, storing and maintaining. The concept is highly focused on reducing inventory levels which can appear in the forms of excess materials, obsolete materials, defective materials, over production, defective production and work in progress. These unnecessary inventories utilize storage space, thus incurring unnecessary storage and maintenance costs and also generate additional risks associated with storage such as those relating to damages and obsolescence. Consequently, capital has to be disbursed earlier which is also unnecessary. In relation to construction sites, where labour, materials and plant are the main types of resources involved, this principle provides the initiative to eliminate idling labour, material wastage, idling plant, and wasting resources on un-related activities.

12.2.2 "Kanban" or Pull System

This principle simply prescribes the need to supply only when there is demand. Low and Choong [4] explained that the essence of this principle is that the flow of materials is "pulled" by the demand side. The supply can be configured in two ways, either in a pull system or in a push system wherein the organization produces on demand or the organization produced based on the demand forecast respectively. "Kanban" is a Japanese term for "card" which was derived from the card system used to signal the delivery of components from one workstation to another [4]. This fundamental concept behind the JIT philosophy is to provide resources only in the required amount at the required time, and therefore plays a major role for all JIT operations. Hence, an established relationship between the supplier and the customer is important to implement this principle. This will be discussed further under the principle of "Supplier relations". The supplier may be required to make frequent deliveries using this principle. In this context, defects may also be identified more readily and where the quality standard of the products may be maintained consistently. This will be discussed further under the principle of "Total quality control" (TQC). In contrast to this principle, the risks associated with unforeseeable conditions, late deliveries and shortages of resources will be discussed as limitations for the implementation of this principle in the construction industry.

12.2.3 Uninterrupted Work Flow

The rationale behind this principle is to keep the production process smooth and the continuous flow of activities as scheduled. Simplifying and rationalizing the processes by reducing unnecessary functions such as inspection and rework would help to facilitate this principle. Similarly, automation and continuous maintenance of machineries or equipment to prevent disruptions arising from breakdowns will also be useful to accomplish this principle. Since the JIT concept encourages the provision of resources only at the required time following the "Kanban" or pull system, any interruption in the production process may impact the entire process and it is, therefore, essential to ensure that the production process is uninterrupted [4].

12.2.4 Total Quality Control

To accomplish the principle of "eliminating waste" which will also occur due to defective components and to maintain "uninterrupted work flow", a TQC system must be incorporated for a successful JIT system to materialize. Inspection will be carried out to first ensure that the product is defect-free before passing the part to the subsequent work station and that every worker is personally responsible for the quality of the work he produces [3]. Hence, this principle is interrelated with the principle of "Employee involvement" which will be discussed later. Quality is one of the three major aspects (the other two being time and cost) of a construction project in which the contractor ensures that all the client requirements are achieved. Hence, this principle is of immense importance for the construction industry.

12.2.5 Employee Involvement

The JIT concept comprises of a set of principles that requires practical implementation throughout the entire production process. The involvement of every worker at all levels in the organization is critical for its success. Low and Ang [3] observed that commitment from the top management is significant as the driving force behind JIT implementation and the involvement of all other employees is necessary for the success of JIT. As a result, the implementation is dependent on the level of their involvement. The awareness of employees, their understanding and knowledge of the JIT concept and their ability to work quickly and as multi-functional workers would consequently be more beneficial.

As described earlier, employee involvement plays a major role for JIT implementation under the principle of "Total quality control". Likewise, the principle of "Continuous improvement" is essentially required as part of employee involvement [4] and will be discussed shortly. This principle underpins the successful completion of construction projects which are highly dependent on human resources throughout the entire project duration.

12.2.6 Supplier Relations

Suppliers are entities external to the organization but are highly related to the production processes. As described under the principle of "Kanban" or pull system, delivering the right items with the required quality at the right time to the right place is of utmost importance in the JIT environment and which is different from that in the traditional processes. Thus, the suppliers play a major role in JIT implementation. Careful selection of a few suppliers who are able to deliver high quality materials at the right time is therefore crucial. Effective communication and coordination with the suppliers which encourage long term agreements founded on mutual trust are also important to ensure proper material flow. Close geographical proximity of the customer and the supplier locally would be more beneficial while reducing the costs of deliveries. Due to the temporary nature of construction projects, implementing this principle may not always be feasible in some situations. On the other hand, the long term relationships between suppliers and building contractors are not entirely new in the construction industry. In this context, Low [2] explained the advantages of a "rim delivery system" under the JIT concept in which the suppliers delivering materials to the customer also call on other suppliers on the way to the customer. This is in contrast to the conventional "spoke delivery system" where the suppliers directly deliver to the customer individually and in the absence of a well coordinated delivery schedule, may lead to traffic congestion in the workplace of the customer.

12.2.7 Continuous Improvement

The JIT philosophy is a never-ending process which provides the means to an end and not an end in itself [3]. The organizations can develop their own internal systems to encourage continuous improvement (or "Kaizen" in Japanese) of the JIT system by motivating and leading the employees to do so. All the employees should collectively implement this continuously and should strive to develop this mindset and habit in their organizations. Low and Choong [4] described employee involvement, which is led and guided by top management commitment, as an essential factor for "continuous improvement". However, the lack of permanency in the construction project teams may appear to be a barrier for implementing this principle in the construction industry. Nevertheless, when considered at the organizational level rather than at the project level, this principle can improve the ways by which projects can likewise benefit in the long term.

12.3 Background of the Construction Industry

It is a well known fact that the JIT concept originated from the manufacturing industry. Since the study discusses the application of the JIT concept in the construction industry, it would be pertinent to first highlight the main differences between the manufacturing industry and the construction industry to better understand the modifications needed for the concept to be applied in the latter.

The construction industry comprises of multi-disciplinary tasks and parties that are inherently complex in nature. The construction industry also produces unique products according to specific client requirements. Normally, the works carried out in the construction industry are project based and where the project teams consist of multi-disciplinary team members (e.g. architects, quantity surveyors, engineers, main contractors, subcontractors, suppliers, etc.) temporarily gathered together to work to achieve common targets within a specific time period and a budget. Activities in a project are implemented in a sequence and many resources (materials, labour and plant) will be utilized for these activities.

Unlike producing for the potential purchasers or consumers in the manufacturing industry, in construction projects, the building owner or property developer is known at the beginning of the project and who initiates the project and assigns other construction team members to the project. The appointed contractor will implement the construction works thereafter. Client involvement is high in construction projects which may lead to variations during the construction process. Construction projects are highly influenced by the external environment such as inclement weather, safety hazards and many other unforeseeable conditions such as shortages of building materials. Due to the often large scope of works operating over an extended time frame, construction projects are highly vulnerable to adverse effects of wide economic fluctuations and unstable political situations.

A typical construction project proceeds through several sequential stages that include inception, feasibility studies, design stage, tendering, production information, construction and completion. The involvement of each stakeholder in these stages is dependent on the type of contractual agreement adopted for the project. As an example, in a traditional contract, the involvement of the main contractor will start after the tendering stage while in Design-and-Build contracts, the main contractor's involvement starts much earlier. The contractual arrangements and the corresponding responsibilities of parties to the contract are also different between different types of building contracts. In traditional contracts, the responsibilities for design and construction are underpinned by separate contractual arrangements. In contrast, in Design-and-Build projects, both design and construction responsibilities are framed in a single contract.

Due to the unique nature of the construction industry, planning and scheduling the work activities would invariably require more efforts. Since the involvement of different stakeholders is a common feature in the construction industry and is highly dependent on human activities, construction projects should be planned and managed with careful attention paid to the day to day details. Conflicts due to different perceptions of members may arise in construction projects. The impacts of legal issues such as building regulations, safety requirements and other general legislation may also be considerable. When the construction related organizations are operating in compliance with other quality assurance requirements, environmental considerations, safety and health issues under the umbrella of either or both local or global institutions, care should be taken in project planning and management to ensure that all regulations have been complied with.

For this reason, different planning and management techniques are used in the international construction industry. However, it is generally considered that the contractor is still responsible for completing the construction project to meet all these regulations as well as other/additional requirements stipulated in the contract.

12.4 Characteristics of Case Study Organization

The case study organization is one of the leading and highly progressive contracting companies in the region, namely Jalab Engineering and Contracting, hereinafter called JEC. (Note: the names of all organizations and projects have been changed to preserve their anonymity.) Certification to the ISO 9001 quality management system forms an integral part of the management process of this organization to ensure consistency and compliance with international standards and best practices. The organization strives to be a leader in environmental management and is ISO 14001 certified. It also actively participates in assisting its key clients to achieve their targeted Leadership in Energy and Environmental Design or LEED ratings for sustainable construction in their projects. The organization is also OHSAS 18001 certified to ensure that specific procedures are correctly implemented to achieve the safest possible working environment for everyone.

The construction site selected for the case study was one of the major construction projects of this organization, namely the "Merdin Shopping Mall Project". The client was the FAM Group of Middle-East and the project was managed by the MacPM Group International with separate services from several consultancy organizations for architectural works, interior design works, design and supervision of mechanical and electrical works, quantity surveying services, structural engineering works and facilities management works.

The development consists of a two-storey shopping mall (accommodating 450 shops), two-storey car park (providing 7,000 parking spaces), Energy Centre and associated external works. Its other facilities include two food courts with about 60 different eating outlets, cafes, restaurants, a hypermarket, a 10-screen cinema complex, an outside cinema, a children's entertainment area, and an indoor skydiving centre. The site was spread over an area of more than 275,000 m^2 and the total gross floor area was 552,000 m^2 including 196,000 m^2 leasable areas. The construction period was 29 months that began in August 2007 and ended in January 2010 to meet the scheduled opening ceremony in March 2010. The project entailed two specific handover dates for the Energy Centre and the mainstream construction completion date respectively.

The Managing Director of JEC leads the organization towards continuous improvement to raise productivity and quality standards. The Project Manager of the Merdin Shopping Mall project was encouraged to proceed accordingly as part of their continuation exercise with the OHSAS 18001, ISO 9001 and ISO 14001 standards and to assist their clients to achieve their targeted LEED ratings as required. The Merdin Shopping Mall project is an important milestone for JEC as the main contractor in their continuous improvement for higher quality and productive construction in the Gulf region. As part of these achievements, this project also set out to achieve a Gold rating for the Leadership in Energy and Environmental Design or LEED rating system.

The project employed a traditional type of contract and re-measurement was carried out to fixed lump sum prices. The application of the JIT principles and potential barriers are examined in this project and presented below.

12.5 Applications of JIT Principles in Case Study

12.5.1 Site Organization

The contractor's project team comprised of several departments overseen by the project manager. The departments consisted of the construction team, contractor's design team, planning team, commercial team, QA/QC team, logistics team, and safety and security teams.

12.5.2 Zoning and Construction Teams

The project comprised three major parts relating to the Main Mall, Car Park and Energy Centre with three area managers assigned to oversee each part by the project manager. In addition, due to the large site area and the large scale project, the construction site was broken down into 13 main work zones with several subdivisions (Fig. 12.1). Each zone had its own dedicated construction team as shown in Table 12.1 and dedicated resources (MM – Main Mall, CP – Car Park, EC – Energy Centre).

These teams comprised of both skilled and unskilled workers, the supervising personnel that included the foreman and site engineers. The allocation of construction activities was carried out based on these teams.

12.5.3 Construction Managers

Apart from physical zoning, separate construction managers for each specific trade were also assigned for proper coordination due to the large scope of works carried out over a widespread area at various levels. Separate construction managers oversee Block work, Finishes to Back of House (BOH) areas, Ceiling and painting to public areas, Granite finishes to public areas, Standard public toilets, Electrical rooms and MEP risers, External works and Landscaping works. As is common in many retail developments, the finishes for the public circulation areas and service areas (BOH areas) were carried out by the main contractor while finishes for the shops were undertaken by the respective tenants.

12.5.4 Site Offices and Other Facilities

Due to the large footprint of the contract, there was a main office which accommodated the senior staff, design staff, administration staff and for document control.

a

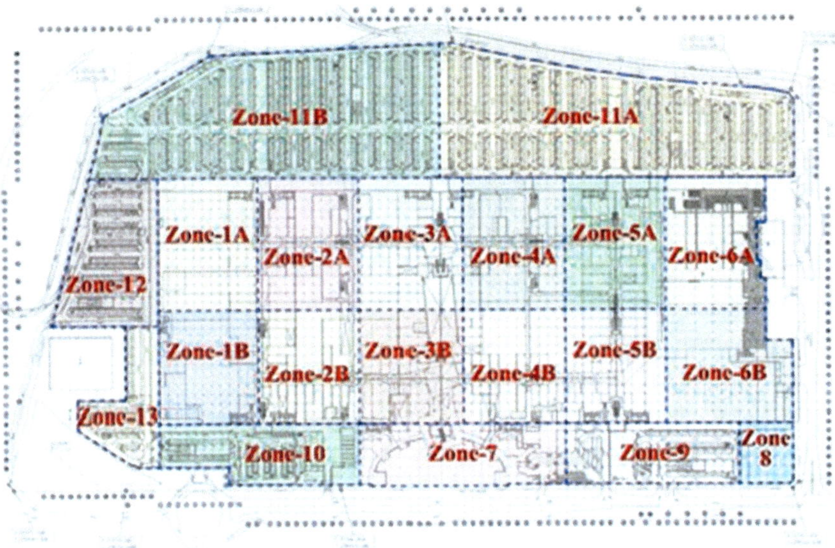

Separation of site area into thirteen main zones and sub-divisions

b

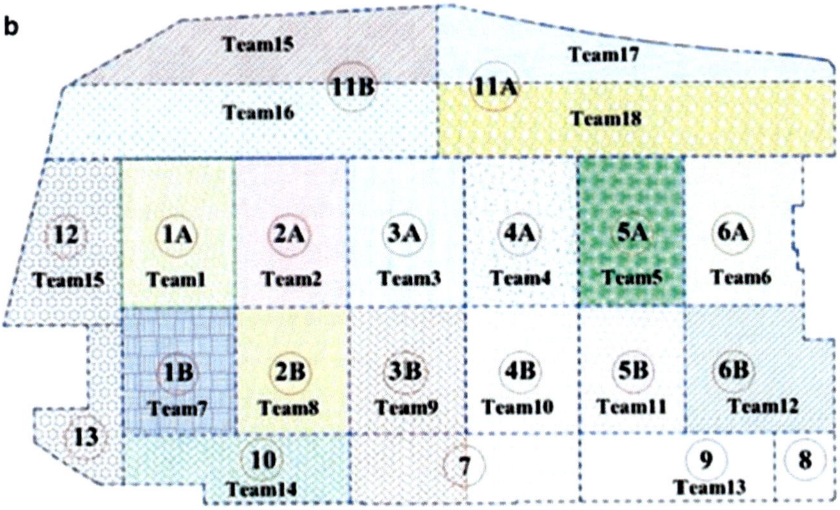

Allocation of construction teams to work zones and sub-divisions

Fig. 12.1: Work zones and sub-divisions

Table 12.1: Zoning and allocation of construction teams

Main zone	Sub-division	Construction team
1 (MM)	1A, 1B	Team 1 and 7
2 (MM)	2A, 2B	Team 2 and 8
3 (MM)	3A, 3B	Team 3 and 9
4 (MM)	4A, 4B	Team 4 and 10
5 (MM)	5A, 5B	Team 5 and 11
6 (MM)	6A, 6B	Team 6 and 12
7 (MM)	–	Team 9 and 10
8 (EC)	–	Team 13
9 (CP)	–	Team 13
10 (CP)	–	Team 14
11 (CP)	11A, 11B	Team 15, 16, 17 and 18
12 (CP)	–	Team 15
13 (CP)	–	Team 15

There was an additional four satellite establishments strategically located around the site which accommodated the construction staff and the subcontractors (Fig. 12.2). Debris clearing, labour facilities and material storage areas were also indicated in the layout. These locations were determined to facilitate proper coordination and easy transmission of information and staff personnel to their respective locations in the site as required and in so doing, not to interrupt nor disrupt the construction activities.

12.5.5 Elimination of Waste

12.5.5.1 Eliminating Idling Time, Traveling Time and Other Human Related Time Wastages

The proper allocation of teams based on zoning was facilitated through ready management of the labour force. Upon the completion of a work activity, the respective supervising personnel of the team can immediately inform the managers to assign the team to another task to eliminate idling time which is considered as a waste under the JIT concept.

Having separate construction managers for different trades was beneficial to add value to the project by eliminating the need for the managers to constantly go through a learning curve for each trade. This eliminated the wastage of time to be familiar with the different disciplines by all area managers in each zone.

For example, due to the criticality of meeting the handover dates for structures such as the Electrical rooms and MEP rooms, one construction manager was

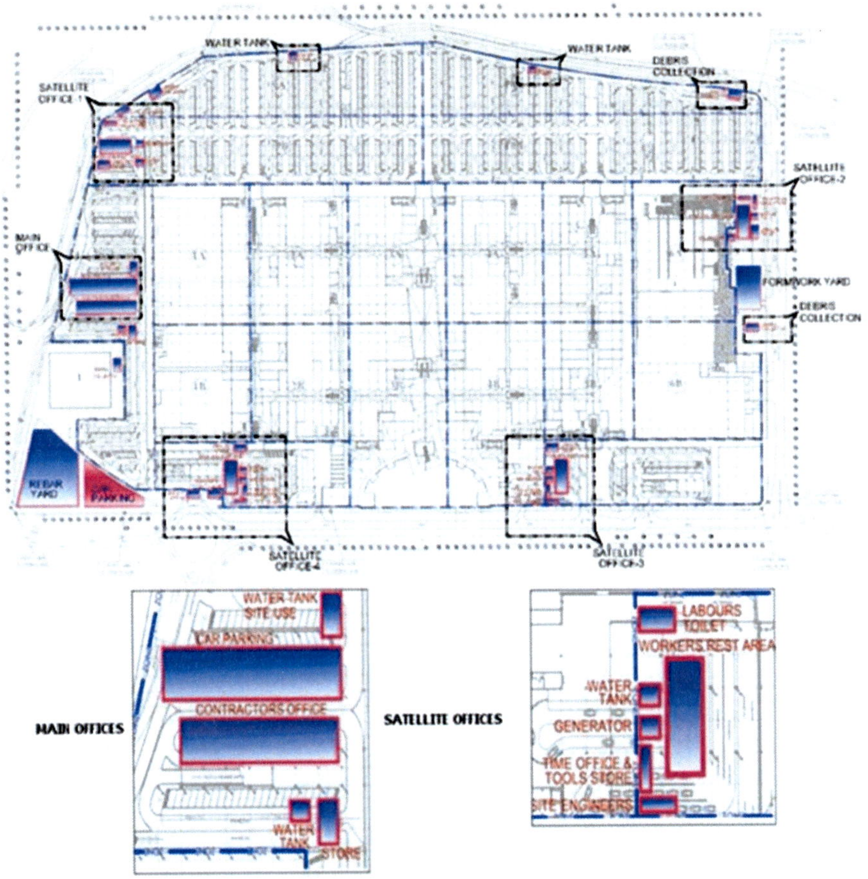

Fig. 12.2: Site establishment layout

dedicated to hand over these rooms within the stipulated time period to give early access to the specialist contractors to these areas to achieve early completion by eliminating unnecessary waiting time.

According to the law of the country, bachelors are allowed to be accommodated only in the industrial areas and therefore, the construction companies must provide suitable accommodation to their workers in worker dormitories which can usually be found in specific industrial areas. As a result, JEC selected the worker dormitories with closer proximity to particular construction sites. In the Merdin Shopping Mall project, all the workers were accommodated in worker dormitories in the Sonanagar area which was just 10 min travelling time from the site. These were the closest worker dormitories for the Merdin site. This, therefore, eliminated unnecessary traveling time to transport workers which would eliminate the time wastage of both the workers and the drivers. Moreover, unnecessary costs of transportation were also eliminated. This is a clear indication of practicing the principle of elimination of waste which reduced the costs and time for transportation that is considered as an activity which does not add value to the product as described by Low and Mok [5].

12.5.5.2 Eliminate Unnecessary Inventories and Waste of Materials

The use of prefabricated components was preferred to eliminate the processing time to achieve target completion dates within the short construction period. Precast concrete slabs were used as part of this strategy where possible. Although these were expensive compared to cast in situ concrete slabs, savings through eliminating many inventories such as form work and reinforcement bars as well as eliminating waste from excess materials such as left over concrete and left over parts from cutting reinforcement bars were able to provide many other benefits of JIT application to overcome the additional cost issues. However, these precast concrete slabs were available only in several standard sizes. Nonetheless, as per the floor slab designs, different shapes and level drops had to be formed for voids, open spaces and steps. Therefore, in such circumstances, cast in situ concrete slabs were used only in such locations where necessary and where a specialist subcontractor prepared the shop drawings together with the design team for approval. These were then fabricated and delivered to site only and strictly in accordance with the agreed delivery schedule to avoid congestion and storage of materials on site.

The contractor was concerned to reduce as much materials on site as possible. However, some materials may have to be stored on site due to inevitable circumstances. About Eighty percent of the total material on site was claimed for monthly payments under the interim progress payment arrangements. Hence, the client has to pay more money each month for materials which do not add value to the construction works yet. On the other hand, the contractor would obtain only 80% of the total amount he spent on these materials. Moreover, due to the summer season throughout six months of each year, special measures have to be taken to protect materials which would incur additional costs which do not add value to the construction works. Capital was, therefore, tied up for purchasing, handling, storing and for securing these materials.

12.5.5.3 Eliminate Waste from Plant and Machineries

Following the collaboration of the construction managers and the area managers, the use of plant and machineries were properly scheduled with up-to-date information according to the agreed program. This was used to eliminate unnecessary waiting time due to the absence of plant and machineries at the required time. Moreover, this helped to eliminate the need for storage area and eliminate motion time for double handling materials as well as eliminate waste of defective materials due to unnecessary waiting time for plant and machineries.

For example, the tower cranes for the erection of precast elements were a case in point. Both the concrete pump and the tower cranes for concreting upon the delivery of ready mix concrete and precast concrete components were allocated on the stipulated dates in the program and as scheduled with the suppliers. The absence of the required plant at the required time may lead to storage space being needed to store the precast components, thus leading to double handling. In addition, the possibility of damages occurring in the precast components would also increase due to double handling. The quality of the ready mix concrete would not be the same after leaving the batching plant for a specific time and therefore needed to be cast in place without undue delay. The timely availability of the concrete pump was therefore essential to avoid deterioration of the concrete quality.

Tower cranes and mobile cranes were placed to cover the entire site as well as the access roads (Fig. 12.3). This was facilitated to place the materials and components only at the point of use from the suppliers' vehicles to avoid double handling. This, therefore, helped to eliminate transportation waste and waste of motion.

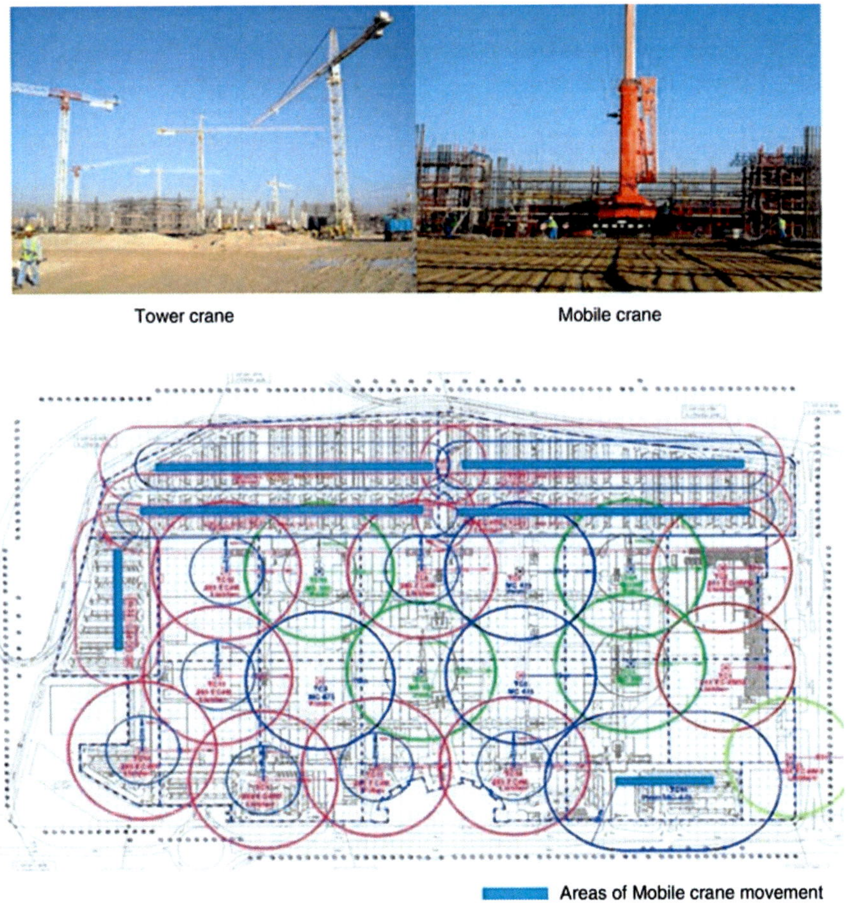

Fig. 12.3: Layout of tower cranes and mobile cranes

The contractor carefully ensured that works were completed correctly, the first time all the time. Therefore, the planning engineers scheduled each and every work activity with the collaboration of the design department and also strove to obtain the maximum possible assistance from the other consultants as well. All the drawings were properly checked to ensure there were no discrepancies leading to reworks. Otherwise, immediate actions were taken to resolve them with the design consultants to eliminate waste from defects and rework.

Although the works were undertaken carefully by the contractor, frequent design changes might still occur after the construction of some elements as well as those elements that were categorized as abortive works and separately priced and claimed by the contractor with higher rates from the client who ordered the change. Although the costs were recovered by the contractor, it appears that wastage of time, resources and interruption to work cannot be completely eliminated altogether in some instances.

12.5.5.4 Eliminate Waste Due to Defective Components and Reworks or Rectification Works

Precast concrete panels for the facade and precast concrete slabs for the upper floors were used as alternative designs as proposed by the contractor. This was for the purpose of simplifying the works to eliminate processing time and to eliminate waste from defects and reworks through minimizing the possibility of mistakes being made.

The volume of concrete required for this project was around 300,000 m^3. Comparatively speaking, this was not considered to be an exceedingly large volume. Furthermore, the project duration was also shorter in comparison with some of the larger building projects elsewhere. Hence, the mobilization of a batching plant was not economical. Using ready mix concrete was much more economical in this case for the project. Moreover, there were some government restrictions against the mobilization of any batching plant, due to the shorter project duration and less quantity of concrete being used. Using ready mix concrete thus reduced unnecessary costs. Proper coordination was carried out to deliver the required volumes of concrete to site only at the required time to eliminate waste due to excess materials. Tests were carried out prior to accepting the truck load of ready mix concrete to ensure that the concrete delivered was of the required strength and quality to eliminate wasting time on future rectification works due to defective materials.

The contractor always strove to use metal form works where possible. All form works were assembled when needed within the site at the nearest available space to the required locations. All the columns and shafts from the footing to the first floor level were cast in one lift to eliminate time wasting for repeating works. Few number of form works were assembled for typical elements and reused for other zones to eliminate waste from overproduction.

A comprehensive data base for the reinforcement bar schedules were used to calculate the exact quantities of reinforcement bars required. These were fabricated in bar bending yards or available free spaces that were closer to the point of use. These schedules facilitated the exact cutting of all required reinforcement bars and gave the exact length for each bar to use the full length of the reinforcement bars delivered to eliminate waste. The reinforcement bars and the fabricated components were marked with different color paints (Fig. 12.4) based on their diameters for easy identification to eliminate waste of motion and reworks due to errors.

Fig. 12.4: Use of color codes to mark different sizes of reinforcement bars and sides of precast concrete components and their intended locations

12.5.5.5 Eliminating Waste Through Proper Organization and Cleanliness of the Site

Locations of site offices and facilities were planned to remain on the same location for as long as possible to minimize the wastage of time in shifting offices which do not add any value to the construction works. In addition to the time for motion, equipment and documents could also be misplaced and damaged due to frequent shifting. Therefore, the offices and facilities were moved only two times within the entire contract period which were inevitable due to the construction works.

In the site, all the columns were marked with grid numbers for easy identification of the locations in the large site (Fig. 12.5) to eliminate transportation waste, waste of motion and to also facilitate the proper coordination with all subcontractors and suppliers.

The area managers and logistics personnel were responsible for assigning the workers to clean debris at the completion of each task and to report to the logistics personnel if special tools and equipment were needed to remove the debris. This practice was followed by all area managers regularly and necessarily to ensure that the site remained clean and clear of debris. Likewise, the site was cleaned daily during the course of the construction works, with construction wastes collected at specified locations (Fig. 12.6). Waste water was removed by tankers while other construction wastes were removed by using trucks and other specialized agencies.

Fig. 12.5: Each and every grid marked on the elements with site cleanliness maintained

Where abortive works occurred due to changes ordered by the clients, these were clearly marked on site for clarity so that everyone knew exactly what needed to be done (Fig. 12.7).

Fig. 12.6: Removing debris

Fig. 12.7: Marked location for abortive works

12.5.6 "Kanban" or Pull System

12.5.6.1 Material Ordering Process

Material ordering was undertaken by the materials section with the collaboration of the commercial department and the design department. Members of the commercial department measured the elemental quantities with the use of the latest shop drawings provided by the design department and inputted these into a Construction Computer Software to transfer these into the required material quantities. All design changes were readily captured, considered and adjusted continuously with the use of this software. The final material quantities at the time of placing the material order would be ordered and the suppliers were usually informed of the required date for the materials. The materials would then be received only two to three days before the activity, depending on the types of materials. This procedure ensured the deliveries of materials only in the required amounts and only at the required time to implement the "Kanban" or pull system.

Although this process by the contractor ensured timely deliveries of materials, the problem associated with this operation was the frequent variations which might ultimately lead to unnecessary deliveries. Due to the large scope of work and frequent variations ordered by the client that were inevitable, all the design changes might not be included in the final quantities ordered. The architectural elements were more vulnerable to changes. In some cases, after ordering the materials, the orders might not be cancelled and the material quantities could not be changed within the short notice given. In such situations, either the contractor or the client has to bear the additional costs of the excess materials and the costs of the new materials although if the changes were ordered by the client, then in all likelihood, the client would need to bear the additional expenses incurred. When the materials were delivered based on drawings approved by the client, the contractor would be able to claim the additional costs under materials redundancy.

For example, three types of blocks were used in this project, namely hollow, solid and AAC (Autoclaved Aerated Concrete) blocks. Many variations occurred from time to time that varied the type of block work used in many locations. The entire construction project comprised of total block work quantities of approximately more than 160,000 m^2. Blocks were ordered in lots considering the sequence of construction, beginning with the basement and then eventually for the upper floors. Standard Public Toilet areas spread over the shopping mall area were considered as a separate package due to its typical design and their block works used were quantified separately. The quantities were based on drawings approved by the client. In this context, AAC blocks of three different thicknesses (100, 150, and 200 mm) were ordered accordingly. Within the period from the date of the purchase order being made to the actual delivery date, the block type for this package was varied when the client finally decided to use solid blocks instead of AAC blocks for the entire package. The construction drawings were issued accordingly. Materials were delivered to the site and the contractor had to provide a larger storage area for them. However, the block work lots of 100 mm thickness were never used until the completion of the project. The approximate total worth of the materials was more than US$ 32,000. Although the contractor claimed this amount under material redundancy by the client, the additional time and labour for related activities such as materials handling, transportation, storing and maintenance were wasted and were not eliminated.

12.5.6.2 Logistics Management

The logistics manager was responsible for planning and monitoring the site logistics. The site logistics plan was circulated to all parties involved in the contract (major material suppliers, subcontractor, etc.) to ensure that all major materials (reinforcement bars, concrete, precast components, etc.) were delivered directly to specific zones and work teams "just in time" for use. The logistics team was responsible for the maintenance of all the internal access roads and the external ring roads (Fig. 12.8). All delivery notes have specified locations and names of persons to be contacted for each of the respective location. All deliveries were reported to the logistics office and directed to specified locations to ensure timely deliveries.

12.5.7 Uninterrupted Work Flow

12.5.7.1 Traffic Management Plan

The site was situated in a residential area with restricted access in two sides and with a highway and a main road in the other two sides as there was an interchange nearby. The traffic management plan was therefore immensely important for the smooth flow of vehicles and construction plant. Figure 12.8 shows the access roads and internal traffic movements. There was a single point of entry and exit to the site for safety and ease of management. The external ring road and the internal access roads helped to ensure the smooth flow of traffic. Logistics supervisors and flag men directed the vehicles to their specified locations. The logistic team was responsible for traffic management and for the maintenance of all these roads to anticipate uninterrupted 24-hour access to the site throughout the contract period. The team helped to ensure that the peak maximum expected number of essential plant and vehicles were able to move freely without any congestion. During the peak time, the maximum expected vehicles were 40 precast concrete trailers, 50 concrete trucks, 15 reinforcement bar trailers, and 25 other vehicles. This was important since the works were progressing concurrently in 13 work zones.

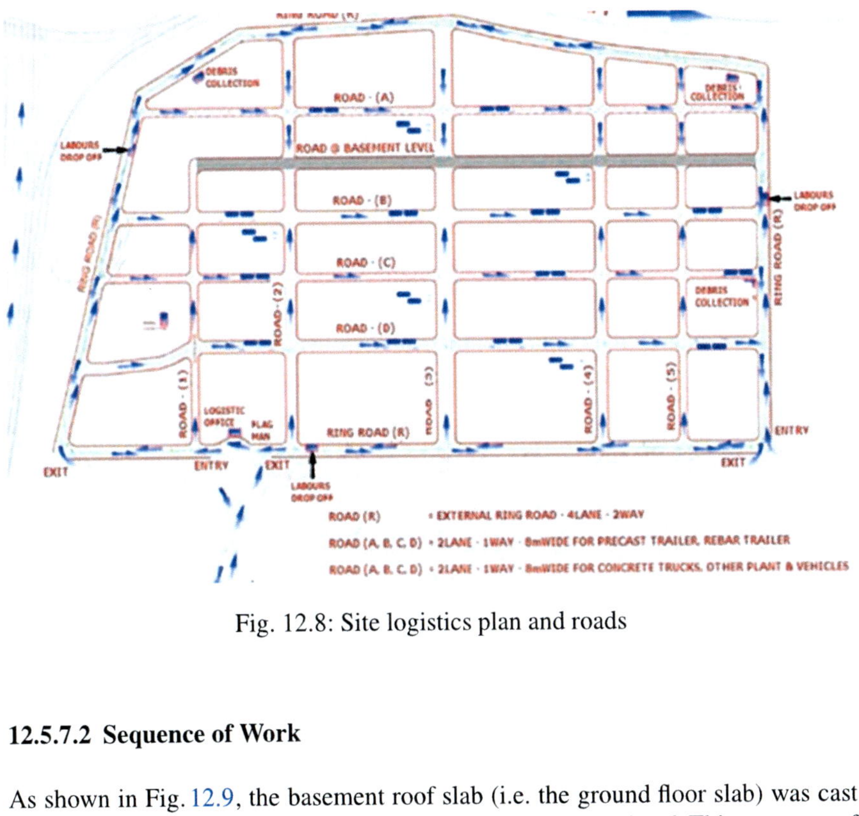

Fig. 12.8: Site logistics plan and roads

12.5.7.2 Sequence of Work

As shown in Fig. 12.9, the basement roof slab (i.e. the ground floor slab) was cast only after the superstructure (columns and beams) were completed. This sequence of works allowed and provided for sufficient time to finalize the underground services and to commence with the roof services as soon as practically possible.

12.5.7.3 Work Shifts

Two work shifts were in place for construction works: the first shift was from 7 am to 5 pm and the second shift was from 5 pm to 7 am on the following day, with

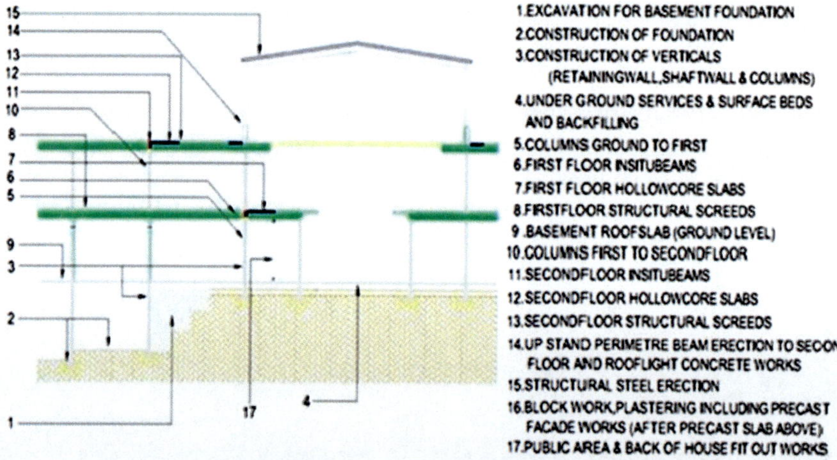

Fig. 12.9: Sequence of construction activities

the work program being developed accordingly. The workers were transported by small buses which were hired on an hourly basis only for the changing shift times (7 am and 5 pm), that is two times per day. A bus would leave from the worker dormitories to the site with workers for the next shift and would bring the workers back from the site to the worker dormitories after they have completed their shift. Logistics personnel coordinated the workers entering and leaving the site so as to avoid any congestion. Moreover, since the bus transporting the workers was the first to reach the site with workers for the following shift, site activities were able to flow uninterruptedly (Fig. 12.10).

Fig. 12.10: Workers coordinated to leave/enter the site without disruption

12.5.8 Total Quality Control

As an ISO 9001 certified organization, quality assurance was an inherent and integrated part of JEC. Accomplishing the works right the first time all the time was a high priority. This helped to ensure that the materials used were of good quality when managed under the TQC concept. Regular cleaning and providing on-the-spot solutions were accomplished in consultation with the area managers with the collaboration of the construction manager for specific trades, all as part of a quality control system.

The lists of the site engineers, supervising personnel and area managers were displayed at several areas in each zone for ease of coordination and for solving problem on-the-spot by discussing with the relevant supervising personnel or with the area manager immediately. This, thus, helped to ensure better quality of work done. This was immensely important because some defects cannot be identified at a later stage and must be solved during the construction stage or just after the completion of the work.

For example, it was difficult to differentiate between the solid and hollow blocks after bonding. No one was able to identify the types of blocks later after these blocks were cemented together to form a wall. Similarly, it was difficult to differentiate between the concrete and block walls after plastering. Any discrepancies in the drawings and specifications have to be solved earlier because it was difficult to identify the building materials later upon covering up.

The allocation of construction managers for specific trades helped to ensure that trades were completed according to the specifications. The construction managers were capable of preventing problems from arising in a particular trade to assure the required quality with their specialized knowledge gained from their past industrial experience.

12.5.9 Employee Involvement

12.5.9.1 Experienced In-House Labour Resource

Since the Merdin Shopping Mall building project was a relatively huge construction project involving a wide range of construction activities spread over a sizable area, a large number of workers were required. JEC had in-house workers who were both skilled and unskilled for this purpose. Many of them were migrants from developing Asian countries with past experiences over many years from working in the Gulf region. Having its own labour resources was a benefit to the organization because the workers were familiar with the organization and had gained much experience with the organization over many years. This pool of experienced in-house labour resource was an asset to JEC for implementing the JIT concept. The valuable labour pool recognized the company's objectives and co-operated to comply with company's policies well. The principle of "employee involvement" was of immense importance and was necessary for successful JIT implementation.

As a result, this helped to eliminate the repetition time for orienting and training new workers from time to time or individually to meet the company's requirements and to eliminate the time wasted due to inexperienced workers. Nevertheless, the existing labour pool was further enhanced by recruiting new workers annually and who were briefed of company's policies and requirements through readily understood awareness programs.

12.5.9.2 Motivating Workers

Employees in all levels were highly motivated to serve according to the organization's values and procedures. Relationships between the management and the employees at each level were strong, helped in no small way through the JEC Annual Day event, among other things. This helped to facilitate the active involvement and engagement of the employees continuously. Employees in all levels were also motivated through training programs and reward systems.

12.5.10 Supplier Relations

The organization placed a high value on establishing supply chains that have the ability to deliver materials and components on time, safely and of the right quality. The organization identified its supply chains as a resource to the company, many of whom were strategic alliance partners, having been involved on repeated occasions in assisting JEC on successful project deliveries in the past.

For example, JEC has a long term relationship with the precast concrete slab manufacturers with many previous and other current projects. With prior notification, this relationship has enabled efficient coordination for the deliveries of slabs only at the required time and with the required quantities for on-site assemblies as planned.

The selection of suppliers was undertaken carefully to ensure that materials and components were delivered in the required quality and the suppliers were in close proximity to the construction site physically. This was because long term supplier relations made it beneficial to develop the "rim delivery system" instead of the conventional "spoke delivery system". When several projects were operating simultaneously by the organization, deliveries of some building materials and components can be made first to one site and further distributions to other sites thereafter. This arrangement helped to reduce unnecessary delivery time and costs to the suppliers and further enhanced the relationship eventually.

For example, in sourcing for ready mix concrete suppliers, a nearby supplier was chosen to ensure that fresh concrete batches of good quality can be delivered in a timely manner.

12.5.11 Continuous Improvement

The in-house labour resource played a key role in continuous improvement which can be readily achieved through the experience and familiarity of the employees. In the process, the efficiency of the construction teams can also be enhanced through the team spirit. Cultivating strong relationships with all levels of employees have made for a better working environment that assisted to spearhead continuous improvement in the organization. Introducing annual and monthly rewards systems have also enabled all employees to align themselves closely with organizational goals and procedures.

For example, rewarding workers with the "Safety Man of the Month" award encouraged all workers to be responsible for their own safety and to ensure the safety of others. This formed an integral part of work activities which supported continuing safety standards under OHSAS 18001. Through annual employee reviews, the skills and knowledge areas to be improved by each employee were identified with the assistance of managers in each department. Employees were encouraged to self-learn and also to attend quarterly training programs held by the organization. Orientation programs and safety induction courses were compulsory for new staff members to ensure that the employees recruited were completely aware of and have the necessary knowledge relating to organization procedures. These collectively helped them to comply with the correct working procedures as an integral part of the organization. Safety induction courses, site awareness programs and meetings were also essential for all subcontractors and their workers.

12.6 Findings and Conclusion

The implementation of JIT principles was found in many activities in the Merdin Shopping Mall project. This provided anecdotal evidence of the successful implementation of JIT principles in a construction site setting. The contractor reached the target completion date as scheduled. When the top management in an organization is committed to implement JIT principles with the collaboration of employees at all levels as well as with the external but highly influential stakeholders such as the subcontractors and suppliers, successful implementation of the JIT concept can be accomplished in the construction industry. The case study presented suggests that it is possible to successfully implement the JIT concept in the construction industry despite the perceived differences between the construction industry and the manufacturing industry where the JIT philosophy first originated. Nevertheless, it should be noted that the implementation of the JIT concept is a continuous process and should not therefore end abruptly upon completion of a building project. Instead, the organization concerned should continue to be imbued with a JIT culture for all its on-going daily operations as well as in other building projects which it undertakes in the future.

Nevertheless, despite the successful implementation of JIT principles, some disruptions may be inevitable from the corresponding barriers created. Because many parties were involved in the Merdin Shopping Mall project, with more than six separate consultant organisations being engaged, isolated efforts made by the contracting organization alone would be difficult for the JIT concept to succeed. Coordination of the works of all these parties was made possible by a separate project management service provider on behalf of the client. In this context, the main contractor may not implement the JIT principles fully. However, this may be made possible with the collaboration of all other parties in the building contract.

Frequent design changes are common in construction projects and which are inevitable for the contractor to manage. There is also the issue of the higher risks associated with the construction industry and the short construction time given for project completion. All these may collectively drive the contractor to resort to delivering building materials and components earlier than required in order not to be burdened by delays caused by their shortages. Eventually, the inventories of building materials and components on site will increase and both the client and the contractor end up incurring unnecessary additional costs and expenses.

The construction industry is unlike other industries where in the latter, the customer is frequently not known during the production stage and where top management spearheads the implementation of the JIT concept single-mindedly. In this context, commitment of the client is of immense importance in the construction industry where the building products are primarily known before-hand and where the construction related organizations (i.e. contractors and design consultants) take instructions from the client.

12.7 Recommendations

It may appear that some management techniques and philosophies may be difficult to implement in the Gulf region because of the high liquidity and strong spending capacities of many local clients who do not necessarily focus too much on economical options and waste minimization. This may seem to be the common situation of many clients in this region over many decades. However, the severe impacts faced by the region arising from the recent global financial crisis in 2008–2009 have forced many organizations to move towards better management tools and to change their attitudes positively towards waste minimization. An increase in awareness of the JIT philosophy and encouragement towards its implementation would appear to be more worthwhile in such a situation.

Since the consultancy organizations play a major role in the traditional contracts, investigating the awareness, perceptions and implementation of the JIT concept in consultancy organizations would help to enhance their collaboration with the construction teams for successful JIT implementation at the project level. Moreover, enhancing the awareness of and conveying the importance of the JIT concept to consultancy organizations and property developers would eliminate the barriers for JIT implementation that may arise from design changes during the construction stage. In the process, this may also facilitate their careful selection of nominated sub contractors and nominated suppliers downstream.

Acknowledgment The authors are grateful to the management and staff involved with the case study project for offering the second author the opportunity to observe how best practices in construction project management may be facilitated through the JIT philosophy. It is hoped that the best practices presented can further demonstrate and showcase the applicability of JIT principles in the construction setting.

References

1. Liker, J.K,: The Toyota Way. McGraw-Hill, New York, United States (2004)
2. Low, S.P.: The just-in-time concept to improving manufacturing productivity: is it applicable to the construction industry? Construction Paper Series **8**, Chartered Institute of Building, 1-8 (1992)
3. Low, S.P., Ang, G.K.: Integrating JIT and 5-S concepts for construction site management: a case study. International Journal of Construction Management **3**(1), 31-47 (2003)
4. Low, S.P., Choong, J.C. : Just-in-time management of precast concrete components. Journal of Construction Engineering and Management **127**(6), 494-501 (2001)
5. Low, S.P. and Mok, S.H.: The application of JIT philosophy to construction: a case study in site layout. Construction Management and Economics **17**(5), 657-668 (1999)
6. Wantuck, K.A.: Just-In-Time for America: A common sense production strategy. The Forum Ltd, Milwaukee, United States (1989)
7. Wu, M., Low, S.P., Xu, M.Z., Shen, Q.P.: The EOQ-JIT Cost Indifference Point Theory and Its Applications in the Ready-mixed Concrete Industry. Hong Kong Polytechnic University Press, Hong Kong SAR, China. (2010)

Chapter 13
Analysis of Transient Throughput Rates of Transfer Lines with Pull Systems

Mahmut Ali Gökçe, M. Cemali Dinçer, and M. Arslan Örnek

Abstract Transfer lines represent the basic manufacturing system of many high volume production systems. Analysis and understanding of transfer lines are of great importance to improve design and operation performance of many manufacturing systems. Majority of research on the throughput of transfer lines concentrate on the steady state results. Due to the changes in manufacturing environment and increasing importance of JIT and pull systems, many transfer lines now have to changeover to different parts' production quickly, probably most of the time, before enough time passes to reach steady state for a specific configuration. In such situations, steady state may never be reached and hence results relating to the steady state do not make sense. In these situations, one would be more interested in transient behavior of the system.

In this study, we offer a novel analytical model for transient throughput analysis of transfer lines. Defining throughput as the number of units produced by a transfer line with buffers per unit time, this chapter shows how to calculate mean and variance of and interval estimates for throughput for a pull type transfer line. Derivation of distribution of transient throughput of transfer lines is presented and sample calculations are provided.

Mahmut Ali Gökçe and M. Arslan Örnek
Department of Industrial Systems Engineering, Izmir University of Economics, Sakarya Cad. No: 156 Balcova Izmir, Turkey, e-mail: {ali.gokce, arslan.ornek}@ieu.edu.tr

M. Cemali Dinçer
Department of Industrial Engineering, Istanbul Bilgi University, Kazim Karabekir Cad. No:2/13, Eyup Istanbul, Turkey, e-mail: cemal.dincer@bilgi.edu.tr

R.Z. Ríos-Mercado and Y.A. Ríos-Solís (eds.), *Just-in-Time Systems*, Springer Optimization and Its Applications 60, DOI 10.1007/978-1-4614-1123-9_13,

13.1 Introduction

Transfer lines are manufacturing systems with special structures. A transfer line is a linear network of machines or workstations $(M_1, M_2, \ldots, M_m)$ separated by buffer storages $(B_0, B_1, \ldots, B_n)$. B_0 is the initial buffer where the material is received from outside of the system, which is fed to M_1, then to B_1 and so forth until material leaves the system from B_m. A typical transfer line is depicted in Fig. 13.1. Buffers are represented by triangles and machines are represented by squares. In a Pull environment, the system will be driven by demand and in a Push environment, the system is driven by forecast.

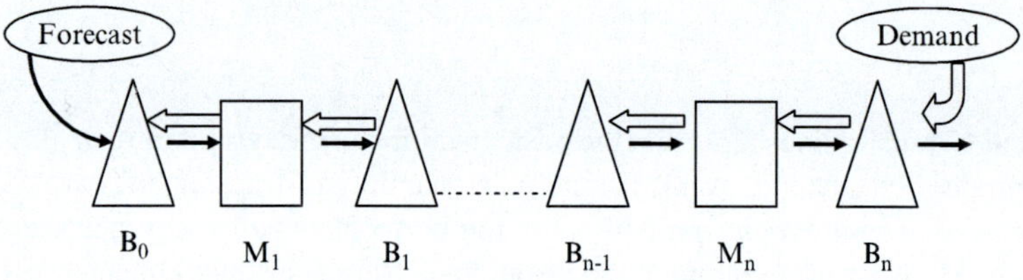

Fig. 13.1: A typical transfer line: push (⟶) vs. pull (⟸)

Transfer lines are important because they represent the basic manufacturing system of many high volume production systems. Analysis and understanding of transfer lines are of great importance to improve design and operation performance of many manufacturing systems. Another important property of transfer lines is that they characterize the relationship of the stages in production with buffers.

Transfer lines were and still are a very integral part of manufacturing. Initially, transfer lines' utilization was to achieve high volumes of production. Last decade or two, world has witnessed a fundamental change in manufacturing. The initial transfer lines did not offer much flexibility. They were made for speed. The new transfer lines now pack flexibility and are able to produce a family of parts, switching from one part to another's production easily with minor setups. This change was brought about with the mass customization requirements. The result is, many transfer lines now have to changeover to different parts' production quickly, probably most of the time, before enough time passes to reach steady state for a specific configuration.

The literature on transfer lines are immense, dealing with almost every imaginable aspect of the transfer lines. These include design, efficiency improvement, reliability analysis etc. Since transfer lines are usually representative of high volume production, throughput of such systems has been researched extensively as well. However, almost all of these studies either concentrated on modeling push systems or dealt with steady state results for such systems. Unfortunately, in many practical situations, mass customization and other factors limit the natural time horizon of the operations. In such situations, steady state may never be reached and hence results relating to the steady state do not make sense. In these situations, one would be more interested in transient behavior of the system.

Transient behavior of such stochastic systems is either hard to obtain owing to the complex calculations required, or is only available for a fairly restricted class of problems.

Mass production systems, with which transfer lines are usually associated, often have excess inventory, higher Work-In-Process (WIP) levels and longer quoted lead-times. The last couple of decades have witnessed a strong and important change in manufacturing paradigm. Just-In-Time (JIT) systems have swept the old systems aside, basically building on the success of Japanese firms, led by Toyota, who successfully implemented this production philosophy. JIT systems are often confusingly understood as just inventory systems. APICS dictionary defines JIT as "a philosophy of manufacturing based on planned elimination of all waste and on continuous improvement of productivity". It also has been described as an approach with the objective of producing the right part in the right place at the right time (in other words, "just in time"). Pull systems are important part of JIT systems. A Pull system's main characteristic is that replenishment occurs only when a pull signal, showing that a part is used downstream, is received.

For Pull Systems, two main types of systems exist. These are Kanban controlled and CONWIP. Kanban, which means "card" or "marker" in Japanese, is the more widely known type of Pull system. Kanban utilizes kanban cards attached to the workstations and bins to tightly control WIP inventory in the system. Total WIP in the system is limited to the (number of cards)*(number of parts per kanban). Production kanbans trigger production and production occurs only when there is a card authorizing production after parts' usage, therefore, pulling the parts through the system. In CONWIP systems, there is a single set of global cards controlling release of work into the system, and total number of cards are fixed, which in turn assures that total number of WIP in the system is constant at all times. In CONWIP, WIP is not controlled at the individual workstation level. As parts leave the system, cards are returned to the start of the system. Once parts are released into the system, they are pushed within the system.

Majority of manufacturing systems configured themselves as a mix of push and pull systems. Great deal of research on how to combine these two exists since both pull and push have their advantages as well as disadvantages [11].

Transfer lines have become more and more suitable for applications of partly or completely pull systems, to benefit from advantages of pull systems such as reduced WIP inventory, increased quality etc. Still, most of the research published so far has dealt exclusively with throughput analysis of push system type transfer lines. To the best of the authors' knowledge, there is a need for analytical models for transient throughput analysis of transfer lines.

In majority of the research so far, the area of the transient behavior for these systems practically remained unexplored. Transient behavior of manufacturing system may have significant implications. If the transient performance of the system is not very good and if steady state is reached after a relatively long time, the loss of production due to bad performance during the transient time can be very important. In many practical (operational) situations where time horizon is too short to reach steady state, it simply does not make sense to deal with steady state results.

In this study, we offer an analytical model for transient throughput analysis of transfer lines. Defining throughput as the number of units produced by a transfer line with buffers per unit time, this chapter shows how to calculate mean and variance of and interval estimates for throughput for a pull type transfer line. This way, it is possible to derive the distribution of throughput from such a system at an arbitrary point in time, which enables us to examine transient behavior of short transfer lines.

The rest of this chapter is organized as follows. Section 13.2 gives a classification for throughput analysis of transfer lines. Section 13.3 presents the model for transient throughput analysis of pull type systems. Some numerical results and discussions are provided in Sect. 13.4. Finally, Sect. 13.5 gives conclusions and directions for future research.

13.2 Literature Review

In this section, we present a short survey for the performance analysis of serial production systems, considering both transient and steady-state analysis.

Since throughput analysis of serial production lines are a popular research area, this review does not claim to be complete or exhaustive, but summarizes current research to the best of the authors' knowledge.

Due to the importance of serial production systems in industry, there are many studies on performance modeling of such systems. There have been a number of good reviews by Koenigsberg [12], Buxey et al. [2], Buzacott and Hanifin [3], Dallery and Gershwin [7], Govil and Fu [10], Papadopoulos et al. [18], and lastly by Li and Meerkov [14]. Throughput analysis has also been studied in a number of important papers and well written books as the subject area. Among these Buzacott and Shantikumar [4], Papadopoulos et al. [18], Gershwin [9], Altiok [1], Tan [21,22,24–26] and Li and Meerkov [14] can be listed. Of these reviews and books, Dallery and Gershwin [7], and Li et al. [13] are the most comprehensive.

Exact analytical results for throughput analysis exist only for one or two machine and one buffer systems depending on the assumptions of the system under study. An important work in this respect was published by Gershwin [9]. In his work, he presents a new approach, which is based on the exact calculation of the production variance for a single machine and then items of a decomposition technique are developed for longer production lines

When analytical results fail for larger size systems, approximation methods, simulation-based methods or heuristic methods are utilized. Approximation methods concentrate either on aggregation or decomposition.

Basic idea of aggregation is to combine the machines and replace them with a single representative machine. Basic idea of decomposition is to decompose the original more complex model into a set of smaller subsystems, for which analysis is possible. Simulation-based methods are useful for analysis of complex systems. But they lack the generality required.

Majority of throughput analysis in literature focus on steady state analysis of production systems. Most of the literature on throughput analysis for steady state concentrate on finding mean throughput rate. Studies on transient throughput and variance of an unreliable serial production system are rather limited. However, transient behavior effects production losses and profitability and it is important to be able to answer questions like what parameters control transient behavior, how long it takes to reach steady state.

The difficulty with transient analysis is that the performance measures are usually very hard to obtain analytically, owing to the fact that the required calculations are complex. The models that can be solved tend to be restricted models of smaller size. Still these restricted models can be useful to provide insight into the inner-workings of the larger models.

Of the research found so far on transient behavior of production system, all that have been found concentrate on a push type serial production system. The research that deal with transient analysis of pull type of serial production systems, until the time of writing of this chapter, mostly employed simulation or approximation-based methods, avoiding analytical approach. In this study, we attempt to fill that void.

A recent comprehensive review of transient behavior serial transfer lines is given in Sariyer [20]. Among the studies for transient analysis are Mitra [16], Narahari and Viswanadham [27], Mocanu [17], and Meerkov and Zhang [15]. We choose to elaborate more on some of these limited number of studies on transient behavior of production systems, which are most relevant to this work.

Dinçer and Deler [8] present an analytical model for n-machine, $n+1$ buffer transfer line with reliable machines. Starting from a small representative model of 1 machine system, they develop mean and distribution of throughput of the system, from which they are able to discern the throughput anytime. They use number of parts leaving the system at anytime to get various moments of the throughput rate. This study is unique in the sense that it allows transient analysis of the throughput of the transfer lines by the analysis of evolution of the stochastic processes under consideration. (From this evolution, they derive the distribution of the throughput of the transfer line.) However, due to the inherent difficulty of the analytical approach, models for two-machine-one-buffer and three-machines-two-buffer cases are explicitly analyzed.

Sader and Sorensen [19], using analogies to electrical systems, provide a model with physically meaningful governing equations that describe both steady state and transient responses of continuous manufacturing systems. This work considers a system with infinite buffers and a push system. They also provide a discrete-event simulation model to compare the results of the proposed model. The authors report on an ongoing study for a similar type model for the pull type serial production lines, which has not been published yet.

Chen et al. [5] present a numerical simulation method based on sample path method for performance measures, such as expected value and variance of the transient throughput, and the probability that measures the delivery in time for serial unreliable work centers. They allow independent and identically distributed (i.i.d.) up and down times for serial production system with no interstage buffer. With this configuration, their model mimics that of a classic push type assembly system.

Ciprut et al. [6] using continuous approximations to flow of material, present a model to calculate variability of serial production line's output. Their study also assumes that the machines have down times. They present a single for non-Markovian cases, which is never starved or blocked and for which the time between failures is known. They then present a network of machines with no buffers. Finally, they present a production dipole with two machines with identical rates and a limited buffer. All of their work, assumes parts are pushed through the system.

Existing research on transfer lines for transient analysis focuses on push systems. To the best knowledge of the authors, only a limited number of studies exist on transient analysis of pull transfer lines. In this work, we try to partly alleviate this deficiency via introducing a novel approach with a simple model for this type of analysis.

13.3 Approach and the Model

As mentioned earlier, transient analysis of any system is challenging. Therefore, we start this section by analytical transient throughput analysis of an atomic pull system, and show proper derivations. We begin by describing the problem on hand, state the assumptions and introduce the notation used.

The system under investigation is an m machine $(m+1)$ buffer transfer line. The first buffer is in front of the first machine. As demand arrives at the last machine, production starts and completion of production on this last machine forms a demand for the previous machine, etc. In short, the system works as a perfect pull system.

The performance measure of interest is the throughput or more specifically in this case, the number of parts leaving the system at any arbitrary time t.

Assumptions of the system are

1. The production line is a serial arrangement of machines. Each machine has capacity of one unit.
2. The processing time for the parts on machine j is an i.i.d. exponential distribution with mean μ_j with density $f_j(t) = \mu e^{-\mu_j t}, \mu > 0, t \leq 0, j = 1, 2, \ldots, m$.
3. The demand is assumed to arrive according to a Poisson Process. Therefore, the interarrival time for the demand for jth machine, is i.i.d. exponential random variable with density function $f_j(t) = \lambda e^{-\lambda t}, \lambda > 0, t \leq 0, j = 1, 2, \ldots, m$. Although interarrival times are for parts, because of the pull system, it is the upstream machine's processing that creates the demand for the machine downstream, the interarrival time is specific to the machine and is therefore referred to with index j.
4. The first machine is never starved and the last machine is never blocked due to the buffers.

5. All machines are assumed to be reliable. They produce 100% good parts all the time. There is also no breakdown for the machines.
6. Machines have no setup times.
7. The production system is assumed to start from an empty and idle state.

Notation used throughout the chapter:

- $N_j(t)$: Number of parts that have left machine j by time t
- m: Number of machines in the system
- μ: Processing time
- λ: Demand rate
- T_μ^i: Processing time of part i
- T_λ^i: Interarrival time for demand of part i
- $f_{\lambda\mu}(t)$: The probability density function of the random variable $T_\lambda + T_\mu$
- $F_{\lambda,\mu}(t)$: Cumulative distribution function (CDF) of the random variable $T_\lambda + T_\mu$
- $f_{\lambda,\mu}^l(t)$: lth convolution of probability density function of the random variable $T_\lambda + T_\mu$
- $E(N_m(t))$: Expected value of the number of parts leaving machine m (mean of throughput) by time t
- $V(N_m(t))$: Variance value of the number of parts leaving machine m (squared deviation of throughput from mean) by time t
- S_n: Completion time of nth part by the system $i = 1, 2, \ldots, n$

13.3.1 The Atomic Pull Model

The atomic pull model corresponds to one machine system. Atomic model is depicted in Fig. 13.2. First we present the evolution of the stochastic process of

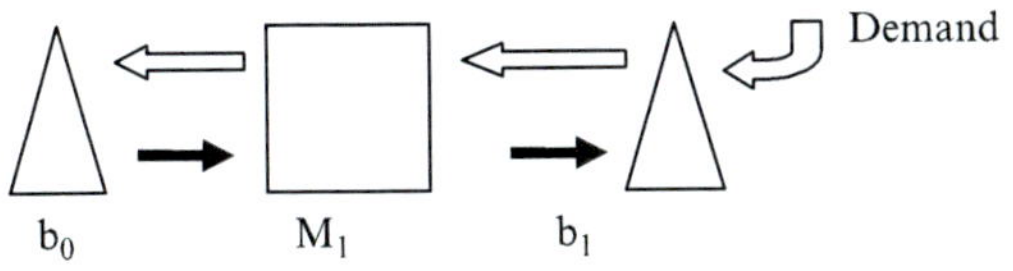

Fig. 13.2: Atomic model: Pull ⟸: Demand Flow, ⟶: Material Flow

random variable $N_1(t)$. We should be interested in $N_1(t)$ because that would give us the number of parts processed by the system. The evolution of the stochastic process adapted from Dinçer and Deler [8] is given in Fig. 13.3. The critical change from their work is that Dinçer and Deler's evolution is based on a push system, whereas. Figure 13.3 shows the evolution of the stochastic process adapted for a pull type system by time t.

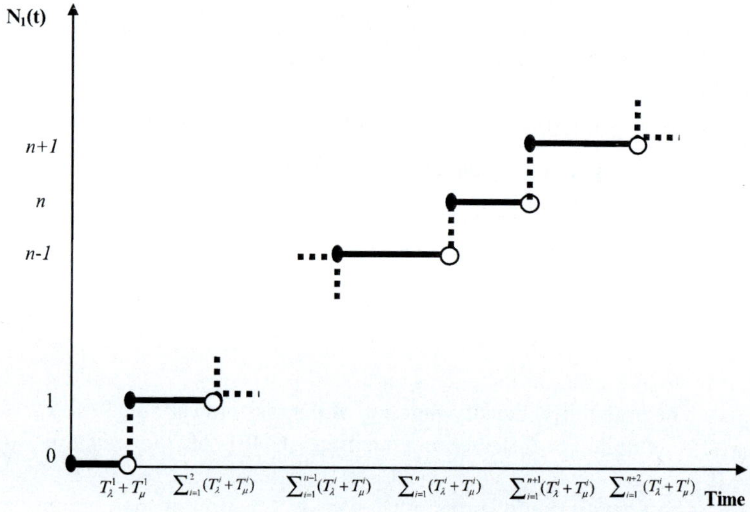

Fig. 13.3: Evolution of the stochastic process

$$
N_1(t) = \begin{cases} 0 & \text{if } 0 < t < T_\lambda^1 + T_\mu^1 \\ 1 & \text{if } T_\lambda^1 + T_\mu^1 < t < \sum_{i=1}^{2}(T_\lambda^i + T_\mu^i) \\ 2 & \text{if } \sum_{i=1}^{2}(T_\lambda^i + T_\mu^i) < t < \sum_{i=1}^{3}(T_\lambda^i + T_\mu^i) \\ \vdots & \vdots \qquad\qquad \vdots \\ n-1 & \text{if } \sum_{i=1}^{n-1}(T_\lambda^i + T_\mu^i) < t < \sum_{i=1}^{n}(T_\lambda^i + T_\mu^i) \\ n & \text{if } \sum_{i=1}^{n}(T_\lambda^i + T_\mu^i) < t < \sum_{i=1}^{n+1}(T_\lambda^i + T_\mu^i) \\ n+1 & \text{if } \sum_{i=1}^{n+1}(T_\lambda^i + T_\mu^i) < t < \sum_{i=1}^{n+2}(T_\lambda^i + T_\mu^i) \\ \vdots & \vdots \qquad\qquad \vdots \end{cases}
$$

From Fig. 13.3 we can easily derive $N_1(t)$, the number of parts that are pulled out of the system by time t as shown above.

For any random variable X with CDF $F_X(x), P(t_1 < X < t_2)$ would be easily calculated as $F_X(t_2) - F_X(t_1)$. The difficulty here is that, we have a constant in the middle and random variables as bounds.

By the assumptions, $T_\lambda^i \sim Exp(\lambda)$ and $T_\mu^i \sim Exp(\mu)$ for $i = 1, 2, ...$, and T_λ^i is independent of T_μ^i for all i. We will therefore use T_λ^i and T_λ, T_μ^i and T_μ interchangeably for the rest of the chapter.

We can therefore define:

S_n = Completion time of nth part by the system and hence $S_n = \sum_{i=1}^{n}(T_\lambda^i + T_\mu^i)$, $S_{n+1} = S_n + T_\lambda^{n+1} + T_\mu^{n+1}$, $S_n = \sum_{i=1}^{n} T_\lambda^i + \sum_{i=1}^{n} T_\mu^i$ which makes S_n sum of two independent Erlang random variables: $S_n = ST_\lambda^n + ST_\mu^n$ where ST_λ^n is then an Erlang (λ, n) and ST_μ^n is an Erlang (μ, n). Therefore,

$$f_{ST_\lambda^n}(X) = \frac{\lambda^n X^{n-1} \mathrm{e}^{-\lambda X}}{(n-1)!}, X \geq 0, \lambda > 0$$

and

$$f_{ST_\mu^n}(X) = \frac{\mu^n X^{n-1} \mathrm{e}^{-\mu X}}{(n-1)!}, X \geq 0, \mu > 0.$$

By using the evolution of the stochastic process one can obtain,

$$P\{N_t = n\} = P\{S_n > t > S_{n+1}\} = P\{(S_n \leq t), (S_{n+1} > t)\}.$$

Using $P(A) = \int P(A|B) f_B(u)\mathrm{d}u$,

$$\begin{aligned}
P\{(S_n < t), (S_{n+1} > t)\} &= P(S_{n+1} > t | S_n < t) P(S_n < t) \\
&= P(S_{n+1} > t | S_n > t) \int_0^t f_{S_n}(\tau)\mathrm{d}\tau \\
&= \int_0^t P(S_{n+1} > t | S_n < t) f_{S_n}(\tau)\mathrm{d}\tau \\
&= \int_0^t [1 - F_{S_{n+1}}(\tau)] f_{S_n}(\tau)\mathrm{d}\tau
\end{aligned}$$

Since $S_{(n+1)t}$ is sum of two Erlang random variables with parameters $(\lambda, n+1)$ and $(\mu, n+1)$ respectively,

$$F_{S_{n+1}}(\tau) = P\{S_{n+1} \leq \tau\} = P\{X + Y \leq \tau\} = \int_0^\infty \int_0^{\tau - X} f_{X,Y}(x,y)\mathrm{d}x\mathrm{d}y.$$

Due to independence of X and Y,

$$\begin{aligned}
F_{S_{n+1}} &= \int_0^\infty \int_0^{\tau - X} \frac{\lambda^{n+1} x^n \mathrm{e}^{-\lambda x}}{n!} \frac{\mu^{n+1} y^n \mathrm{e}^{-\mu y}}{n!} \mathrm{d}x\mathrm{d}y \\
&= \int_0^\infty \left[\frac{\lambda^{n+1} x^n \mathrm{e}^{-\lambda x}}{n!} \int_0^{\tau - X} \frac{\mu^{n+1} y^n \mathrm{e}^{-\mu y}}{n!} dy \right] \mathrm{d}x.
\end{aligned}$$

If $X \sim$Erlang$(\lambda, n+1)$ then $F_X(x) = 1 - \sum_{k=0}^n \frac{\mathrm{e}^{\lambda x}(\lambda x)^k}{k!}$.

$$\begin{aligned}
F_{S_{n+1}}(\tau) &= \int_0^\infty \frac{\lambda^{n+1} x^n \mathrm{e}^{-\lambda x}}{n!} \left[1 - \sum_{k=0}^n \frac{\mathrm{e}^{(-\mu(\tau - x))} [\mu(\tau - x)]^k}{k!} \right] \mathrm{d}x \\
&= 1 - \int_0^\infty \frac{\lambda^{n+1} x^n \mathrm{e}^{-\lambda x}}{n!} \sum_{k=0}^n \frac{\mathrm{e}^{(-\mu(\tau - x))} [\mu(\tau - x)]^k}{k!} \mathrm{d}x
\end{aligned}$$

$$f_{S_n}(\tau) = \int_0^\tau \frac{\lambda^n(\tau-y)^{n-1}e^{\lambda(\tau-y)}}{(n-1)!}\frac{\mu^n y^{n-1}e^{-\mu y}}{(n-1)!}dy$$

$$= \frac{(\lambda\mu)^n}{[(n-1)!]^2}e^{-\lambda\tau}\int_0^\tau [y(\tau-y)]^{n-1}e^{-(\lambda-\mu)y}dy.$$

Therefore,

$$g_{S_{n+1}}(\tau) = \frac{(\lambda\mu)^{n+1}}{[(n)!]^2}e^{-\lambda\tau}\int_0^t [y(\tau-y)]^n e^{-(\lambda-\mu)y}dt \text{ and}$$

$$G_{S_{n+1}}(\tau) = \int_0^\tau g_{S_{n+1}}(u)du$$

$$P\{N(t)=n\} = \int_0^t [1-G(\tau)]f_{S_n}(\tau)d\tau \text{ for } n \geq 1$$

$$\sum_{n=0}^{\infty} P\{N(t)=n\} = 1 \rightarrow P\{N(t)=0\} = 1 - \sum_{n=1}^{\infty} P\{N(t)=n\}$$

$$E[N(t)] = \sum_{n=1}^{\infty} nP\{N(t)=n\} \text{ and } E[(N(t))^2] = \sum_{n=1}^{\infty} n^2 P\{N(t)=n\}$$

$$Var(N(t)) = E[(N(t))^2] + (E[N(t)])^2.$$

This way, we can actually calculate expected value and variance of the production system working under a pull approach, for any arbitrary time t. The derivation of the distribution of transient behavior follows a similar approach to Dinçer and Deler [8]. The fundamental difference between derivations of the two approaches stems from the following: pull is demand driven whereas push is forecast driven.

Therefore, the steady state expected number of parts is obtained by $E[N] = \lim_{t\to\infty} E[N(t)]$, and the steady state variance is given by $Var(N) = \lim_{t\to\infty} Var(N(t))$. Unfortunately, even for small sized problems, calculating a closed form for the distribution of transient throughput of transfer line, $(N(t))$ is impossible for the time being. Nevertheless, it is possible to numerically calculate related integrals to find appropriate probability values $(P\{N(t)=n\})$ for a given demand and service rates at an arbitrary, that is, transient time t.

13.4 Experiments and Numerical Results

This part of the study presents numerical results as part of experimentation from the proposed method given in Sect. 13.3. The implementation of the method was achieved using Maple, for which the maple code is provided in Appendix.

First case considered is (λ,μ) = (Demand rate, Processing rate) = (4,2). Figure 13.4 shows the plot of number of parts produced during time $t = 1$ through

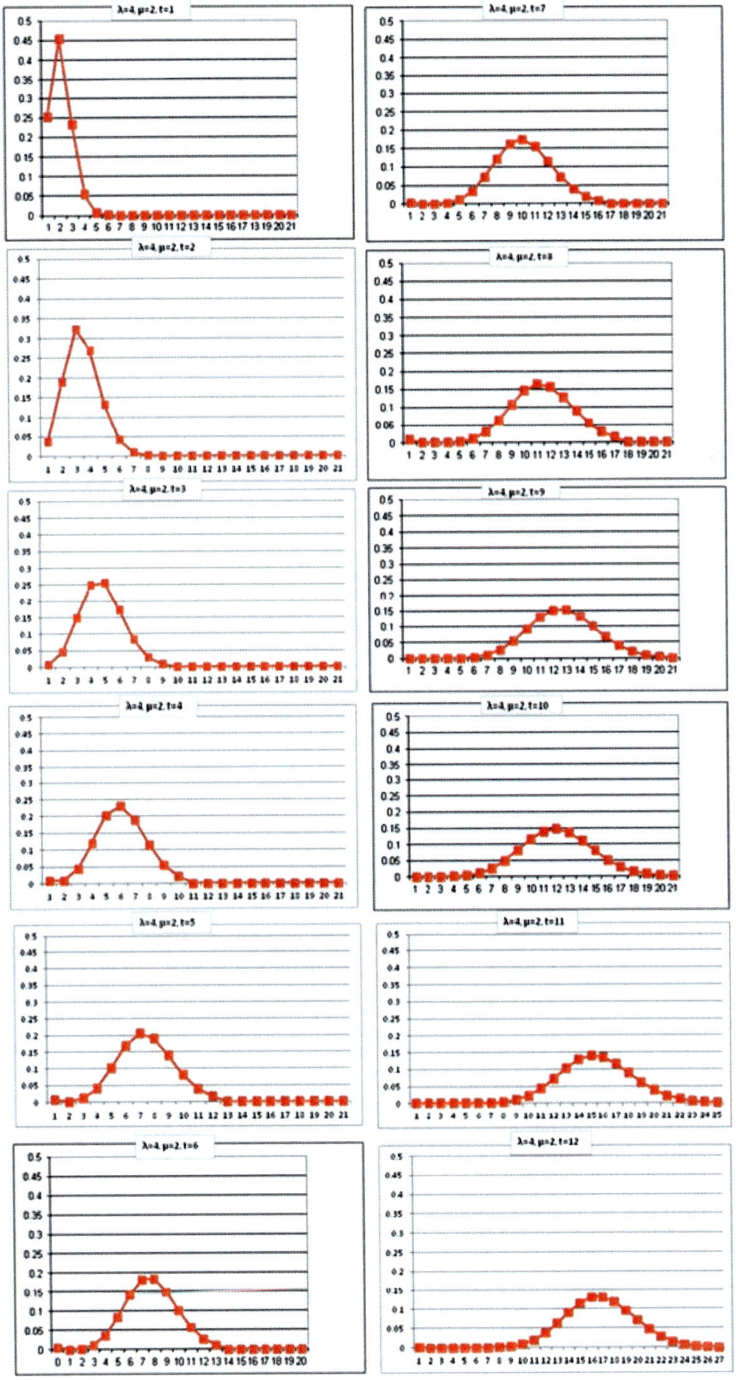

Fig. 13.4: Transient throughput of model transfer line, $\lambda = 4$, $\mu = 2$

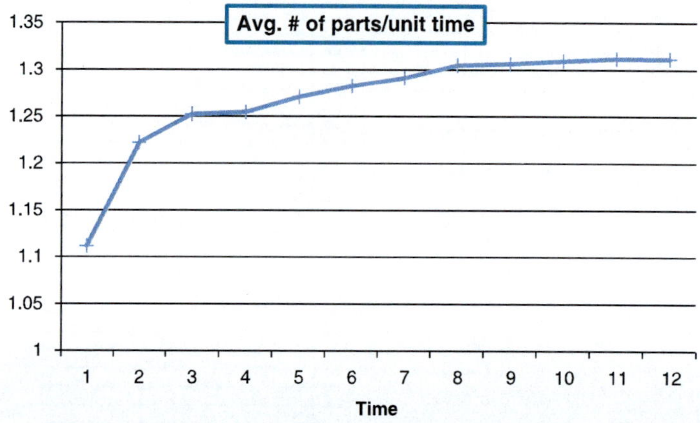

Fig. 13.5: Average number of parts produced / unit time

Table 13.1: Summary results for $\lambda = 4$, $\mu = 2$

t	Mean throughput	Std. Dev. Of Throughput
1	1.1114	0.6708
2	2.4443	1.1351
3	3.7571	1.4527
4	5.0186	1.6507
5	6.3554	1.8628
6	7.6965	2.0539
7	9.0395	2.2303
8	10.4400	2.4345
9	11.7600	2.5714
10	13.0960	2.7121
11	14.4300	2.8478
12	15.7400	2.9645

12. Time dependent average parts produced per unit time (Fig. 13.5) shows that after $t = 12$, the system reaches steady state. The reason for this is initially because the system is empty, the average parts per unit time increases. As time progresses, since demand rate is higher than the production rate, blockage occurs and average parts per unit time settles. Table 13.1 displays the mean throughput and variance of throughput for each transient time.

Additionally, we present results for cases, when λ is fixed and μ varies, and when μ is fixed and λ varies. For fixed demand rate, we present the changes in mean throughput and standard deviation of throughput for varying service rates. Figures 13.6 and 13.7 present these instances when $\lambda = 4$ for $\mu = 2$ and $\mu = 8$, respectively.

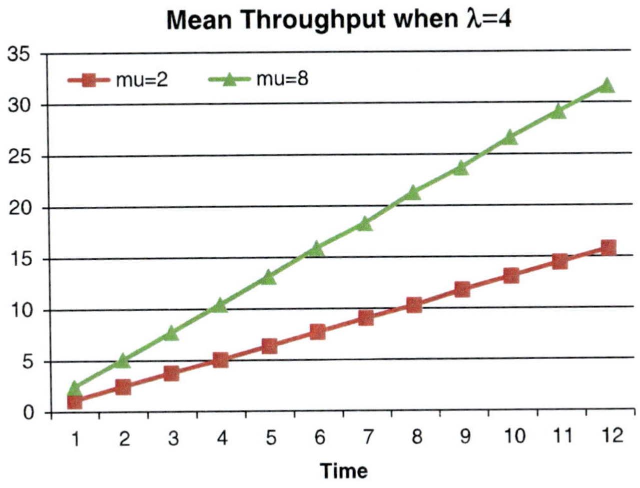

Fig. 13.6: Mean throughput for fixed demand $\lambda = 4$

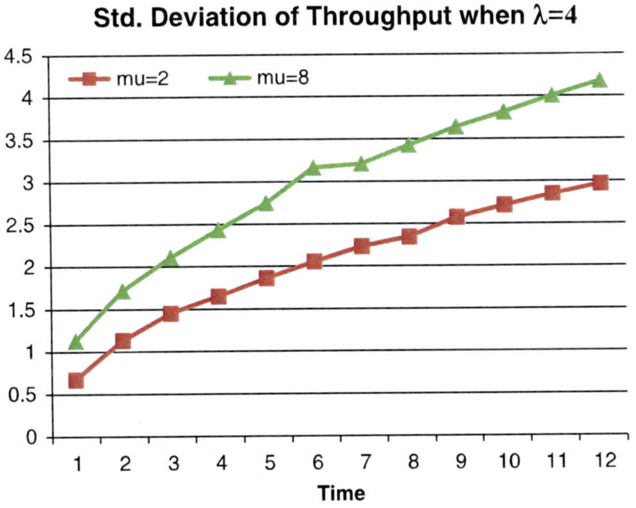

Fig. 13.7: Standard deviation of throughput for fixed demand $\lambda = 4$

For fixed service rate, we also present the changes in mean throughput and standard deviation of throughput for varying demand rates. Figures 13.8 and 13.9 present these instances when $\mu = 2$ for $\lambda = 4$ and $\lambda = 8$. Finally, Figs. 13.10 and 13.11 are for instances when $\mu = 4$ for $\lambda = 2$ and $\lambda = 8$.

Figures 13.6–13.11, give us the chance to uniquely observe exactly what happens to a system like the one under review in transient time, measured in mean throughput and standard deviation of the throughput.

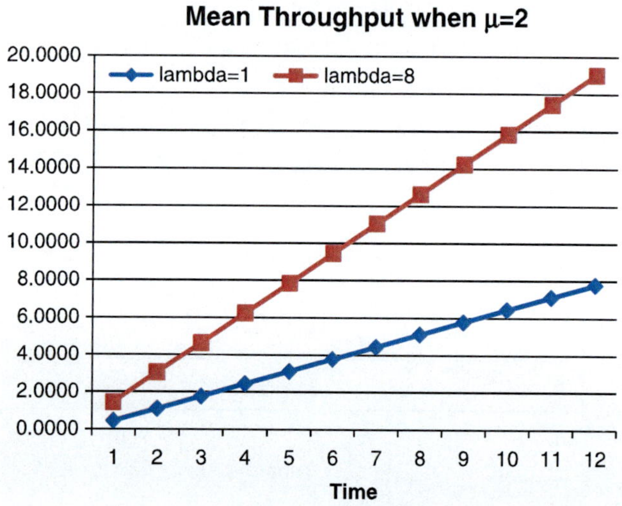

Fig. 13.8: Mean throughput for fixed service rate $\mu = 2$

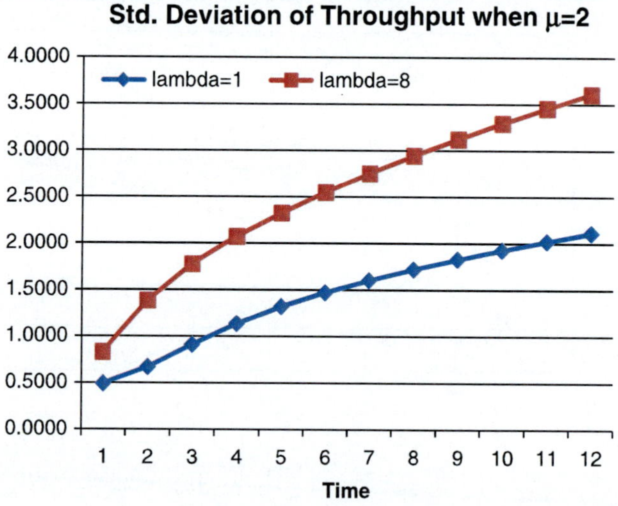

Fig. 13.9: Standard deviation of throughput for fixed service rate $\mu = 2$

13.5 Conclusions and Future Research

This chapter presents a method to calculate transient throughput of a pull type system. Majority of research on the throughput of transfer lines, concentrate on the steady state results. However, ever increasing mass customization and number of models produced from a transfer line, require many quick changeovers from one part/model to the other. In these kind of situations, there is usually not enough

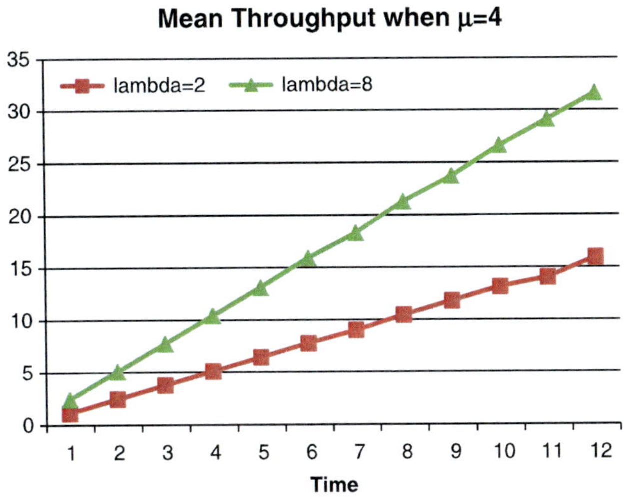

Fig. 13.10: Mean throughput for fixed service rate $\mu = 4$

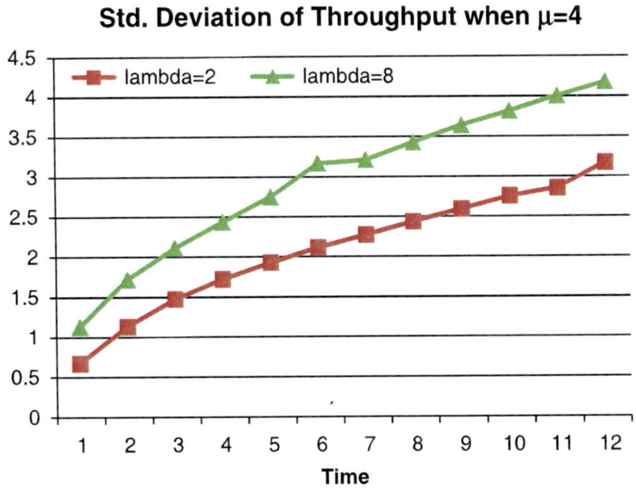

Fig. 13.11: Standard deviation of throughput for fixed service rate $\mu = 4$

time for the system to reach steady state. The relevant performance measures are those during the transient time, until the next changeover. Although there are studies on transient behavior of manufacturing systems, these are mostly based on push type systems. JIT and pull type systems, which have been very successful and are widely implemented, have been almost completely ignored for transient behavior studies.

The proposed model in this chapter, is not only able to calculate the transient mean throughput, but also allows calculation of variance and standard deviation since the whole distribution of the transient throughput is known. Derivation of distribution of transient throughput of transfer lines are presented and sample calculations are provided.

The work in this chapter by no means suggests that either a pure pull type or a pure push type system is the correct choice. The correct choice is probably a hybrid of the two. But this study will form the basis of analytical study of such hybrid systems as well.

A natural extension to this work, would be to enlarge the current model by adding multiple machines and inserting reliability. As the model enlarges, calculations would be increasingly hard, therefore if one can figure out, based on the smaller examples, using certain rules of thumbs to approximate the transient behavior would be a valuable contribution.

Appendix: Sample Maple Code ($\lambda = 4, \mu = 2, t = 1$)

```
> restart;
> fd:=fopen("c:\\mag\\dene.out", APPEND);
fprintf(fd,"\n");fprintf(fd,"\n");
> printlevel:=0; #  So that no output is printed on screen
 unless specifically printed by command
> mean_thruput:=0; printlevel:=0;
> tot:=0;
> epsilon:=0.001;
> mu[1]:=4;
> lambda:=2; t:=1;
> fprintf(fd,"epsilon= %10.5f\n",epsilon);
> fprintf(fd,"lambda= %d\nmu= %d\ntime= %d\n", lambda, mu[1],t);
> P(N(t)=0):= evalf((1/(mu[1]-lambda))*(mu[1]*
 exp(-lambda*t)-lambda*exp(-mu[1]*t)));
> tot:=tot+P(N(t)=0);
> for n from 1 by 1 to 200 while ((1-tot)>epsilon) do
> F_Hypo(lambda, mu[1],u):= 1 -   (1/(mu[1]-lambda))*(mu[1]*
 exp(-lambda*(u-t))-lambda*exp(-mu[1]*(u-t)));
> G_Hypo(lambda, mu[1],u):= 1- F_Hypo(lambda, mu[1],u);
> F_S[n](u):=int((((lambda^n)*(z^(n-1))*exp(-lambda*z))/(n-1)!)*
 int((((mu[1]^n)*(w^(n-1))*exp(-mu[1]*w))/(n-1)!),w=0..u-z),
   z=0..u);
> f_S[n](u):= diff(F_S[n](u),u);
> P(N(t)=n):= int(G_Hypo(lambda, mu[1],u)*
 f_S[n](u),u=t..infinity);
> last_prob:=P(N(t)=n);
> last_prob:=evalf(last_prob);
> v1[n]:=evalf(P(N(t)=n));
> v2[n]:=n;
> tot:=tot+P(N(t)=n);
> tot:=evalf(%);
> end do;
> v1[0]:=P(N(t)=0);
> for i from 0 to n do print(v1[i]);
 fprintf(fd,"%d    %10.7f\n",i,v1[i]); end do;
```

```
> mean_thruput:=sum(m*P(N(t)=m), m=1..(n-1));
> evalf(mean_thruput);
> fprintf(fd,"\n");
 fprintf(fd,"mean_thruput=%10.7f\n",mean_thruput);
> variance_thruput:= sum(((m-mean_thruput)^2)*
 P(N(t)=m), m=1..(n-1));
> evalf(variance_thruput);
> fprintf(fd,"variance=%10.7f\n",variance_thruput);
> std_dev_thruput:=sqrt(variance_thruput);
> evalf(std_dev_thruput);
> fprintf(fd,"std_deviation=%10.7f\n",std_dev_thruput);
> fclose(fd);
```

References

1. Altiok, T.: Performance analysis of manufacturing systems. Springer, Belin (1997)
2. Buxey, G.M., Slack, N.D., Wild, R.: Production flow line systems design–a review. AIIE Transactions **5**, 37–48 (1973)
3. Buzacott, J.A., Hanifin, L.E.: Models of automatic transfer lines with inventory banks: a review and comparison. AIIE Transactions **10**, 197–207 (1978)
4. Buzacott, J.A., Shantikumar, J.G.: Stochastic models of manufacturing systems. Prentice Hall, Engelwood Cliffs, NJ (1993)
5. Chen, C.T., Yuan, J., Lin M.-H.: Transient throughput analysis using sample path method for serial unreliable work centers. International Journal of The Computer, The Internet and Management **11**(1), 30–41 (2003)
6. Ciprut, P., Hongler, M.-O., Salama Y.: On the variance of the production output of transfer lines. IEEE Transactions on Robotics and Automation **15**(1), 33–43 (1999)
7. Dallery, Y., Gershwin, S.B.: Manufacturing flow line systems: a review of models and analytical results. Queuing Systems **2**, 3–94 (1992)
8. Dinçer, C., Deler, B.: On the distribution of throughput of transfer lines. Journal of the Operational Research Society **51**, 1170–1178 (2000)
9. Gershwin, S.B.: Manufacturing Systems Engineering. Prentice Hall, New Jersey (1994)
10. Govil, M.C., Fu, M.C.: Queueing theory in manufacturing: a survey. Journal of Manufacturing Systems, **18**, 214–240 (1996)
11. Hopp, W., Spearman, M.: Factory Physics. McGraw-Hill, Irwin (2000)
12. Koenigsberg, E.: Production lines and internal storage–a review. Management Sciences **5**, 410–433 (1959)
13. Li, J., Blumenfeld, D.E., Huang, N., Alden, J.M: Throughput Analysis of production systems: Recent Advances and Future Topics. International Journal of Production Research **47**, 3823–3851 (2009)
14. Li, J., Meerkov, S.M.: Production systems engineering. Wingspan Press, Livermore, CA (2007)
15. Meerkov, S.M., Zhang, L.: Transient behavior of serial production lines with Bernoulli machines. IIE Transactions **40**, 297–312 (2008)
16. Mitra, D.: Stochastic theory of a fluid model of producers and consumers coupled by a buffer. Advances in Applied Probability **20**(3), 646–676 (1998)
17. Mocanu, S.: Numerical algorithms for transient analysis of fluid queues. In Fifth International Conference on the Analysis of Manufacturing Systems, Zakymthos, Greece (2005)
18. Papadopoulos, H.T., Heavy, C., Browne, J.: Queueing Theory in Manufacturing Systems Analysis and Design. Chapman & Hill, London, UK (1993)

19. Sader, B.H., Sorensen, C.D.: A new technique for modelling production control schemes in manufacturing systems. International Journal of Production Research **48**(23), 7127–7157 (2010)
20. Sariyer, G.: On the transient analysis of transfer lines M.Sc. Thesis, Graduate School of Sciences, Izmir University of Economics (2009)
21. Tan, B.: An efficient method for variance of the output in a production line with a finite buffer. In Proceeding of International Workshop on Performance Evaluation and Optimization of Production Lines, Samos, Greece, 135–157 (1997)
22. Tan, B.: Variance of the throughput of an N-station production line with no intermediate buffers and time dependent failures. European Journal of Operational Research **101**(3), 560–576 (1997)
23. Tan, B.: An analytic formula for variance of output from a series-parallel production system with no interstation buffers and time-dependent failures. Mathematical and Computer Modelling **27**(6), 95–112 (1998)
24. Tan, B.: Asymptotic variance rate of the output of a transfer line with no buffer storage and cycle-dependent failures. Mathematical and Computer Modeling **29**, 97–112 (1999)
25. Tan, B.: Variance of the output as a function of time: Production line dynamics. European Journal of Operational Research **117**, 470–484 (1999)
26. Tan, B.: Asymptotic variance rate of the output in production lines with finite buffers. Annals of Operations Research **93**, 385–403 (2000)
27. Viswanadham, N., Narahari, Y.: Transient analysis of manufacturing systems performance. IEEE Transactions on Robotics and Automation **10**, 330–345 (1994)